SCIENTISTS

SCIENTISTS

EXTRAORDINARY PEOPLE WHO ALTERED THE COURSE OF HISTORY

EMMA CAMPBELL, NICOLA CHALTON, JASMINE FARSARAKIS, SUSAN BOYD LEES,
ELIZABETH MILES, SIMON RICHES, VANESSA SCHNEIDER, HUGO SIMMS

EDITOR:
MEREDITH MACARDLE

METRO BOOKS
NEW YORK

Designed and produced by Basement Press
61 Ivydale Road
London SE15 3DS, United Kingdom
www.basementpress.com
Design & picture research: Pascal Thivillon
Project editor: Nicola Chalton
Editorial: Meredith MacArdle and Nicola Chalton
Science consultant: David Hawksett
Glossary: Simon Riches

Metro Books
122 Fifth Avenue
New York, NY 10011

ISBN-13: 978-1-4351-1018-2

Printed and bound in Malaysia

10 9 8 7 6 5 4 3 2 1

Picture credits Heritage Image Partnership/© British Library Cover. Book Builder All portraits. History of Science Collections, University of Oklahoma Libraries; © the Board of Regents of the University of Oklahoma pages 13, 63, 67, 87, 99, 111. Basement Press pages 19, 35, 51, 53 (left), 97 (from a photo Renee Comet - National Cancer Institute). National Library of Medicine (NLM) pages 25, 29. Library of Congress, Prints & Photographs Division pages 39, 55, 59, 61, 69, 73, 85, 113, 117, 133, 162. Visipix.com pages 57, 93. Photo © Luc Viatour page 41. Photo © Thomas Claveirole - Creative Commons page 53 (right). Photo © C. E. Ford page 65. GNU Free Documentation License pages 77, 83, 91, 115. US National Oceanic and Atmospheric Administration Central Library page 79. Rocky Mountain Laboratories, National Institute of Allergy and Infectious Diseases, National Institutes of Health page 101. Centers for Disease Control and Prevention page 107. Photo © Falcorian - creative commons page 123. Photo © Konstantin Binder - GNU Free Documentation License page 127. U.S. Geological Survey Photographic Library page 135. New York Times archives page 138. Photo © Mila Zinkova - GNU Free Documentation License page 141. Centers for Disease Control and Prevention page 143. Photo © Jack Aeby - United States Department of Energy page 145. NASA page 146. NASA Jet Propulsion Laboratory (NASA-JPL) page 149. NASA/ESA/AURA/Caltech page 151. NIST/JILA/CU-Boulder page 153. U.S. Air Force page 155. National Archives and Records Administration page 157. Photo © Andrew Dunn - GNU Free Documentation License page 165. Smithsonian Institution - National Museum of American History page 169. Otisarchives - creative commons page 171. NASA Marshall Space Flight Center (NASA-MSFC) page 173. Photo © Matt Crypto page 175. CDC/ Mary Hilpertshauser page 177. © Michael Ströck - GNU Free Documentation License page 179. © V8rik - GNU Free Documentation License page 183. OAR/National Undersea Research Program (NURP) page 185. Photo © Sherseydc - creative commons page 189. Photo © Ute Kraus - GNU Free Documentation License page 191.

Publisher's note Every effort has been made to ensure the accuracy of the information presented in this book. The publisher will not assume liability for damages caused by inaccuracies in the data and makes no warranty whatsoever expressed or implied. The publisher welcomes comments and corrections from readers, emailed to info@basementpress.com, which will be considered for incorporation in future editions. Likewise, every effort has been made to trace copyright holders and seek permission to use illustrative and other material. The publisher wishes to apologize for any inadvertent errors or omissions and would be glad to rectify these in future editions.

Contents

World map and timeline

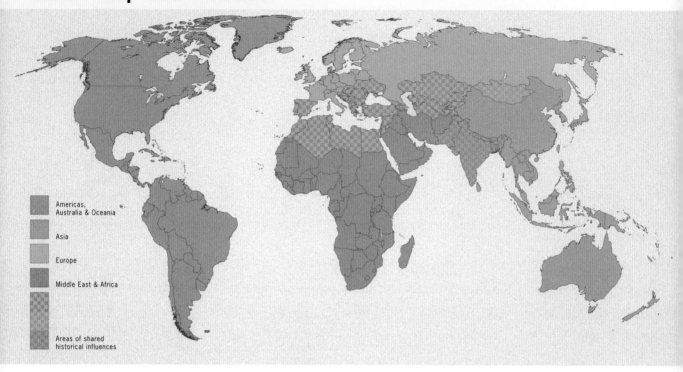

- Americas, Australia & Oceania
- Asia
- Europe
- Middle East & Africa
- Areas of shared historical influences

This history of scientists and their world-changing discoveries begins with a map and timeline. These are designed to help locate scientists both geographically and within a general temporal framework.

World map

The map is divided into four major areas: the Americas, Australia, Oceania (all colored purple); Asia (blue); Europe (green); and the Middle East and Africa (brown). The demarcations provide a rough guide to each person's birth place and sphere of influence and the same color-coded theme is carried through the rest of the book. Of course divisions of this nature must reflect a certain arbitrariness, and cultural influences are rarely so geographically bound. It will be noted that some areas on the map overlap, representing approximate

regions of shared historical influence. For example, the Iberian Peninsula – now modern Spain and Portugal, and part of Europe – is shown to overlap with North Africa to indicate the spread of Arab culture during the Golden Age of Islam. Likewise, the classical Greeks born in what is now modern Turkey, part of the Middle East, are located in an area sharing historical links with Europe, indicating the vast reach of Greek culture during ancient times.

Timeline

The timeline highlights periods of intense scientific activity, and locates scientists in a stream of time running from the fifth century BCE to the present day. The numbers in the colored dots refer to the list of scientists on the page opposite; their positions on the timeline are determined according to birth dates.

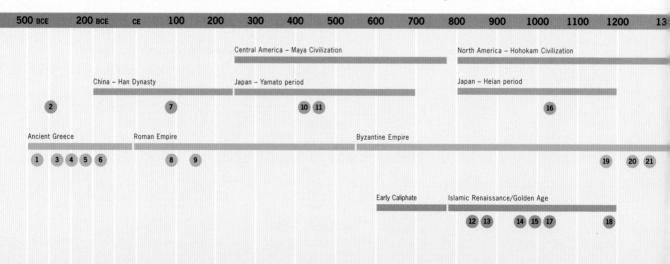

Americas, Australia, & Oceania

35	Benjamin Franklin 1706–90
51	James Dwight Dana 1813–95
61	Thomas Edison 1847–1931
64	William Morris Davis 1850–1934
67	Albert Abraham Michelson 1852–1931
73	Florence Bascom 1862–1945
80	Edwin Hubble 1889–1953
86	Linus Pauling 1901–94
90	Gregory Goodwin Pincus 1903–67
91	Barbara McClintock 1902–92
92	Rachel Carson 1907–64
95	Jonas Salk 1914–95
97	Richard Feynman 1918–88
98	Paul Berg b. 1926
99	Elias James Corey b. 1928
100	Sylvia Earle b. 1935
102	Sidney Altman b. 1939
106	Craig Venter b. 1946

Asia

2	Gan De c.400–c.340 BCE
7	Zhang Heng 78–139 CE
10	Zu Chongzhi 429–500
11	Aryabhata 476–550
16	Shen Kuo 1031–95
27	Xu Guangqi 1562–1633
71	Jagadis Chandra Bose 1858–1937
81	Megh Nad Saha 1893–1956
82	Satyendra Nath Bose 1894–1974
89	Zhang Yuzhe 1902–86
93	Subrahmanyan Chandrasekhar 1910–95

Europe

1	Hippocrates c.460–c.370 BCE
3	Aristotle 384–322 BCE
4	Euclid c.325–c.265 BCE
5	Archimedes c.287–212 BCE
6	Hipparchus 190–120 BCE
8	Ptolemy c.83–c.161 CE
9	Galen 129–c.216
19	Leonardo Fibonacci 1170–1250
20	Roger Bacon c.1214–c.1292
21	Jacob ben Machir ibn Tibbon c.1236–c.1305
22	Abraham Zacuto c.1450–1510
23	Leonardo da Vinci 1452–1519
24	Nicolaus Copernicus 1473–1543
25	Georgius Agricola 1494–1555
26	Garcia de Orta 1501–68
28	Galileo Galilei 1564–1642
29	Johannes Kepler 1571–1630
30	Blaise Pascal 1623–62
31	Robert Boyle 1627–91
33	Antony van Leeuwenhoek 1632–1723
34	Isaac Newton 1642–1727
36	Carolus Linnaeus 1707–78
37	James Hutton 1726–97
38	Antoine Lavoisier 1743–94
39	Joseph Banks 1743–1820
40	Alessandro Volta 1745–1827
41	Edward Jenner 1749–1823
42	John Dalton 1766–1844
43	Amedeo Avogadro 1776–1856
44	Carl Friedrich Gauss 1777–1855
45	Michael Faraday 1791–1867
46	Roderick Impey Murchison 1792–1871
47	Jean-Baptiste Dumas 1800–84
48	Justus von Liebig 1803–73
49	Charles Darwin 1809–82
50	Robert Wilhelm Bunsen 1811–99
52	Rudolf Virchow 1821–1902
53	Gregor Johann Mendel 1822–84
54	Louis Pasteur 1822–95
55	Ferdinand Cohn 1828–98
56	James Clerk Maxwell 1831–79
57	Alfred Bernhard Nobel 1833–96
58	Robert Koch 1843–1910
59	Ludwig Boltzmann 1844–1906
60	Wilhelm Conrad Röntgen 1843–1923
62	Ivan Petrovich Pavlov 1849–1936
63a	Karl Ferdinand Braun 1850–1918
63b	Guglielmo Marconi 1874–1937
65	Emil Fischer 1852–1919
66	Henri Becquerel 1852–1908
68	Santiago Ramón y Cajal 1852–1934
69	Sigmund Freud 1856–1939
70	Heinrich Hertz 1857–94
72	Max Planck 1858–1947
74	Marie Curie 1867–1934
75	Albert Einstein 1879–1955
76	Alfred Wegener 1880–1930
77	Alexander Fleming 1881–1955
78	Niels Bohr 1885–1962
79	Erwin Schrödinger 1887–1961
83	Leo Szilard 1898–1964
84	Enrico Fermi 1901–54
85	Werner Heisenberg 1901–76
87	Kurt Alder 1902–58
88	Paul Dirac 1902–84
94	Alan Turing 1912–54
96a	Francis Crick 1916–2004
96b	James D. Watson b. 1928 [US]
101	Günter Blobel b. 1936
103	Richard Dawkins b. 1941
104	Stephen Hawking b. 1942
107	Tim Berners-Lee b. 1955

Middle East & Africa

12	Al–Battani c.850–c.929
13	Rhazes 860–925
14	Ibn al–Haytham (Alhazen) c.965–c.1040
15	Avicenna (Ibn Sina) c.980–1037
17	Al–Zarqali (Azarquiel) 1028–87
18	Ibn al–Baitar 1190–1248
105	Ahmed H. Zewail b. 1946

Timeline axis: 00 1400 1500 1600 1700 1800 1900 2000

South America – Inca Empire

Age of Revolutions

World War I & World War II

35

51

61 67 80 86 91 95 97 99 106
64 73 90 92 100
71 81 89 93 98 102
82

China – Ming Dynasty

27

European Renaissance

Enlightenment

22 23 24 25 26   28 29   30 31 32 33 34   36 37 38 40 42 43 45 49 52 58 62 72 79 84 94 101 107
39 41   44 46 50 53 59 63a 63b 83 85 96a 103
47 54 60 65 74 87 96b 104
48 55 66 75 88
56 68 76 105
57 69 77
70 78

Mughal Empire

7

Introduction

"The history of science and technology has consistently taught us that scientific advances in basic understanding have sooner or later led to technical and industrial applications that have revolutionized our way of life . . . What is less certain, and what we all fervently hope, is that man will soon grow sufficiently adult to make good use of the powers that he acquires over nature."

Enrico Fermi, "The Future of Nuclear Physics", unpublished address (1953)

The polymath and influential nineteenth-century figure William Whewell coined the word "scientist" in 1833 to describe a person who uses methods of observation and experiment to discover truths about the natural world. Before this, scientists were called "natural philosophers".

The natural philosophers of ancient Greece rarely relied upon experimental results in their investigations, preferring pure argument. However, the work of some of these early thinkers can be seen to represent the first flowerings of scientific knowledge that would eventually transform our world, such as the mathematical models tested through experimentation by Archimedes (born c.287 BCE), and the systems set up by Hippocrates (born c.460 BCE), the "father of medicine".

By the time of the Islamic Golden Age, the early methods emphasizing experimental verification used by the Arab physicist and polymath Ibn al-Haytham in his *Book of Optics* (c.1021 CE) were approaching the scientific standard of today, though evidence was still thin on the ground. The men of learning of the European Renaissance faced other difficulties: Copernicus (born 1473), a Polish priest and astronomer; Galileo (born 1564), an Italian astronomer; and Roger Bacon (born c.1214), a humble and devout English friar, worked courageously in the shadow of a terrifyingly repressive religious censorship, risking their lives by challenging the orthodoxy of their day.

The advances of Isaac Newton, who explained how the universe is physically held together, dominated the seventeenth century. His treatise published in 1687 described universal gravitation and the three laws of motion, laying the groundwork for classical mechanics, which were to dominate the scientific view of the physical universe for the next three centuries and significantly advance the Scientific Revolution.

In the nineteenth century, the Austrian monk Gregor Mendel studied traits of inheritance and founded modern genetics, a discipline that helped to explain Charles Darwin's discoveries in evolution recorded in his 1859 book, *On the Origin of Species*. Louis Pasteur found that microorganisms could cause disease, paving the way for the flowering of molecular biology in the twentieth century – the century that also witnessed the genius of Albert Einstein and his many contributions to physics.

Not all scientific contributions can be said to have advanced humanity. The arrival of Enrico Fermi, a physicist who had escaped from fascist Italy in 1938, was the help America needed in the race with Germany to possess the atomic bomb. Fermi and his fellow physicists produced the first controlled atomic chain reaction, and Hiroshima and Nagasaki graphically illustrate how the United States won the race to be the first to deploy the bomb. But as Fermi and many others soon realized, responsible science that recognizes how its power must be harnessed to do good, and not to spread chaos, is the key to its future.

Although the selection of scientists in this book is, by its nature, somewhat arbitrary, as so many influential scientists have lived and could have been included, those chosen are generally acknowledged as leaders in their field or representative of a particular movement. All can be said to have left indelible marks on the world.

On using the book

■ The entries on each scientist are arranged chronologically, roughly according to birth dates. Entries are color-coded to match the geographical regions described on pages 6–7.

■ Each entry includes a short description of the life and work of the scientist, essential points about their work, key dates in their biography, and an analysis of their legacy and contribution.

■ Cross-references to other scientists in this book are given in grey bold type (e.g. **Galileo Galilei**) on their first appearance in an entry. These cross-references allow the reader to track scientific developments by turning to all the pages that are linked in this way.

■ Scientific terms are in dark bold type (e.g. **quantum mechanics**) the first time they appear in each entry, and are explained in the glossary on pages 198–204.

■ Generally, scientists are referred to by their names in common usage, although full and alternative names are also given.

■ Arabic names are used unless a Latinized version is more well known in Western culture.

■ Chinese names – with the surname first – are given in the modern Pinyin transliteration, although the older Wade-Giles version is also mentioned.

Hippocrates

Known as the "father of medicine", Hippocrates was the first to dissociate the symptoms of illness with religion. As well as describing several illnesses, his guidelines for physicians' conduct, professionalism, and responsibility in preserving life are the basis of the Hippocratic Oath, the oath medical students still swear to today before entering medical practice.

Born on the island of Kos in ancient Greece, Hippocrates was a descendant of two legendary ancient Greek figures, Asclepius and Hercules. Little is known about exactly what Hippocrates wrote and did, since much of his work is indistinguishable from other practitioners of Hippocratic medicine. However, it is recorded that he traveled throughout ancient Greece practicing medicine and teaching his own theories, and treating and diagnosing many of the important political and academic figures of his time. By the end of his life he was renowned as the greatest physician of the time. He was also known to be kind, calm, and practical.

> *It is thus with regard to the disease called Sacred ... Men regard its nature and cause as divine from ignorance and wonder ...*
>
> From *On the Sacred Disease*, part of the *Hippocratic Corpus* (c.300 BCE)

Essential science

The Hippocratic or Koan school of medicine

Medicine in ancient Greece was divided into the Knidian and Koan schools. The Knidian school's focus on diagnosis was based on many erroneous assumptions about the human body (the Greek taboo forbidding the dissection of humans meant that there was virtually no knowledge of human anatomy and physiology). By contrast, the Hippocratic or Koan medical strategy focused on patient care and prognosis, and achieved greater success by applying general diagnoses and non-invasive treatments.

People in Hippocrates' time believed that sickness was a form of punishment from the gods, and Hippocrates was the first to disagree. He taught that illness was a result of an imbalance of the fluids known as the four humors, which were blood, black bile, yellow bile, and phlegm. He also believed that the body had certain points of "crisis", which were time points during disease progression where the patient would either improve by the healing power of nature, or possibly suffer a relapse.

The Hippocratic Oath

Hippocrates' work *On the Physician* instructs physicians to be clean, well kept, honest, kind to their patients, calm, understanding, and

Legacy, truth, consequence

- The Hippocratic school of medicine established medicine as a distinct profession.
- The Hippocratic Oath in modern form still governs physicians' conduct and responsibility today.
- Hippocrates first described the symptoms of "Sacred", the ancient term for epilepsy; clubbing fingers (also known as "Hippocratic fingers"); haemorrhoids; and the "Hippocratic face", which is the gradual change in the face as illness progresses or after death.

Key dates

c.460 BCE Born on the Greek island of Kos.

c.300 BCE Production of the *Hippocratic Corpus*, a collection of approximately 70 early medical works from ancient Greece, is believed to have been compiled by Hippocrates' students and followers. This collection contains textbooks, lectures, research notes, and philosophical essays on medicine, and includes the Hippocratic Oath.

c.370 BCE Dies, probably in Larissa, Greece.

serious. He was insistent that physicians' practices should follow specific guidelines on lighting, personnel, instruments, and techniques when handling patients. It was also important for physicians to keep clear and accurate records that could be used by other physicians later on. The records were to list details of symptoms including complexion, pulse, fever, pains, mobility, and any excretions, as well as family history and details of the patient's environmental exposure. These are founding traits of modern clinical medicine.

The Hippocratic Oath, a modern version of which is still taken today by medical students when they enter medical practice, is included in the work the *Hippocratic Corpus*. The oath highlights the most important aspects of a physician's professional responsibility: to preserve life, to share experience and teachings, and to be morally uncorrupt.

Gan De

An ancient Chinese astronomer, Gan De and his contemporary Shi Shen were the authors of the world's earliest known star catalogs. He cataloged thousands of stars, and recognized more than 100 constellations. He also made the first known record in the world of seeing a satellite of Jupiter.

Gan De (Kan Te) lived during the turbulent **Warring States period** of ancient China, when, as the name suggests, there was no unified government and rival states fought for supremacy. Given the troubles of the time it is not surprising that there are no surviving records about his life, but it was later reported that he lived in the state of Qi. According to the traditional Chinese naming system, Gan was his family name, and De was his personal name.

We do know that from even earlier times, Chinese emperors and kings always had a group of **astronomers** attached to their courts to fulfil several important functions: maintaining clocks to measure time; keeping the calendars which determined the dates of the annual festivals and religious ceremonies; keeping astronomical records; and predicting **eclipses** and other heavenly events. A court astronomer would have been a respected official and very likely a wise, learned man.

We do not even have surviving copies of Gan De's works: all we have are later accounts of them and copies of parts of his writings incorporated into other books, particularly the *Treatise on Astrology of the Kaiyuan Era*, compiled many centuries later, from about 718 to 726 CE, during the reign of the emperor of the Tang dynasty Xuan Zong. This book includes Gan's eyewitness account of observing one of the satellites of Jupiter – the first known such record.

During the Warring States period, the regular 12-year passage across the skies of the bright, visible light of Jupiter was used to count years, so it was the focus of concentrated observations and predictions. Without telescopes, Gan and his colleagues had to rely on the naked eye, but they made acute calculations to guide them to the best times to make celestial observations.

Gan De became an expert on the movements of Jupiter, and among his lost books was a monograph solely on this planet, *The Book of the Annual Star*, in which he first recorded the existence of an "alliance" or satellite.

Legacy, truth, consequence

■ Gan De was one of the founders of the enduring science of astronomy of China. He helped lay the basis for the scholarly, detailed, and empirical tradition that lasted there for centuries.

■ His observation of what was almost certainly one of Jupiter's four large moons was performed with the naked eye, and was long before **Galileo Galilei** officially "discovered" the satellites in 1610 using his newly-developed telescope.

■ In 1981 Chinese astronomers proved that during a narrow window when Jupiter is closest to the earth, at least two satellites can be seen with the naked eye by people with excellent eyesight.

Key dates

c.400 BCE	Born in China.
364 BCE	Year when it is thought that Gan De observed one of Jupiter's satellites.
c.340 BCE	Dies in China.

Essential science

Astronomy

Gan De's star catalog was more comprehensive than the first-known Western star catalog, drawn up by the Greek astronomer **Hipparchus** in about 129 BCE, who listed about 800 stars.

Together with Shi Shen (*fl.* fourth century BCE) Gan De was among the first early astronomers to approach an accurate measurement of the year, since they calculated it as 365 and one-quarter days. They also made reasonably accurate observations of the five major planets.

Jupiter ... rose in the morning and went under in the evening together with the Lunar Mansions [lunar divisions] Xunu, Xu and Wei. It was very large and bright ... there was a small reddish star appended to its side. This is called an alliance.

Gan De quoted in the *Treatise on Astrology of the Kaiyuan Era* (c.718–726 CE)

Aristotle

One of the giants of the classical Greek intellectual world, Aristotle is probably best known as a philosopher, but he was interested in every aspect of the natural world, and his ideas on subjects as varied as biology and cosmology had a lasting influence in the West.

Born to a Macedonian medical family, Aristotle was one of the stars of Plato's school in Athens. There he studied philosophy and astronomy, as well as mathematics, which Plato thought was a suitable discipline to sharpen the mind.

He left Athens possibly because he was not appointed head of the Academy after Plato's death, but perhaps because Philip of Macedon's expansionist wars made Macedonians unpopular. Aristotle only returned after Alexander the Great (356–323 BCE) – Philip's son and Aristotle's pupil – had conquered all of Greece.

While he was running his own school, the Lyceum, named after its district in Athens, he continued extensive studies into almost every subject then defined, including politics, physics, philosophy, biology, mathematics, logic, and poetry. Because his method of teaching and debate was to walk around discussing topics, Aristotelians are often called Peripatetics.

After Alexander's death ill-feeling towards Macedonians flared up again, and Aristotle fled, supposedly stating in a reference to Socrates' execution: "*I will not allow the Athenians to sin twice against philosophy.*"

> ## In all things of nature there is something of the marvelous.
> On the Parts of Animals (c.350 BCE)

Legacy, truth, consequence

■ Aristotle's ideas on physical sciences and cosmology were accepted in the Arab world, and later in Christian countries once they had been reintroduced to Europe. He dominated intellectual life for centuries, to the point when if a scientific idea wasn't in his writings, it wasn't considered important. Ironically, this stifled the very exploration of the world he advocated.

■ Some of his zoological observations were confirmed to be accurate in the nineteenth century.

■ French schools are called lycées after his Lyceum.

Key dates

384 BCE	Born in Stagira in northern Greece. Trains in medicine.
c.367 BCE	Joins Plato's philosophical Academy in Athens.
347 BCE	After Plato's death, travels around Asia Minor (modern Turkey) and Greek islands, studying wildlife and biology.
c.334 BCE	Founds his own school, the Lyceum, in Athens.
323 BCE	Flees to the Greek island of Euboea.
322 BCE	Dies in Chalcis, Euboea.

Essential science

Aristotle believed that reality is expressed in physical objects (not "ideas", as Plato thought), and that the ultimate essential form of things could be determined through detailed examination followed by inductive reasoning. He put this into practice, observing and commenting on most of the natural world, from the laws of motion (stones are drawn to their natural home of earth, while fire is drawn upwards to its natural home in the heavens), to a spherical earth as shown by the earth's arc-like shadow on the moon during an eclipse. Some of his most enduring theories were:

The earth and the heavens

According to Aristotle, everything on earth was made up of four elements – earth, air, fire, and water – and was always changing as things came into being, then decayed and died. The heavens, on the other hand, consisted of a fifth element, aether, and were perfect and unchanging. There was no room for a vacuum.

As a refinement of existing **geocentric** and spherical ideas, his universe consisted of concentric aetheric shells around the earth, each containing one of the heavenly bodies, while at the edge were the stars, fixed within the outermost shell. Heavenly bodies moved at a uniform pace in perfect circles.

Biology

The world's first great biologist, he amassed huge amounts of data on the behavior and structures of animals and plants. He classified more than 500 different species using a logical system based on the means of reproduction, which enabled him to make some correct analyses, such as that dolphins are mammals.

Euclid

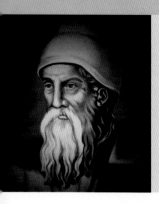

Euclid of Alexandria has been called the "father of geometry". He was the most significant mathematician of the ancient world, best known for his 13-book manual on mathematics, the *Elements of Geometry*. This became the most successful textbook in the history of mathematics and is considered to be one of the world's great classic books. It was the main source for mathematics in Europe and the Middle East until the nineteenth century when non-Euclidian geometry appeared.

In the Middle Ages Euclid was sometimes confused with the philosopher of the same name, Eukleides, who came from Megara and lived about 100 years earlier than Euclid, at about the same time as Plato (c.428–347 BCE). So Euclid is sometimes erroneously referred to as Megarensis (from Megara).

The only information we have on Euclid's life is from the brief biography written many centuries later by the Greek philosopher Proclus (c.410–485 CE) as part of his commentary on the first book of Euclid's *Elements of Geometry*. Proclus included a summary of all known important Greek mathematicians, a useful historical source, but all that he recorded is that during the reign of the Egyptian pharaoh Ptolemy I Soter, who ruled from 323 to 285 BCE, Euclid was teaching in Alexandria, which in those days was a famous center of learning because of its magnificent library. Proclus wrote that, when consulted by the pharaoh for a quick way to understand **geometry**, the mathematician replied that there was no "*royal road to geometry*".

Euclid's proper name, Eukleides, reveals his Greek origins. Egypt had been conquered by Alexander the Great of Macedon in 332 BCE, and in Euclid's time Egypt was still ruled by Greeks. It is now generally accepted that Euclid predated **Archimedes** (c.287–212 BCE).

Essential science

Elements of Geometry

Although its full title implies that it only covers geometry, the *Elements* is actually a long study in 13 books of all known mathematics, presented in a well-organized, well-written, logical framework that is easy to follow. It contains definitions, proofs of mathematical theorems, axioms or unproved general assumptions that he called "postulates", and other unproved assumptions that he called "common notions".

Geometry

Plane geometry (with fewer than three dimensions) is covered in Books One to Six, with a solid foundation supplied first in the form of 23 definitions such as "*A point is that which has no part*" or "*A line is a length without breadth*", which cover the basic properties of triangles, parallels, parallelograms, rectangles, and squares. The properties of circles and problems concerning circles merit two whole books, Three and Four, and these early books also cover what has been called geometric **algebra**, which looks at algebraic theorems to do with equivalent geometric shapes.

As in all Euclid's material, the geometry section covers both theory and practical application. It includes the important theorem known as **Pythagoras' theorem**: in a right-angled triangle, the square of the hypotenuse (the line opposite the right angle) is equal to the sum of the squares of the other two sides.

The Golden Section

In Book Two Euclid discusses what he called "the section," dividing a line into two parts so that the ratio of the larger part to the smaller part is the same as the ratio of the original line to the larger segment. In the European Renaissance, when rediscovered Greek works were widely discussed and followed, architects and artists realized that its proportions are aesthetically pleasing, providing both beauty and balance, and they renamed it the golden section. It is also known as the divine proportion and is equal to around 1:618.

Proportions

Book Five contains Euclid's clear exposition of the theory of proportions as well as an important solution to the problem of **irrational numbers** that is needed before the remaining books can be followed.

Number theory

Books Seven to Nine explore **number theory**, or the mathematics of whole numbers greater than one. Euclid provides 22 new definitions in this field, including concepts such as even and odd numbers, unity, and prime numbers, which he defines as whole numbers that can only be divided by themselves and one. He also proves that there is an infinite number of primes.

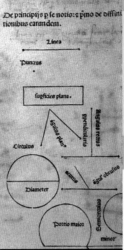

Detail of an 1842 edition of the *Elements of Geometry*.

Key dates

c.325 BCE   Born in Alexandria, Egypt.
before 285 BCE   Teaches mathematics in Alexandria and presumed to be compiling his *Elements of Geometry*.
c.265 BCE   Dies in Alexandria.

Legacy, truth, consequence

■ Comprehensive, concise, and easy to follow, Euclid's *Elements* is the world's oldest continuously used math textbook. Many of his explanations and theorems cannot be better described even today. The earlier works that he incorporated and explained became obsolete and many of them have not survived.

■ From the ninth century, Arabic translations were being studied in the Middle East, and the twelfth-century Latin translations galvanized Christian Europe. The *Elements* was the main source of geometric reasoning, theorems, and methods for about 2,000 years.

■ The geometrical system that Euclid described and that was named after him was thought to be the only possible geometry until the nineteenth century when mathematicians finally demonstrated forms of non-Euclidean geometrics. The art of M. C. Escher (1898–1972) is often based on non-Euclidean geometry.

Euclid's most famous work, commonly called the *Elements*, included a compilation of mathematical knowledge from a number of sources, including Hippocrates of Chios who worked at around 460 BCE (a different man from the doctor Hippocrates of Cos), and Theudius (fourth century BCE), whose writings would have been studied by **Aristotle** at Plato's Academy. However, the bulk of the mathematical proofs was Euclid's original work.

> *Things which are equal to the same thing are equal to each other.*
>
> Elements of Geometry

Platonic solids

Solid geometry, or the study of three-dimensional figures, is dealt with in Books 11 to 13, in which Euclid explores the intersections of lines, planes, and some parallelograms. He proves that the relationship between the areas of circles is the same as the relationship between the squares of their diameters, and that the volumes of spheres have the same relationship to each other as do the cubes of their diameters. He does this using the **method of exhaustion**.

The *Elements* concludes with instructions on how to construct and circumscribe five regular solids that are now known as the **Platonic solids** after the philosopher Plato who extensively discussed their nature. They are a pyramid (four faces), cube (six faces), octahedron (eight), dodecahedron (12), and icosahedron (20). It also supplies a proof that there are only five of these congruent shapes.

Euclidean geometrics

The study of points and lines, and planes and other figures is called **Euclidean geometry** in honor of the ancient Greek. One of mathematics' enduring questions was posed by his fifth or "parallel postulate", which says that there can only be one line that contains a given point and is parallel to another line, but it took more than 2,000 years before anyone was able to disprove this theory and present a valid argument for "**non-Euclidian geometrics**". (See **Carl Friedrich Gauss**, pages 78–9.)

Some of his other postulates seem to be obvious, such as "It is possible to draw a straight line between any two points", but might have hidden implications, such as "All right angles are equal", which requires the shape to be independent of the position in space in which it is drawn.

Other works

Euclid wrote some works that have not survived, but apart from the *Elements*, his known books are: *Data*, a collection of 94 advanced geometric propositions; *Optics*, the first Greek treatise on perspective; *On Divisions of Figures*, which looks at how to divide a figure into two parts; the *Phaenomena*, an introduction to mathematical astronomy; and *Catoptrics*, about the mathematics of mirror images. All of these works are shorter than the *Elements*, but follow its basic logical structure with definitions and proved propositions.

Archimedes

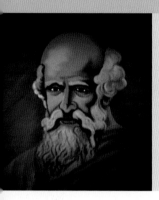

Considered to be one of the greatest mathematicians of all times, Archimedes of Syracuse stood out as an intellectual figure even among the other great minds of ancient Greece. Although he is now appreciated for his original mathematical insights, during his lifetime he became famous mainly as an inventor of impressive mechanical devices. In popular culture he is also legendary for jumping out of the bath naked.

Archimedes was born in the independent Greek city-state of Syracuse on the island of Sicily. During his lifetime Syracuse became one of the battlegrounds in the Second Punic War (218–201 BCE) between Rome and Carthage and its allies (including Syracuse) – the same war in which Hannibal of Carthage crossed the Alps with the help of elephants to invade Rome.

The young Archimedes was probably exposed to mathematics early on, because in his treatise *The Sand Reckoner* he mentions that his father, Phidias, was an **astronomer**. As a young man he might have studied in Alexandria in Egypt, home of the ancient world's great library, because he later corresponded with several scholars who lived there. He was on good terms with King Hieron II of Syracuse and his son Gelon, and one historian claimed they were related. This is about the sum of our knowledge of Archimedes the man, but although a biography of him by a contemporary is lost, he gained such a powerful reputation in his own lifetime that other scientists and historians relayed stories about him.

Ironically, his reputation grew not because of his original advanced mathematics – which he considered to be the only worthy study and which he described clearly and eloquently – but because of the war-machines and other mechanical devices he invented. Although he helped keep the invading Romans at bay

Essential science

Archimedes' principle

Archimedes' most famous theorem, which he supposedly discovered while in the bath, works out the weight or volume of a body immersed in a liquid. According to tradition, King Hieron suspected that the goldsmith who had made him a new crown was dishonest and had adulterated the gold with silver. He asked Archimedes to investigate without destroying the crown, and after puzzling over the problem for some time, the answer came to Archimedes while he was in the public baths. (His servants had to regularly drag him away from his work to make sure he washed.) Archimedes was so excited by his insight that he jumped out of the baths and, forgetting to get dressed, ran home naked through the streets shouting *"Eureka!"* (*"I have found it!"*)

If he had been able to melt the crown down into a cube he could have easily measured its density, which would have been lower than expected if cheap and less dense metals than gold had been used. But instead he realized when he got in the bath that he had displaced a certain amount of water, and since water cannot be compressed, the volume that he displaced would be the same volume as the body put in it. So he immersed the crown in water, measured the amount of water it displaced, divided that by the weight of the crown, and arrived at the crown's density.

In his treatise *On Floating Bodies* Archimedes does not relate the story of the crown, but it is there that he states his principle that a body immersed in a fluid experiences a buoyant force equal to the weight of the displaced fluid, thereby giving rise to the science of **hydrostatics**, or the study of the mechanical properties of liquids at rest.

Method of exhaustion and pi (π)

Archimedes used a technique called the **method of exhaustion** to calculate the areas, volumes, and other properties of circles. This involves drawing a straight-sided polygon around the outside of the circle and another inside the circle, then adding sides to the polygons until they approximated the area of the circle. Since the area and other properties can be more easily worked out for a polygon than for a circle, this method allowed him to find the properties of all circular bodies, as well as to discover a near approximation of the value of pi, which he estimated as between 3.1429 and 3.1408. Considered to be an early form of **integral calculus**, this method also helped him prove that the area of a circle is pi multiplied by the square of the circle's radius.

Spheres and cylinders

Archimedes considered that his discoveries concerning the relationships between a sphere and a cylinder circumscribing the sphere to be his most important – and most beautiful – achievements.

Archimedes' screw as pictured in a fourteenth-century woodcut

c.287 BCE	Born in Syracuse, Sicily.
c.260 BCE	By now is writing his treatises on mathematics and science.
213 BCE	Archimedes' war machines beat off a Roman invasion. The Roman siege and blockade begins.
212 BCE	Killed in the Roman conquest of Syracuse.

Legacy, truth, consequence

for a while, eventually they did overrun Syracuse, and Archimedes was killed in the conquest despite orders from the Roman general Marcellus that the great scientist was to be left unharmed. Centuries later, the Greek historian Plutarch (46–120 CE) reported different versions of the murder. In one story, Archimedes was carrying scientific instruments to Marcellus' headquarters when Roman soldiers, thinking he had valuables, killed him for the loot. In another version, Archimedes had drawn a geometrical diagram in the sand, and was so involved in his puzzle that he ignored a Roman soldier's orders to come to Marcellus, crying out "*Do not step on my circles!*" Impatiently, the soldier simply stabbed the old man.

- Originally more famous for his machines than for his mathematics, his theorems and equations were forgotten for centuries until they were rediscovered in the sixth century CE and recognized for great, original works.
- He is held to be the "father" of disciplines as varied as hydrostatics and integral calculus.
- Modern experiments have shown that the claw could be a functioning defense, but a mirror weapon would work only in very narrow weather conditions.

He proved that the sphere has two-thirds of the volume and surface area of the cylinder, including its bases.

Law of the lever
Although he did not invent levers, Archimedes supplied the first rigorous explanation for how they work in his two-volume *On the Equilibrium of Planes*, stating "*Equal weights at equal distances are in equilibrium, and equal weights at unequal distances are not in equilibrium but incline towards the weight which is at the greater distance.*"

Other mathematics
Among the many topics he explored were the values of **square roots**; calculating the arc of a **parabola** or the areas of sections of geometrical shapes; and the properties of "Archimedean solids" (symmetric, semi-regular **polyhedron** forms).

Archimedes' screw
Archimedes may have invented the water pump known as the "Archimedes' screw", which consists of a long screw enclosed in a cylinder, tilted so that its bottom is placed in a water source. When the handle is turned the screw draws water up the cylinder. Archimedes supposedly built it to empty bilge water from the huge ship he designed, the *Syracusia*. Archimedes' screws are still widely used in countries such as Egypt for irrigation.

... certain things first became clear to me by a mechanical method, although they had to be proved by geometry afterwards because their investigation ... did not furnish an actual proof. But it is of course easier, when we have previously acquired, by the method, some knowledge of the questions, to supply the proof than it is to find it without any previous knowledge.

The Method (c.250 BCE)

Other machines
Many of his machines were built to help defend Syracuse from the Roman assault, but Archimedes also developed them as a pleasing scientific exercise. He designed a compound pulley, giant catapults, and the "claw of Archimedes" or the "ship shaker". This was a crane with a large metal grappling hook or claw that could either smash on to invading ships in an attempt to sink them, or could snatch them and lift them out of the water. He is also supposed to have created a mirror weapon that would focus sunlight to burn up a wooden ship.

Hipparchus

Often referred to as one the greatest astronomical observers of all time, Hipparchus is credited with many fundamental advances in astronomy and with important contributions to the foundation of trigonometry.

Born in Nicaea (now Iznik, Turkey), Hipparchus conducted much of his work in Rhodes and probably some in Alexandria. Little is known about his life and only one of his works (a minor one) survives: the "Commentary on Aratus and Eudoxus".

In the second century CE, **Ptolemy** viewed Hipparchus as his most important predecessor and often quoted him in his work. His depiction on coins and on astronomical works for several centuries after his death shows that Hipparchus was highly influential.

> *... conditions ... forced [Hipparchus] ... to conjecture rather than to predict, since he had found very few observations of fixed stars before his own time ...*
>
> Ptolemy, discussing Hipparchus in the *Almagest* (second century CE)

Essential science

Astronomy

The best-known discovery that Hipparchus made was the **precession of the equinoxes**. This refers to the rotational axis of the earth moving in a cone-shaped path (imagine the slow wobble of a spinning top, with its axis tracing a circular path), which causes the angle of the earth's axis to vary relative to the sun and the stars. One such circuit takes about 26,000 years, and this was calculated very accurately by Hipparchus.

Through comparison of his own careful measurements with previous records (for example by Timocharis of Alexandria about 150 years earlier), Hipparchus determined that the **equinoxes** were occurring slightly later than expected if calculated according to the positions of the stars. This difference is explained by the two definitions that exist for a year: the time it takes for the sun to return to the same position relative to the stars (**sidereal year**), or the length of time between equinoxes, i.e. repetition of seasons (**tropical year**). The tropical year is slightly longer than the sidereal year so the equinoxes gradually occur later and later as time passes until, after about 26,000 years, the equinoxes return to their original times of year. Hipparchus used Babylonian data to calculate the lengths of the sidereal and tropical years with great accuracy.

Hipparchus produced one of the earliest catalogs of stars, based on his painstaking observations of their positions. From observations

Legacy, truth, consequence

■ Even Ptolemy's estimation of the annual precession of the equinoxes, about 300 years later, was not as accurate as the value calculated by Hipparchus. This is just one of many examples of how far ahead of his time Hipparchus was.

■ Hipparchus' star catalog recorded about 850 stellar positions with greater accuracy than ever before, and his observations were used by Ptolemy and even later by Edmond Halley (1656–1742).

■ Believed by some to have invented trigonometry, Hipparchus at least created an early version of a trigonometric table (a calculation aid used in trigonometry). He may thus have made the first step in turning Greek astronomy into a practical rather than a theoretical science.

Key dates

c.190 BCE	Born in Nicaea, Bithynia (now Iznik, Turkey).
147 BCE	First recorded observations are made by Hipparchus.
134 BCE	Observes a new star (stars were previously thought to be fixed in number).
c.129 BCE	Completes his star catalog.
127 BCE	The last year he is reported to have worked.
c.120 BCE	Dies, probably in Rhodes, Greece.

of a solar **eclipse** from two different locations (the moon obscured the entire sun at Syene but only four-fifths of it at Alexandria), Hipparchus worked out the approximate distance of the moon from the earth. The actual distance (on average 238,855 miles, or 384,400 kilometers) lies within the range that he calculated.

Trigonometry

Hipparchus is also the first person in the West documented to have used **trigonometry** systematically. He may have written one or more books on the subject and Indian trigonometric tables are thought to be based on his work.

Zhang Heng

A Chinese inventor, mathematician, astronomer, and poet, Zhang Heng is best known for building the world's first seismograph to record earthquakes. He also invented many other mechanical devices, improved the calendar, and made a more accurate calculation of the value of pi (π).

Zhang (Chang) Heng lived during the prosperous and relatively peaceful Eastern Han Dynasty in ancient China. He was brought up in the Confucian tradition that stressed ethics and scholarship, and as a young man he became well known for his poetry and literary works. He only took up the study of **astronomy** when he was 30, entering government service a few years later.

In Imperial China, government officials were also scholars who were trained in Confucian philosophy, and most scientists would seek a government position of some sort. Zhang was humble and self-effacing, and although he did eventually rise to become a minister, he never accepted the senior promotions that his work deserved.

> ## The steps of heaven [the risings, settings, and movements of constellations] follow unvarying rules.
>
> Attributed comment in Fan Ye's *History of the Later Han Dynasty* (c.450)

Legacy, truth, consequence

- Zhang's machine could not predict earthquakes, but it did allow the government to prepare disaster relief more efficiently.
- Doubts were cast on his machine when a ball was triggered although no one had felt an earthquake shock. However, a few days later news arrived of a devastating earthquake about 400 miles away, in the direction indicated by the dragon.
- In 2005 Chinese scientists reconstructed Zhang's seismograph and proved that it did successfully record earthquakes.
- Zhang's complex gears and improved use of water-power inspired future Chinese inventors and mechanical engineers.

Key dates

78	Born in Nanyang, China.
c.116	Becomes an official at the Emperor's court in Luoyang. Goes on to become chief astrologer.
132	Builds his seismograph.
139	Dies in Luoyang, China.

Essential science

The seismograph

"In the first year of the Yang-Chia reign-period (c.132) Zhang Heng also invented an earthquake weathercock …"
(*The History of the Later Han Dynasty*, c.450)

Zhang believed that earthquakes were caused by air that had become trapped and compressed so that when it broke free it would batter the earth. He called his machine an instrument for measuring the seasonal winds and the movements of the earth.

His original seismograph was lost, but a detailed record of it survived. It was a large bronze jar, about six feet (1.8 m) in diameter, with eight dragon heads outside the jar, each holding a bronze ball in its mouth. Around the base of the vessel were eight bronze open-mouthed frogs, one beneath each dragon, and when an earthquake vibration was felt, the dragon facing the shock would open its mouth. The ball would then fall into the mouth of the frog below it, alerting the observers. Inside the jar were complex mechanisms including a suspended pendulum, cranks, pivots, and sliding levers that would unhinge the dragon's jaw while locking all the other dragons in place. Like most Chinese instruments, the whole construction was elaborately decorated.

Mechanics

Zhang was an acknowledged expert on geared machines, and created an odometer or "recording carriage"; the first known rotating **armillary sphere** or celestial globe, powered by water; and an improved water clock, among other instruments.

Astronomy and mathematics

His other achievements, which tend to be overshadowed by the magnificent seismograph, include his proposal that pi is the square root of 10, or 3.162; an improved calendar; and a detailed star map showing 124 constellations and a total of 2,500 stars. His theory of the universe was less successful: he believed that the sky was like an egg, and the earth was like the egg yolk, lying in its center.

Ptolemy

The last of the great ancient Greek astronomers, Ptolemy created the first mathematical model of the universe that explained the movements of the sun and planets. Based on the theory that the earth is the center of the cosmos, his model provided Europe's view of astronomy for around 1,500 years.

We know nothing about Ptolemy's personal life apart from the fact that he lived in Egypt when it was a province of the Roman Empire. Although his first name, Claudius, is Roman, his proper surname, Ptolemaeus, suggests Greek heritage, and he used Greek for his writings. Modern astronomers have confirmed that he made his observations of the heavens from the town of Alexandria, whose magnificent library was a magnet for ancient scholars of all disciplines.

Until Ptolemy began writing, there had been a gap in Greek **astronomy** since the days of **Hipparchus** (c.190–c.120 BCE), and it is only thanks to Ptolemy that we know about Hipparchus' work, such as his systematic observations of celestial bodies which

Ptolemy combined with his own data. Ptolemy was a great synthesizer, and acknowledged using earlier theories in his explanation for how the universe moved.

Ptolemy's incredible influence is obvious by the fact that his **geocentric** model was followed for around 1,500 years, first in the Middle East and then in Western Europe. It aligned with religious belief, and scholars who dared dispute it faced a death sentence from the rigid and repressive Catholic Church. However, in the light of today's emphasis on scientific honesty, he is a controversial figure. By 1008, Arab astronomers were questioning his data and his ideas, and centuries later it was clear that at least some of his recorded observations were falsified to match his theories.

Essential science

The *Almagest*

Ptolemy's masterpiece was his thirteen-volume *Almagest*, named from the Arab translation (*al-Majisti* or *Great Work*) of his title *Mathematike Syntaxis* or *Mathematical Composition*. In this he laid out his understanding of the structure of the cosmos and his new mathematical systems that explained and predicted the movements of the planets. From earlier Greek natural philosophers/scientists, such as **Aristotle**, he had inherited the firm beliefs that the earth was a perfect sphere at the center of the universe, with the other known heavenly bodies orbiting around the earth, and that the movement of the sun and planets was at a uniform speed and in a perfect circle. Some of his own basic theories were that the moon was the closest heavenly body to the earth, followed by the planets, and then the stars, which were fixed points of light in a rotating sphere.

Despite these fundamental beliefs, in order to mathematically explain the movements of heavenly bodies Ptolemy had to violate his own rules by assuming that the earth is actually *not* the exact center of the planetary orbits. Pragmatically, he and his followers accepted this displacement, known as the "eccentric", as just a minor blip in the essential geocentric theory.

Ptolemy used a combination of three geometric constructs. The first, the eccentric, was not new, nor was his second construct, the **epicycle**. This proposes that planets do not actually move in

large circles, but instead move around small circles or epicycles which in turn revolve around the circumference of a larger sphere focused (eccentrically) on the earth. Progress along the epicycle explains why planets sometimes appear to move backwards, or "retrograde".

His third construction – the equant – was revolutionary, and Ptolemy invented it to explain why the planets sometimes seem to move faster or slower, rather than uniformly. He suggested that the epicycle's center of motion on the circumference of its larger circle is not aligned with either the earth or the actual eccentric center of the larger circle, but with a third point, the equant, which is situated opposite the earth and at the same distance from the true center of the sphere as is the earth. It is only from the point of view of the equant that the planet would appear to be in uniform motion.

These three mathematical constructions, epicycle, eccentric, and equant, were complex and unsatisfactory to purists, but they worked. They explained some puzzling aspects of astronomy, such as why planets sometimes appear to move backwards in the night sky and why they appear brighter, and therefore closer, at different times. Together they actually approximated the modern view of the universe, in which the planets orbit the sun in elliptical paths. So Ptolemy's work allowed for reasonably accurate predictions of planetary positions for many years.

Legacy, truth, consequence

■ The *Almagest* defined astronomy up till the sixteenth century, when **Copernicus** presented his model of a **heliocentric** universe. While Ptolemy's geocentric model is wrong, it was in keeping with religious belief, so it became a paradigm. And, although his framework was incorrect, his mathematical models actually worked, and could be used to gain sufficiently accurate predictions and explanations of the behavior of the sun, moon, and planets.

■ Many of our modern constellations derive from Ptolemy's classifications.

■ Ptolemy treated **astrology** as a serious science and generations of "natural philosophers" followed in his footsteps. His *Tetrabiblos* was the standard astrological textbook for centuries.

■ His *Geographia* had a huge impact when it was reintroduced to Europe centuries later, galvanizing interest in the rest of the world. Unfortunately, for the far-flung areas he relied on hearsay, and since he also underestimated the earth's circumference, parts of his maps were wildly wrong. In particular, he greatly exaggerated the width of the European-Asian continent, helping to convince the explorer Christopher Columbus (1451–1506) that a westwards route to Asia across the Atlantic would be shorter. On this quest he found America, instead of Asia.

c.83   Born, possibly in the town of Ptolemais in Egypt.
c.127–c.150   Makes astronomical observations from Alexandria.
c.161   Dies, probably in Alexandria.

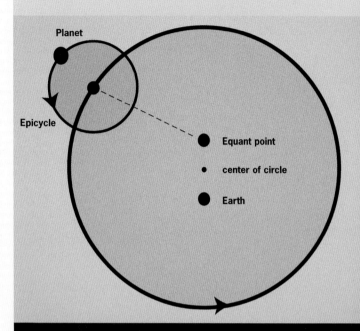

Ptolemy's geocentric model placed the earth not at the exact center of the planetary orbits.

[After the mathematical concepts] we have to go through the motions of the sun and of the moon, and the phenomena accompanying these motions; for it would be impossible to examine the theory of the stars thoroughly without first having a grasp of these matters.

The Almagest (second century CE)

The *Almagest* also contained a star catalog of more than 1,000 stars and 48 constellations, listing the stars' longitude, latitude, and magnitude (brightness). He included tables of observations of the sun, moon, and the five known lesser planets, Venus, Mercury, Mars, Jupiter, and Saturn, together with instructions for using the tables to predict celestial movements.

Geography

Ptolemy's other major work was his eight-volume *Geographia*, the oldest surviving atlas or collection of maps, in which he provided a compilation of geographical knowledge of the known world. It included an improved method of projecting maps, discussed **latitude** and **longitude**, and gave coordinates for some 8,000 places. From north to south, his maps covered from the Shetland Islands (northern Scotland) to anti-Meroe in the upper Nile valley, Africa, and from west to east he covered from the Cape Verde islands in the Atlantic Ocean to the middle of China.

Other publications

Ptolemy summed up his astronomical tables in a separate *Handy Tables*, which became the model for later quick reference tables. This supplied known positions of the sun, moon, and planets, and gave instructions on how to use these to' make predictions or for navigation. He also wrote the textbook on **astrology**, *Tetrabiblos* (*Four Books*), and a book on **optics**. In this he foreshadowed the modern scientific method by combining experiments, observations, and reason to correctly conclude that rays of light are emitted from objects, not from the eyes. He therefore disproved various optical theories of such giants as **Euclid** and Aristotle.

Galen

An ancient Greek physician and prolific writer, Galen carried out revolutionary dissections and experiments to reach his theories of anatomy and medicine. His overall view of the body was considered practically infallible in the Middle East and Europe until the seventeenth century.

Galen or Galenos of Pergamon was mistakenly referred to by Renaissance scholars as Claudius Galenus. In Arabic he is called Jalinos.

His home, Pergamon (now Bergama in Turkey), was an ancient Greek city, although in his day it was part of the Roman Empire. The son of a well-off architect, Galen had a good education and chose to specialize in medicine, a popular topic in Pergamon, where there was a famous temple to the god of healing, Asclepius.

After studying, Galen worked as a doctor in a gladiator school, and learnt a great deal about open wounds and physical trauma.

Ambitious and clever, Galen first moved to the empire's capital, Rome, in 162, working his way to becoming a physician to the emperors Marcus Aurelius, Commodius, and Septimius Severus. Through lectures, public demonstrations, and prolific writings (about 300 treatises), he became popular and successful in his own lifetime.

Legacy, truth, consequence

■ His anatomy, systematic methods, and theories became the standard in the Roman world.

■ In about 850 many of his books were translated into Arabic and influenced the development of Arab medicine. And by 1200 there were Latin translations of Arabic versions that became hugely influential in Europe for centuries.

■ Only in the later Renaissance were some of his ideas challenged and disproved.

Key dates

129	Born in Pergamon (now in modern Turkey).
148–9	Studies medicine in Alexandria, Corinth, and Smyrna.
162	Moves to Rome.
166	Returns to Pergamon after plague hits Rome.
169	Moves back to Rome.
191	Fire destroys many of his records.
c.216	Dies in Rome.

[securing the ureters with ligatures] shows the bladder empty and the ureters quite full and distended ... on removing the ligature ... one then plainly sees the bladder becoming filled with urine.

On the Natural Faculties (c.170)

Essential science

Dissection and anatomy

Galen thought that anatomy was the basis of medical knowledge, but Roman law banned the dissection of human corpses, so he had to work on animals. This occasionally led him into mistakes, for example, his description of the uterus is only relevant to dogs.

His many experiments included severing pigs' spinal cords and tying up the nerve of the larynx to show paralysis. He also tied off the ureters to examine the functions of the bladder and kidneys.

Galen pioneered scientific medical investigations. He also introduced the still-standard taking of the pulse, and his emphasis on bloodletting as a cure-all was continued until the 1800s.

Discoveries

Galen made several important discoveries – sometimes disproving earlier theories – including:
• urine is formed in the kidneys, not the bladder

• veins and arteries are different
• arteries carry blood, not air
• a description of heart valves

Operations

Never afraid of trying, Galen carried out some operations on the brain and eye, such as cataract removals using a long needle, that were not successfully repeated for nearly 2,000 years.

Bodily theory

Galen followed **Hippocrates** in thinking there were three connected bodily systems: brain and nerves; heart and arteries; liver and veins, each responsible for different features such as thought or growth. He also believed, after **Aristotle**, that the body was governed by four humors, or fluids, representing different qualities, and he drew up a list linking the humors to specific organs.

Zu Chongzhi

An important Chinese mathematician, astronomer, and engineer, Zu Chongzhi was the first person in the world to calculate the mathematical constant pi (π) to the seventh decimal place. He produced an advanced calendar calculating the exact length of a year to 365.24281481 days – only 50 seconds different from the modern estimate – and built amazing mechanical devices.

Zu Chongzhi (Tsu Ch'ung-chih) came from a family of distinguished government officials and **astronomers** who originally lived near modern Beijing in the north of China. Like many hundreds of thousands of others, they fled south to the Yangtze valley in the fourth century to escape raiders from outside China.

Zu followed the family tradition by entering Imperial service, working as a civil and military administrator. But his main interest was in calculating calendar dates. He created an improved calendar, but died before it was adopted.

Although his advanced mathematics would only have been acknowledged by fellow scholars, he became famous for his mechanical creations, a south-pointing chariot and a "self-powered" boat, thought to be a paddle-wheel.

> **[my calendar is] ... not from spirits or from ghosts, but from careful observations and accurate mathematical calculations ...**
>
> Quoted in *History of the Southern Qi Dynasty* (c.520)

Legacy, truth, consequence

■ Ten years after Zu's death, his son convinced the emperor to adopt his calendar, which was named the Da Ming (Taming) or Great Brightness calendar.

■ In 1084 Zu's mathematical book *Method of Interpolation* became one of the classical texts that were part of the examinations all Imperial scholar-officials had to pass. However, it was eventually dropped from the syllabus because it was considered to be too difficult for most students.

Key dates

429	Born in Jiankang, China.
436–59	Correctly predicts four eclipses.
462	Completes his improved calendar.
478	Reinvents the south-pointing chariot.
500	Dies in China.

Essential science

Calendar

The ancient Chinese lunar calendar was a crucial prop for the Imperial government since it determined the official dates of important festivals. At the time, the calendar was based on a 600-year cycle with an extra month inserted every 221 years. Zu was the first Chinese calendar-maker to take into account the **precession of the equinoxes**, which makes the tropical year (time between spring equinoxes) about 21 minutes shorter than the **sidereal year** (time for the sun to return to the same place against the background stars). He realized that the calendar could be improved if it was based on a cycle of 391 years, adding an extra month in the 144th year.

Zu's calculations were astonishingly precise: as well as calculating the tropical year, he also identified the draconic or nodal month (time for the moon to return to the same orbital node or point where it crosses the earth's orbit) as 27.21223 days (practically the same as today's figure of 27.21222 days). Knowledge of this helped him to predict lunar **eclipses**. He also estimated Jupiter's year or orbital period as equivalent in time to 11.858 earth years, close to the modern value of 11.862.

Mathematics

With his son Gengzhi, Zu wrote a book on mathematics called *Zhui Shu* (*Method of Interpolation*), containing formulas for the volume of a sphere, cubic equations, and a value of pi between 3.1415926 and 3.1415927, a more accurate approximation than Europeans would achieve for another 1,000 years.

South-pointing chariot

A mechanical compass that always points to its original direction, this device was, in Zu's time, known only from legend. Zu built a version in bronze using a system of complex differential gears. His machine had a figure of an official with one arm pointing south, mounted on a swiveling base fixed to a chariot. No matter how the carriage moved, the arm always pointed south.

Aryabhata

Considered to be the father of the Hindu–Arabic number system that is used today, Aryabhata's one surviving book promoted the ideas of zero and of numerical place-values, along with his own original concepts such as trigonometry, to Arab scientists and from there to Europe.

Aryabhata was born in the region of ancient India known as Ashmaka, and all we know about him is that he studied at Kusumapura, a famous center of learning (probably modern Patna). Sometimes confused with a later mathematician, he is also referred to as Aryabhata I or the Elder.

Aryabhata was in the vanguard of the revival of science after centuries of decline. Most of his texts are lost and are only known from later commentators. Obviously a youthful genius, he says in his great work, the *Aryabhatiya*, that he was 23 when he wrote it in 499.

Comprising 118 poetic verses, this book summarized current Indian **astronomical** and mathematical rules, and presented his own original work. Sometimes cryptic, the *Aryabhatiya* was probably meant to be explained by a teacher, not studied on one's own.

Add four to 100, multiply by eight and then add 62,000. By this rule the circumference of a circle of diameter 20,000 can be approached.

Aryabhatiya (499)

Legacy, truth, consequence

■ The *Aryabhatiya* influenced later Indian mathematicians as well as Arab scholars, being one of the works which introduced zero and place-values to the Middle East. From there, his ideas filtered through to Europe, helping bring about today's Hindu-Arabic numerals.

■ The words "sine" and "cosine" derive from mistranslations of Aryabhata's terms "jya" and "kojya".

Key dates

476	Born in the Ashmaka region of India.
499	Only 23, writes his masterpiece, later called by other mathematicians the *Aryabhatiya*.
550	Dies in India.

Essential science

Mathematics

Aryabhata used the concept of zero and the **place-value** system. He was also one of the world's earliest mathematicians to give a close value of pi (π) as 3.1416, and was the first person to acknowledge that this was only an approximation. He also knew that it is an irrational number – one that cannot be written as a simple fraction because the decimal goes on infinitely.

He was the first mathematician to describe the basics of **trigonometry**, later called the tables of sines, and the *Aryabhatiya* is the oldest surviving work to give a method for finding **square roots**.

Aryabhata presented an unusual method of expressing extremely large numbers, using letters of the alphabet as mnemonics instead of using numerals. And in **algebra**, he introduced a method of breaking factors down into smaller numbers by means of an **algorithm**.

Astronomy

In Aryabhata's time **cosmology** was often explained through mythology, suggesting that gods and demi-gods moved celestial bodies or swallowed the sun to create **eclipses**. He, however, proposed only scientific theories, helping move Indian astronomy to a more rational basis.

Among his correct statements are that the earth rotates on its axis and that the moon reflects light rather than emitting it. He also provided a scientific explanation of eclipses – that they are due to shadows cast by the earth – and he supplied an accurate way of calculating these events.

Aryabhata calculated the earth's circumference as 24,835 miles (39,968 km), a very close approximation to the modern value of 24,902 miles (40,076 km). He also worked out that the **sidereal rotation** period, or time for the earth to rotate relative to the fixed stars, was 23 hours, 56 minutes, and 4.1 seconds – only a fraction of a second out from today's value of 23 hours, 56 minutes, and 4.091 seconds. His calculation for the length of the **sidereal year** was also remarkable: only slightly more than three minutes too long. His were probably the most accurate values in the world at the time.

Al-Battani

One of the greatest of the early Arab astronomers, al-Battani was also a mathematician. He made many significant astronomical discoveries, including the amazingly accurate determination that our solar year is 365 days, 5 hours, 46 minutes, and 24 seconds long. He also improved on some of Ptolemy's ideas and drew up an influential astronomical textbook and set of tables.

Abu Abdallah Mohammad ibn Jabir ibn Sinan al-Raqqi al-Harrani al-Sabi al-Battani was Muslim, as indicated by his first names, but the "al-Sabi" part of his surname suggests that his ancestors may have been members of the star-worshipping Sabian sect. The Sabians produced many great astronomers and mathematicians, and al-Battani's father is reported to have been a well-known instrument maker and astronomer. Al-Battani lived at a time when the Muslim empires encouraged learning and kept alive the science and philosophy of ancient Greece and Rome. At the crossroads between East and West, Muslim scholars also received ideas from the Asian civilizations of China and India, and incorporated them with their own discoveries into a body of knowledge that was later passed on to Europe. Al-Battani was probably born in 850, and received his early education from his father, but he moved to the town of al-Raqqah, on the Euphrates. It was there that he made his astronomical observations, using some massive instruments.

> ... a more accurate description [of the sun and moon's] motions than that given in Ptolemy's Almagest.
>
> A description of al-Battani's work in Ibn al-Nadim's *Fihrist* (*Index*, 988)

Legacy, truth, consequence

- Al-Battani was one of the early medieval astronomers who established Arabic as the principal language of astronomy for centuries. The few Christian and Jewish scholars in the ninth to eleventh centuries followed this standard.
- His *Zij* or tables were translated into Latin in 1116 and widely used. For example, **Nicolaus Copernicus** mentioned al-Battani in his groundbreaking work on **heliocentrism**.

Key dates

c.850	Born in Battan, Harran, Mesopotamia (now Turkey).
877	Begins to makes astronomical observations from the town of al-Raqqah on the Euphrates river, and later from Damascus, Syria.
918	Ends his period of observations.
929	Dies in either Damascus or Samarra (Iraq).

Essential science

Astronomy

Based on his observations of the sky, al-Battani produced a set of elaborate astronomical tables recording positions of the sun, moon, and planets, and explaining how these can be used to anticipate future positions. These tables, the *al-Zij al-Sabi*, or *Sabian Tables*, were more accurate than any others available at the time.

Accompanying this he wrote an astronomical treatise, *Movements of the Stars*, covering the timing of new moons, predictions of eclipses, and calculations of the length of the **solar** and **sidereal years**.

Al-Battani was able to calculate the solar or **tropical year** as 365 days, 5 hours, 46 minutes, and 24 seconds, only a couple of minutes off today's reckoning of 365 days, 5 hours, 48 minutes, 45 seconds. He also gave a close calculation of the **precession of the equinoxes** as 54.5 arcseconds per year (equivalent to 1 degree in 66 years).

Correcting Ptolemy

Al-Battani made a discovery that had eluded **Ptolemy**, that the farthest distance of the sun from the earth varies. As a result he correctly predicted **annular eclipses**, in which the moon does not quite cover the whole face of the sun. He also improved Ptolemy's models of the orbits of the sun and the moon.

Mathematics

Al-Battani was the first writer to use **trigonometry** for astronomical calculations. This replaced the ancient Greeks' "chords", which involved dividing a circle into smaller, manageable parts.

He may have independently developed some trigonometic notions, such as sines. More importantly, al-Battani took care to explain his mathematical steps in the hope that others would be able to build on his work and expand his ideas.

Rhazes

An Iranian Muslim philosopher and doctor, Rhazes is considered to be the father of pediatrics for writing the first book on childhood illnesses as a separate field of medicine. He wrote several other influential medical texts, including the first distinction between smallpox and measles, and was one of the Islamic doctors who helped forge modern medicine by beginning to identify the organic causes of diseases.

Rhazes' Persian and Arab contemporaries would have known at once that he came from Rayy (Rai), an ancient town near modern Tehran in Iran, for the ar-Razi at the end of his full name, Abu Bakr Muhammad ibn Zakariya ar-Razi, simply means "from Rayy".

The few records of his early life suggest Rhazes started out as either a jeweler or a money-changer. He was also a musician and an **alchemist**, and he only became interested in medicine when an alchemical experiment blew up in his face and damaged his eyesight. He was at least 30 when he began medical and philosophical studies in Baghdad, then a center of Islamic science, but he soon became a famous doctor, writing more than 100 medical texts as well as at least that many on general science and philosophy. His continuing interest in alchemy, which in those days was considered to be just another natural science, contributed to his skill as a doctor, since it was from that practice that he first learnt the **empiricism** he brought to medicine.

He was a popular teacher, drawing crowds to his lectures, and filling his hospitals with students. Patients – or enquirers with a scientific question – were first seen by the newer students, or first circle, and then were passed on to the next circle if necessary, until, if no one else could help, they reached Rhazes himself.

Rhazes lived a simple life, and although he earned a good living he also treated the poor for free, and gave so much to charity that in the end he died poor himself. According to legend he suffered from an eye cataract towards the end of his life, but he refused to have any treatment for it, saying that he had seen so much of the world he was tired of it.

Essential science

Rhazes had a very modern steady, methodical way of researching cures and a clinical approach to studying illnesses. He considered himself to be the Islamic equivalent of **Hippocrates**, the great ancient Greek physician, and he suggested that medicine should be seen as a philosophical discipline, since it required independent thinking.

The Comprehensive Book
Among Rhazes' most significant books were two medical encyclopedias. The largest, the *Kitab al-Hawi* (*The Virtuous Life*), which was translated into Latin after his death, in 1279, as *The Comprehensive Book*, was a nine-volume compilation of his notebooks. It included knowledge from Greek, Syrian, and Arabic sources, some Indian ideas, and Rhazes' own many discoveries and commentaries.

Treatise on the Small Pox and Measles
Razes was the first known doctor to discover that smallpox and measles are different diseases, and to identify the differences between them. In his treatise on them he wrote:
"*The eruption of smallpox is preceded by a continued fever, pain in the back, itching in the nose and nightmares during sleep … A swelling of the face appears, which comes and goes, and one notices an overall inflammatory color noticeable as a strong redness on both cheeks and around both eyes … There is a pain in the throat and chest and one finds it difficult to breath and cough. Additional symptoms are: dryness of breath, thick spittle, hoarseness of the voice, pain and heaviness of the head, restlessness, nausea and anxiety. (Note the difference: restlessness, nausea and anxiety occur more frequently with 'measles' than with smallpox. On the other hand, pain in the back is more apparent with smallpox than with measles.)*"

Other works
Among his many "firsts" were his study of childhood illnesses, *The Diseases of Children*, and papers on allergies and hay fever (his splendidly titled *Article on the Reason why Abou Zayd Balkhi Suffers from Rhinitis when Smelling Roses in Spring*). He had the insight that some fevers are the body's defense mechanism to fight infection, recorded the first known use of animal gut for sutures, and was the first to use Plaster of Paris for casts. He was also one of the first practicing doctors to discuss **medical ethics** and the reasons why people choose to put their trust in a particular doctor.

In addition to his writings, Rhazes used several medical instruments that later became standard, and in the course of his alchemical work he discovered the medicinal value of ethanol. He did not hesitate to challenge wrong ideas; in his paper titled

A 1667 copy of the chapter on anatomy from Rhazes' *Book on Medicine*.

Legacy, truth, consequence

- A freethinker who was convinced of the value of experiment, Rhazes tried to show that the authority of the ancients could be challenged. He encouraged research in several fields – science and technology as well as medicine.
- From the twelfth century onwards, his books were translated from Arabic to Latin, particularly by Gerard of Cremona (c.1114–87). Arabs, Jews, and Christians considered him to be one of the greatest medical authorities ever known, and *The Comprehensive Book* became a standard text for Islamic and European medical students for centuries.
- His rational philosophy and outright attacks on religion did not win much support in Islamic countries, and many of his philosophical works are now known only through fragments.
- Rhazes' work contributed to the slow understanding that diseases have organic causes and are not due to magic, fate, or supernatural powers.

The doctor's aim is to do good, even to our enemies, so much more to our friends, and my profession forbids us to do harm to our kindred, as it is instituted for the benefit and welfare of the human race, and God imposed on physicians the oath not to compose dangerous remedies.

Quoted in FSTC Ltd article "Islamic Science, the Scholar and Ethics" (2006)

"Doubts about **Galen**" he showed how his clinical experiences sometimes clashed with the statements of the great Greek authority, and he wrote several attacks on medical charlatans or quacks. One of his particularly popular works was a sort of first-aid book, a general medical manual for use at home or while traveling, which he dedicated to the poor or to anyone who was not able to find a doctor.

Humors

Although Rhazes discovered many cures through observation and experiment, he lived at a time in history when the explanations for disease were still uncertain. According to one story, he was called to treat an emir who was so crippled by arthritis that he could not walk. Rhazes ordered the man's best horse to be brought to the door, before treating the patient with hot showers and a potion. Then he pulled out a knife, swore at the man, and threatened to kill him. The emir jumped to his feet and charged at the doctor, who fled for his life to the waiting horse. When he was sure he was safe he wrote to the emir, explaining that his treatment had softened the humors (elemental liquids thought to govern the body) and he had left it to the patient's own temper to finish dissolving them.

Key dates

860	Born in Rayy, Iran.
c.890	Studies medicine in Baghdad before returning to Rayy to run a hospital.
c.901	Moves to Baghdad and directs a large hospital there. Also becomes court physician.
902–8	Writes his general textbook of medicine while in Baghdad.
c.907	Returns to Rayy and teaches.
c.925	Dies in Rayy.

Ibn al-Haytham (Alhazen)

Often known in the West by his Latinized name Alhazen or Alhacen, the Arab scholar and Muslim polymath Abu Ali al-Hasan ibn al-Haytham is considered to be the father of optics. He was also an early pioneer of the scientific method, and he made many other important contributions in fields as varied as mathematics, physics, anatomy, and the philosophy of science.

Ibn al-Haytham was sometimes called al-Basri, meaning from the city of Basra in modern Iraq, or al-Misri, meaning from Egypt, where he worked. He trained originally as a public administrator, but also studied religion. At the time, the Islamic world was involved in religious civil wars centering on the true heir of the Prophet Muhammad, and Ibn al-Haytham concluded that none of the current religious groups offered the truth. Inspired by the writings of **Aristotle**, he turned instead to the study of science and mathematics, although he remained a devout Muslim all his life.

In 996 the cruel and ruthless al-Hakim became caliph (leader of Islam). Based in Egypt, he was a patron of the sciences and founded a great library in Cairo, drawing scholars such as Ibn al-Haytham from all over the Islamic world.

He was hired by the caliph to regulate the flow of the Nile by building large-scale engineering works such as a network of dams,

but as he journeyed up the Nile on an exploratory survey he realized the plan would not work. At first al-Hakim accepted the news calmly, but the caliph was known to be mercurial and violent, and to escape certain punishment Ibn al-Haytham pretended to go mad.

As a result he was confined to his house, where he busied himself with scientific studies, including writing his masterpiece, *Kitab al Manazir* (*The Book of Optics*). Upon al-Hakim's death in 1021, Ibn al-Haytham was able to prove he was only pretending to be mad, and left his "prison". There are conflicting accounts of this misadventure: one report says instead that he fled to Syria to avoid al-Hakim's anger.

The rest of his life was devoted to science and mathematics. He supported himself by teaching and copying texts, and wrote more than 200 works, most of which are now lost. He was probably associated with the University of al-Azhar at the Azhar Mosque in Cairo.

Essential science

Optics

Apart from his classic *Book of Optics*, Ibn al-Haytham wrote several other papers on **optical** subjects. Using experiments, tests, and observations on the properties of light and the eye, he presented the first coherent, modern theory of light and vision, particularly the basic idea that rays of light are reflected from objects to the eye. Until his work, most scholars followed the theory of emission, put forward by the Greek scientists **Euclid** and **Ptolemy**. This proposes that we see things because the eye emits rays of light that sense an object and bounce back to the eye. The other main alternative theory was that of **Aristotle**, that physical forms enter the eye from an object.

Ibn al-Haytham struck down these theories through observations of everyday events, for example that the eye is dazzled by looking at a bright light, so the rays cannot be coming from the eye itself, and through logic, arguing that we see distant objects as soon as we open our eyes, whereas if it was a physical form of that object entering the eye we would have to wait for it to arrive.

His many explorations of light included: experiments with lenses, prisms, and mirrors to discover that light travels in straight lines; a discussion that all light is the same, regardless of its source,

so sunlight, light reflected from a mirror, and firelight all have the same properties; studies of the geometrical relationships between angles of reflection or **refraction** and the important discovery that the ratio between the **angle of incidence** and refraction does not remain constant; the first full and correct analysis of a camera obscura or pinhole camera, which he used to observe colors in light; and the discovery that the curve of a lens allows it to focus light. He was the first scientist to explore the separate vertical and horizontal components of reflected and refracted light.

All the experiments described in *The Book of Optics* used just a few simple household objects such as a screen, a lamp, or even a blank wall, and are mainly designed to be carried out by one person. Ibn al-Haytham presumably conducted them while locked up alone in his house.

Ibn al-Haytham also studied refraction in the atmosphere; from his observation that twilight occurs when the sun is 19 degrees below the horizon, he calculated the height of the atmosphere, concluding that it stretched for 9.3 miles (15 km). Today it is estimated that three-quarter of the atmosphere's mass is within 7 miles (11 km) of the earth. Beyond this, the atmosphere becomes thinner until it fades into space at around 62 miles (100 km) from the earth.

> ***I constantly sought knowledge and truth, and it became my belief that for gaining access to the effulgence and closeness to God, there is no better way than that of searching for truth and knowledge***
>
> Ibn al-Haytham, "The Winding Motion" (eleventh century)

Frontispiece from a 1572 Latin edition of *The Book of Optics*.

Legacy, truth, consequence

■ *The Book of Optics* was translated into Latin as *Opticae Theosaurus* and became available to European scientists. It completely changed the way optics and sight were perceived, and was the most important work on optics for centuries. His discoveries paved the way for the development of optical aids such as the telescope and the microscope.

■ Ibn al-Haytham also made a major impact with his general approach to science. His works introduced the fundamental elements of the modern scientific method to European scholars.

■ Apart from the field of optics his work on **number theory**, algebra, and geometry also contributed to later scientific developments.

Key dates

c.965	Born, possibly in Basra, modern Iraq.
c.996	Goes to Cairo, a major center of learning.
c.1011	Pretends to be mad to avoid punishment by Caliph al-Hakim. Confined to house arrest for ten years. Works on his theories of light.
1021	On the death of al-Hakim, leaves confinement and publishes *The Book of Optics*.
1027	Writes an autobiography which ignores his personal life and concentrates on the development of his mathematics and science.
c.1040	Dies, possibly in Cairo, Egypt

Psychology of vision

Ibn al-Haytham took his interest in light a stage further and wrote a pioneering study of the psychology of sight. He was the first person to believe that vision perception actually occurs in the brain, and that the eyes only register the sense of sight. He studied optical illusions, examined the conditions needed for good vision, and argued that perception can be subjective: he pointed out that when people report what they have seen they can often be affected by mood, illness, or other aspects of their circumstances. However, his analysis of the structure of the eye is flawed because this otherwise expert on lenses overlooked the function of the eye's lens.

Scientific method

In reaching his theory of vision Ibn al-Haytham pioneered the use of the modern scientific method, combining experimental work, observation, and logical arguments. In his paper "The Winding Motion" he argues that scientists should test their theories, and that blind faith should be reserved for true religious teachers only:

> *"That is how experts in the prophetic tradition have faith in Prophets, may the blessing of God be upon them. But it is not the way that mathematicians have faith in specialists in the demonstrative sciences."*

Very much like the modern method, he organized his investigations in several stages:

• Stating the problem
• Formulating a hypothesis
• Testing the hypothesis through verifiable experiments
• Interpreting the data and formulating a conclusion
• Publishing his findings

Mathematics

Ibn al-Haytham produced important papers linking algebra and geometry, and presented solutions to several mathematical problems. A question he himself posed became known as Alhazen's problem:

> *"Given a light source and a spherical mirror, at what point on the mirror will the light be reflected to the eye of an observer?"*

His solution used sophisticated geometry and covered not only spherical mirrors but also cylindrical and conical ones.

Other works

Ibn al-Haytham was an all-round scientist. He also wrote on anatomy, astronomy, engineering, mathematics, medicine, philosophy, physics, and psychology.

Avicenna (Ibn Sina)

An Iranian/Persian polymath, Avicenna was probably the most famous of all the great philosopher–scientists of the medieval Islamic world. As both a doctor and as a philosopher he became well known in his own lifetime, and he had a major impact after his death on the development of philosophy and the teaching of medicine in Europe and in the Middle East.

A thoroughly precocious child, Avicenna, the Latinized name for Abu Ali al-Husayn ibn Abd Allah ibn Sina, had memorized the Qur'an and other classic Islamic texts by the time he was ten. Soon after that he surpassed his teachers, and began his own course of study in Islamic law, philosophy (logic and metaphysics), and medicine.

His father, a governor of a royal estate, was himself a scholar, and had many learned friends, so the young Avicenna benefited from intellectual discussion and encouragement. He sought wisdom anywhere he could find it, learning arithmetic from an Indian greengrocer and some science from a wandering scholar. He was still young when he successfully treated the Sultan of Bukhara, Nuh ibn Mansur, and was partially rewarded with access to the many manuscripts in the royal library, a new source of knowledge that helped him finish his studies by the time he was 21.

Avicenna lived in a time of turmoil. Not only were Turkish tribes displacing Iranian rulers in central Asia, but in Iran local leaders were throwing off the central control of the Abbasid caliphate based in Baghdad. His life was profoundly affected by these political uncertainties.

In 999 the Bukharan ruling family was overthrown by Turkish invaders led by Mahmud of Ghazna, and rather than stay and either be killed or forced to work for them, Avicenna embarked on a long period of wandering around Iran. He had several adventures: escaping a kidnap attempt and going into hiding; arrest and imprisonment; breakneck flight in disguise. During all this he was writing his scientific and philosophical papers and, whenever he settled in one place long enough, practicing as a doctor.

In about 1024 he finally found refuge as doctor and advisor to the ruler of Isfahan, Ala al-Dawla, remaining in his service until he died.

Essential science

In Avicenna's own words *"Medicine is no hard and thorny science, like mathematics and metaphysics, so I soon made great progress; I became an excellent doctor and began to treat patients, using approved remedies."* From the beginning of his medical studies as a youth, Avicenna both read books and attended at patients' bedsides. Throughout his career he stressed the need for **empirical** medicine: examining, testing, not taking anyone's theory for granted without proof.

As a young medical prodigy Avicenna became famous, even more so when he cured the Sultan of Bukhara. Although he had many demands made on him, he was also known for his compassion and for treating the poor without expecting any payment.

The Canon of Medicine

About 40 of Avicenna's known works are on medicine, but he also wrote on subjects as varied as physics, engineering, and mathematics, as well as expressing some of his ideas in poetry. His most famous works were *The Book of Healing*, a huge encyclopedia of science and philosophy, and the 14-volume *Qanun fi-l-Tibb* or *The Canon of Medicine*, which quickly became the standard medical

textbook in the Middle East and Europe. While this mostly presented his own discoveries and experiences, he also included knowledge that he found valuable from the ancient Greeks (**Hippocrates** and **Galen**), and from Persia, Mesopotamia, and even India.

Avicenna took medicine an important step forward. Among his many discoveries were:
- the contagious nature of some infectious diseases
- the impact of environment and diet on health
- spread of diseases in water or soil
- **neuropsychiatry**

And in the *Canon* he discussed many innovations such as:
- random, controlled clinical trials of medicines
- quarantine to contain the spread of infections
- speculation on the existence of microorganisms

His detailed descriptions of medical matters included sexual diseases (and sexual perversions), skin diseases, the anatomy of the human eye, facial paralysis, and diabetes. *The Canon of Medicine* was organized logically into separate parts, covering physiology, hygiene, illnesses, treatments, and medicines. And, perhaps because of his philosophical training, Avicenna was particularly interested

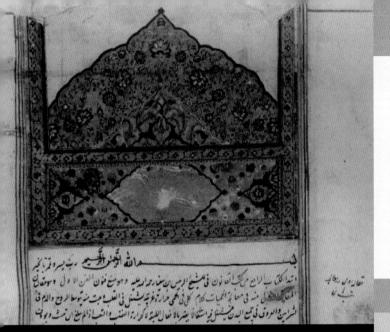

Beginning of the fourth book of *The Canon of Medicine*, from a copy made in Iran (fifteenth century).

- Avicenna's important medical work, *The Canon of Medicine*, became the most widely used textbook in medical schools in both the Islamic and the Christian worlds. It was appreciated for its logical approach to science and encyclopedic coverage of medical matters and it was not surpassed for centuries.

- He was also well known as a philosopher whose writings had a significant effect on European scholastic thinkers, especially Thomas Aquinas (c.1225–74).

- Within Iran, Avicenna is a national icon, thought to be one of the greatest Iranians ever.

Key dates

c.980	Born near Bukhara – then in Iran (Persia), now in Uzbekistan.
c.996	Begins to study medicine while still a boy.
997	Successfully treats the Sultan of Bukhara.
c.999	Leaves Bukhara after Ghaznid invasion; wanders around Iran and central Asia, staying in Jurjan, Rayy, Qazvin, and Hamadan.
c.1001	Only 21, writes his major medical work, *The Canon of Medicine*.
c.1024	Settles in Isfahan, Iran, as a doctor and advisor to local ruler Ala al-Dawla.
1037	Dies at Hamadan, northern Iran.

After such an eventful life Avicenna's friends suggested he should slow down and live quietly for a while, but Avicenna replied: "*I prefer a short life with width to a narrow one with length.*" One biographer recorded disapprovingly that Avicenna was the first philosopher who liked wine and women, but when he realized he was probably dying from colic, or intestinal problems, he freed his slaves, gave away his money to the poor, and asked for the Qur'an to be read to him.

in psychology and in exploring the effect that the mind can have on the body.

Avicenna was a very practical doctor; he is known to have carried out at least one operation, on a friend's gall bladder.

Drug tests
His rules for effective drug trials covered much the same ground as those used today, such as ensuring that there is no extra material in the drug that could interfere with the test; using it only on a "simple" illness, not a "composite" disease; gathering together a large enough test group that the results could not just be accidental; and, in a nod to modern sensibilities about animal testing, arguing that experiments must be carried out on humans, because "*testing a drug on a lion or a horse might not prove anything about its effect on man*".

Philosophy of science
Although Avicenna was effectively an Aristotelian, he criticized **Aristotle**'s method of **induction** for reaching the basic hypotheses of a science, and instead replaced it with experiments and examination.

The knowledge of anything, since all things have causes, is not acquired or complete unless it is known by its causes. Therefore in medicine we ought to know the causes of sickness and health. And because health and sickness and their causes are sometimes manifest, and sometimes hidden and not to be comprehended except by the study of symptoms, we must also study the symptoms of health and disease.

On Medicine (c.1020)

Shen Kuo

A talented polymath, Shen Kuo was a Chinese government official as well as a mathematician, astronomer, geographer, geologist, and historian of science. He made careful notes of his own extensive inquiries and also created a vast encyclopedia of all current scientific and technical knowledge, leaving a wonderful legacy to historians of science. In this book he described many of China's inventions such as the magnetic compass and moveable type.

Shen Kuo (Kua) was the son of a minor provincial official during the Song dynasty. The family moved around as Shen's father accepted different posts, and even as a youth Shen made copious notes on the topography and wildlife of the places they visited.

Shen followed in his father's footsteps by entering government service, where his energy and administrative abilities were soon noted, and a place at court was found for him. He served in many roles: a diplomat, settling a border dispute; an administrator in charge of large-scale engineering works; the senior finance minister; university chancellor; director of **astronomy**.

He also became involved in political in-fighting, since he supported the reformist party of the Minister Wang Anshi (1021–86). Wang was unpopular with wealthy landowners who lost income because of his reforms, and when he fell from power Shen became politically

vulnerable. His enemies seized their chance in 1081 when Shen was commanding military forces against invaders. The Chinese forces were badly defeated in one battle, and although it was not Shen's fault, he was blamed for the disaster and ousted from office.

From then on Shen devoted himself to exploring the natural world and recording experiments. He was curious about everything: wildlife, technology, astronomy, **geology**, engineering, math, medicine, and recording scientific knowledge. In the 1070s Shen had bought a garden estate near modern-day Zhenjiang, Jiangsu province. He was convinced he had already seen its beauty in his dreams, so he named it "Dream Brook". It was there that he completed his encyclopedic record of science and technology commonly called the *Dream Pool Essays*, although he called it *Brush Talks from Dream Brook*. He must have felt lonely and isolated

Essential science

Apart from the 507 essays in *Dream Pool Essays*, Shen wrote several other books (most of which are lost), including two geographical atlases, notes on calendar calculations, a pharmaceutical treatise, a mathematical approach to music, art criticism, and on divination.

Moveable type

In the *Dream Pool Essays* Shen was the first person in the world to describe moveable printing type, which had recently been invented by the artisan Bi Sheng (990–1051), who used ceramic pieces. Shen pointed out that the system was fast and efficient for printing hundreds of copies, but incredibly tedious for just a handful of copies. With Shen's records to hand, Chinese inventors were able to increasingly refine the printing process.

Magnetic compass

Shen also wrote the first report on another important invention, the magnetic needle compass, which later proved to be so important for navigation, although it is not clear whether Chinese sailors were actually using it for navigation at the time he was writing. Shen was the first known person to discover that needles point to the magnetic north, not to true north and south: *"[The*

magnetic needles] are always displaced slightly east rather than pointing due south". He concluded this after:

1. first making improved astronomical instruments, such as a wider sighting tube for observing the stars;
2. spending five years with his colleague Wei Pu (*fl.* 1075) measuring the position of the pole star every night, as well as making other observations;
3. using these measurements to improve calculations for the **meridian**;
4. experimenting with suspended magnetic needles to determine their direction.

From studying the waxing and waning of the moon and comparing it to a ball covered with white powder viewed from the side in full light, at which point it appears like a crescent, Shen proposed that the moon, and other celestial bodies, were spherical in shape and that the moon does not produce its own light, but simply reflects it. This confirmed the ideas put forward earlier by **Zhang Heng** and others. Shen was one of the first astronomers to form a hypothesis of the retrograde motion of planets (see also **Nicolaus Copernicus**), and, unlike the ancient Greeks and contemporaneous European and Arab scientists, he did not believe

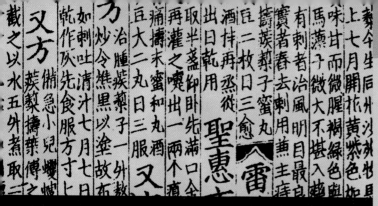

Detail of an early Chinese woodblock book, c.1250. Shen's records helped to refine the printing process.

I mixed sawdust with wheat flour paste to represent the configuration of the terrain upon a wooden base ... The emperor later gave orders that similar wooden maps should be prepared by all prefects.

Dream Pool Essays (1088)

Legacy, truth, consequence

■ Shen's scientific approach to his observations and experiments helped move Chinese science away from mysticism and superstition.

■ From as early as 1095 other Chinese scholars were referring to the *Dream Pool Essays* and praising Shen's work. Later inventors relied on his records for their developments.

■ Shen and Wei Pu's five-year astronomical observation program was not rivaled until the Danish astronomer Tycho Brahe in sixteenth-century Europe.

in his old age, for he is reported to have said: *"Because I had only my writing brush and ink slab to converse with, I call it* Brush Talks.*"*

Shen's second wife, Lady Zhang, was domineering and aggressive, alienating his children from his first marriage. Nevertheless, when she died, in 1094, he tried and failed to drown himself in the Yangtze River. He died a year later.

that planetary orbits were circular, but proposed they were shaped like a willow leaf – an extremely accentuated ellipse.

Mathematics

Although he was a highly practical man, Shen was unlike many Chinese mathematicians in that he excelled in abstract math, particularly spherical **geometry**, where, in exploring the lengths of the arcs of circles, he came close to developing **trigonometry** (sines, cosines, etc).

Geology

Shen Kuo was at his best when making detailed observations of the natural world and then trying to work out an explanation. As a result, he wrote the first known hypothesis of geomorphology (the way that land structures are formed) and of **palaeoclimatology** (the study of fossil plants to indicate climate changes). In the Taihang and Yandang Mountains he noted that although they were hundreds of miles from the ocean, fossil shells could be found in a certain geological stratum or level of the mountains. He concluded that at one time the area must have been either seashore or actually under water, but since then the sea had shifted. He made the leap

to hypothesizing that the continent must have been formed over an immense time-span for all the sediments to build up.

As well as on mountains, Shen found fossils underground. About 1080 he saw that a landslip on the bank of a river near modern Yan'an, northern China, had opened up a large underground cavern in which were hundreds of fossilized bamboo plants, even though bamboo no longer grew in that region. He concluded that local climates must change over time. As it happens, the fossils were not actually bamboo, but his science was still sound.

Anatomy

The traditional Chinese theory was that human beings had three throat valves. Shen dissected the bodies of executed bandits to discover only two valves, the larynx, which he thought distributed *qi* (life force) from air throughout the body, and the esophagus for food.

Other ideas

One of Shen's innovations was a three-dimensional map with a wax model on a wooden base. He produced an improved ink from petroleum soot, and left history with a description of an early camera obscura, dry docks, and canal locks.

Al-Zarqali (Azarquiel)

Known in Europe by his Latinized name Azarquiel or Arzachel, the Spanish Arab al-Zarqali was the foremost astronomer of his time. He prepared the influential *Toledan Tables*, continued the Arab tradition of correcting and improving classical sources of knowledge, and was renowned for making accurate instruments, including famous water clocks.

Abu Ishaq Ibrahim ibn Yahya al-Zarqali came from a family of artisans, and it was as a maker of delicate but accurate scientific instruments that he earned the nickname "al-Nekkach", or "engraver of metals". He supplied his technical skills to **astronomers** of the great Spanish Muslim center of learning in Toledo, and it was they who persuaded him to read their books and get an education although he was already a grown man.

He quickly absorbed mathematics and astronomy, becoming a rising star in the scientific community, but he never forgot his first skill, and continued to make precise astronomical instruments.

Al-Zarqali lived in uncertain times. The Muslim kingdoms of Spain were constantly under attack by Spanish Christians. Toledo became insecure, so al-Zarqali fled.

> *The clocks consisted of two basins, which filled with water or emptied according to the increasing or waning of the moon.*
>
> Al-Zarqali quoted in *Islam in Andalus*, A. Thomson and M. A. Rahim (1996)

Essential science

Toledan tables and astronomy

Al-Zarqali's compilation, based mainly on his own observations, was widely accepted as the most accurate yet produced. These *Toledan Tables* listed the times and positions of the settings of planets, compared the months of several different calendars, including the lunar cycle, and helped users predict solar and lunar eclipses. He also compiled valuable tables of **latitude** and **longitude**.

Al-Zarqali was the first astronomer to prove conclusively that the apogee of the sun relative to the fixed stars (the most distant point in its orbit) is not fixed, but moves slowly, at a rate he calculated of 12.04 seconds per year, close to today's measurement of 11.8 seconds. He was also the first to realize that in **Ptolemy**'s complex model for the planet Mercury, involving two **epicycles**, the primary epicycle is not a circle, as it is for all other planets, but an oval shape.

On a geographical point, al-Zarqali improved on the work of earlier Arab astronomers by making a more accurate calculation of the length of the Mediterranean Sea: 42 degrees longitude rather than Ptolemy's original exaggerated estimate of 62 degrees.

Legacy, truth, consequence

■ Al-Zarqali wasn't always correct. He believed the mistaken theory of trepidation of the **equinoxes** – that the sphere of fixed stars oscillates slightly – and erroneously thought that his observations of the changes in obliquity of the earth's ecliptic proved this.
■ Despite this error, in general his work endured because it was accurate. His astronomical treatises were translated into Latin very early on, and his *Toledan Tables* were adapted to create tables for different locations in the Christian West. The later, important *Alfonsine Tables* (c.1252–70) were based on his work and endured in Europe late into the sixteenth century.
■ Along with other Arab scientists he played a part in inspiring the future mathematically-based astronomy in Europe. Both **Abraham Zacuto** and **Nicolaus Copernicus** acknowledged their debts to him.

Key dates

1028	Born in Toledo, Spain.
c.1058	Begins to study astronomy.
1062	Builds famous water clocks in Toledo.
c.1085	Flees Toledo in face of Christian advance.
1087	Dies, probably outside Spain.

Instruments

Al-Zarqali invented a new accurate **astrolabe**, an instrument used then for astronomical observations and to help calculate time. Unlike previous versions, his was universal and could be used at any latitude.

Water clocks

Al-Zarqali's water clocks not only showed the hours, but also indicated the days of the lunar months; water would start to flow at the start of a new moon, and the clock's basins would start to empty the day after the full moon. Unfortunately, they were so intricate that when they were taken apart in 1133 for examination, no one could work out how to put them back together again.

Ibn al-Baitar

One of the most important botanists and pharmacists of the Middle Ages, the Arab doctor al-Baitar produced an extensive encyclopedia of the medical uses and properties of plants that was unmatched in Europe or the Middle East for centuries.

Abu Muhammad Abdallah ibn Ahmad ibn al-Baitar Diya al-din al-Malaqi was born in Arab Spain to a well-known family; the al-Baitars had been veterinary surgeons for generations, and their very name derived from the Arabic name for their profession – "baitarah" (baytarah).

Abu Muhammad was more interested in human patients than animals. He studied in Seville under the leading physicians or "herbalists" of the time, as well as under specialist **botanists**. In those days most medicines came from plant sources, so all doctors were called herbalists, but al-Baitar became a genuine botanist as well as a doctor.

During al-Baitar's lifetime the Christian reconquest of Iberia disrupted the centers of learning in Arab Spain. Like thousands of other Muslims he emigrated, but in his case he turned his flight into an opportunity to further his knowledge of plants and local plantlore across northern Africa. Some time soon after 1224 he settled in Egypt, where the ruler Ayyubid al-Malik al-Kamil encouraged scholarship and learning, as chief herbalist.

Al-Baitar continued to collect plants and study their properties, and usually took an artist with him on his travels to draw his specimens. He explored Palestine, Arabia, Greece, Turkey, and Armenia, and when al-Kamil took over Syria in 1227 he was able to add the plants of that region to his collections.

> *... I have only given my authorship to what belongs to me alone, and have only written what I am certain is correct and that can be relied upon with great confidence.*
>
> Preface, Book of Simple [Herbal] Remedies

Legacy, truth, consequence

■ Al-Baitar introduced many new remedies and helped transform **pharmaceutics** from folklore into a modern discipline.

■ Both his botanical and his pharmacological classifications were accepted as the authoritative versions. Although his works were not immediately translated in full, sections of them filtered though to Europe.

■ His *Book of Simple (Herbal) Remedies* was finally translated in full into Latin in the fifteenth century as *Simplicibus*, and later became a standard text for centuries.

Key dates

1190	Born near Malaga, Spain.
1216	Begins plant-collecting expeditions.
1219	Leaves Spain and travels along the northern coast of Africa as far as Asia Minor (modern Turkey).
c.1224	Becomes chief herbalist in Egypt. Continues to travel widely.
1227	Visits Syria to research and collect plants.
1248	Dies in Damascus, Syria.

Essential science

According to his student Ibn Abi Usaybi'a, al-Baitar had an incredible memory for plants and their medical uses, and could also remember exactly what other herbalists, including the ancient Greeks **Galen** (129–c.200 CE) and Pedanius Dioscorides (c.40–c.90 CE) had written. His facts were *"obtained through experimentation and observation"*, and he extensively tested his drugs.

He gathered his knowledge into two important encyclopedias. The *Kitab al-Jami fi al-Adwiya al-Mufrada* (*Book of Simple [Herbal] Remedies*) was a systematic compilation of the medicinal and general properties of plants and vegetables. It incorporated previous useful work both Arab and Greek, but the bulk of the book came from his own studies. In all he included 1,400 different plants, about 200 of which had not previously been recorded. It was the greatest medieval treatise on plants.

His second monumental work, *Kitab al-Mughni fi al-Adwiya al-Mufrada* (translated as *The Sufficient*), was solely to do with medicinal cures, listing plants according to the diseases of the body that they can help treat: ear, head, eye, and so on. In addition to Arab names, al-Baitar added the Greek and Roman names of the plants, and even included surgical advice.

Leonardo Fibonacci

A sophisticated mathematician who found practical applications for abstract theorems, Fibonacci helped revive the science of mathematics in late medieval Europe by introducing the decimal system he learnt from the Arabs. He also made many of his own original contributions to mathematical theory, and he has been called the first great mathematician of Christian Europe.

During his lifetime Leonardo Pisano (Leonardo of Pisa) used several names, including Bigollo (the Traveler or possibly the Vagrant), but he is now most commonly known by a nickname meaning son of Bonacci, Fibonacci.

Although born in Italy, when he was a child he moved to North Africa to join his father, Guilielmo Bonacci, who worked as a representative for merchants of the Republic of Pisa in modern Bejaia, northeastern Algeria. Fibonacci accompanied his father on his business trips around the Mediterranean, visiting Egypt, Syria, and Sicily, among other countries.

Guilielmo recognized that the Arabs had preserved the classical Greek and Roman knowledge and had also received newer ideas from India and China, so their mathematics was way in advance of Europe. In the hope that Leonardo would be able to help with business accounts, he had the boy tutored by Arab scholars. With a natural mathematical mind – he considered mathematics to be an "art" – Fibonacci soaked up learning in all the countries they visited.

In about 1200 Fibonacci returned to Pisa and began to write books (some of which are lost to us) introducing to Europe several important concepts in **algebra** and arithmetic. In particular, he was responsible for Europe adopting our present-day numbering system, the Hindu-Arabic system with ten digits, including zero (0), and the decimal point. A number sequence he used in his work is now named after him, the Fibonacci sequence. He also made many personal contributions, especially offering practical applications of mathematics for business, accounting, and surveying.

Fibonacci was particularly drawn to practical matters, but at the same time he explored theoretical algebra and **geometry**. Although at the time his work on **number theory** was not recognized for the great achievement it is, he still gained a reputation for wisdom. He was rewarded with a salary by the City of Pisa for his advice and teachings, and successfully solving mathematical problems set for him as a challenge by scholars at the court of the Holy Roman Empire.

Essential science

Arab numerals

When Fibonacci was born, Christian Europe was still using Roman numerals – I, II, III, IV, V, VI, VII, VIII, IX, X, D, C, M, and so on. In his book *Liber Abaci*, published in 1202, Fibonacci argued persuasively for the adoption of the Hindu-Arabic decimal system including 0, a decimal point, and place-values (the position indicating whether a number is a tenth, hundredth, etc).

Liber Abaci contained several basic examples of adding, subtracting, multiplying, and dividing using the Hindu-Arab system: examples of standard introductory arithmetic similar to those any young child would be taught at school today. The book also studied problems involving **simultaneous linear equations**.

Practical applications

One of the skills Fibonacci learnt on his travels around North Africa was how to quickly convert between different currencies in order to calculate prices, profits, and losses. In *Liber Abaci* he provided basic lessons and examples for merchants covering a range of arithmetic problems including the price of goods, how to calculate profit on transactions, how to convert between the various currencies in use in Mediterranean countries, and theoretical problems to do with remainders which had originated in China.

In his book *Practica Geometriae*, written in 1220, Fibonacci explored problems of geometry, and included valuable advice on calculations for surveyors.

The Fibonacci sequence

A mathematical problem Fibonacci posed in the third section of *Liber Abaci* is solved by the sequence of numbers now known as the Fibonacci sequence, for which he is best remembered today:

> "A certain man put a pair of rabbits in a place surrounded on all sides by a wall. How many pairs of rabbits can be produced from that pair in a year if it is supposed that every month each pair begets a new pair which from the second month on becomes productive?"

The resulting sequence, which can be applied to many situations, starts as 1, 1, 2, 3, 5, 8, 13, 21, 34, 55 ... (Fibonacci omitted the first

■ **Hindu-Arabic numerals:** Fibonacci was responsible for convincing Christian Europe to adopt Arabic numerals, the mathematical concept of zero (0), and the decimal place system that is used today. In his books and teachings, he argued vehemently that these are much simpler than Roman numerals and provide a straightforward way to perform calculations. Although cumbersome, Roman numerals do allow simple addition and subtraction, but are very complex for multiplication and division, and at the time most accountants would have used an abacus instead. Fibonacci was not the first mathematician to propose Arabic numerals, but earlier texts, such as works by the Arab mathematician Al-Khwarizmi (c.780–c.850), were too academic to be widely read. With its everyday examples, Fibonacci's book was soon circulating among professional men as well as scholars. Europeans quickly realized that this proposed system allowed basic arithmetic to be performed swiftly and simply, and was not only more efficient and effective, but was also far more elegant for written calculations. Fibonacci's ideas were soon adopted and had a profound impact on European systems and thought.

■ **The Fibonacci sequence:** Fibonacci lived at a time when there was little interest in scholarship for its own sake, so the significance of the Fibonacci sequence was not recognized during his lifetime. But in recent times it has been accepted as a major contribution to number theory, with applications in many varied areas of science and mathematics, ranging from botany to psychology, music, and astronomy. For example, Fibonacci numbers apply to the distances between the sun and its planets in our solar system, and to the conversion of miles to kilometers. The Fibonacci sequence has even made its mark on popular culture.

> ## These are the nine figures of the Indians: 9 8 7 6 5 4 3 2 1. With these nine figures, and with this sign 0 which in Arabic is called zephirum, any number can be written, as will be demonstrated.
>
> *Liber Abaci* (1202)

Key dates

1170	Born, probably in Pisa, Italy.
1202	Writes *Liber Abaci* (*The Book of the Abacus* or *The Book of Calculating*), a groundbreaking mathematical text for Europe.
1220	Writes *Practica Geometriae* (*The Practice of Geometry*).
1225	Writes *Flos* (*The Flower*), a collection of solutions to problems set for him as a challenge by scholars at the court of Frederick II, the Holy Roman Emperor.
1225	Writes *Liber Quadratorum* (*The Book of Squares*), containing many impressive calculations mainly on quadratic equations.
1250	Dies, possibly in Pisa.

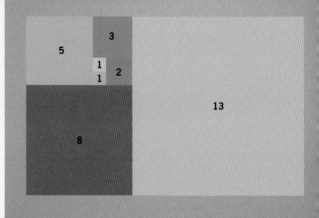

A tiling of squares showing the Fibonacci number sequence.

term in *Liber abaci*) with each number the sum of the two preceding numbers.

Number theory

Apart from the Fibonacci sequence, he made several other significant contributions to number theory, particularly in his 1225 book *Liber Quadratorum* (*The Book of Squares*). Not until the seventeenth century did any other Christian European offer so many original mathematical ideas.

One of the ideas he first presented is that square numbers can be constructed as sums of odd numbers:

> *"I thought about the origin of all square numbers and discovered that they arose from the regular ascent of odd numbers. For unity is a square and from it is produced the first square, namely 1; adding 3 to this makes the second square, namely 4, whose root is 2; if to this sum is added a third odd number, namely 5, the third square will be produced, namely 9 …"*
>
> *Liber Quadratorum* (1225)

Roger Bacon

Roger Bacon is sometimes seen as a pioneer of the modern scientific method. Perhaps his most important contribution to medieval science was his interpretation of early scientific works – most notably on optics – from their original languages.

Born near Ilchester in South West England, Bacon was educated at the University of Oxford and, soon after, in Paris – at the time a center of excellence. Some say that he earned a doctorate in theology, and he reputedly became known as *"doctor mirabilis"* ("wonderful teacher"). Rather than follow the traditional **scholastic** path, he immersed himself in experimental scientific research, and made his own interpretations of scholarly documents from their original languages, including Greek, Hebrew, and Arabic.

About 1250, he returned to Oxford, and joined the Franciscan Order. His orthodoxy was doubted by some and suspicions spread about his involvement in practices such as **alchemy** and **astrology**. Probably as a result of these rumors, Bacon was sent back to Paris in the late 1250s. His academic activities were restricted and he was prohibited from publishing his work.

In 1266 an outlet for Bacon's academic work appeared. He had become acquainted with Cardinal Guy le Gros de Foulques, who was now Pope Clement IV. The pope commanded Bacon to disregard the restrictions imposed on him, and in 1267 Bacon dispatched in secret to the pope his *Opus Majus* (*Greater Work*), followed by his *Opus Minus* (*Minor Work*), and before the pope's death in 1268, his *Opus Tertium* (*Third Work*). These were later rejected by the Minister General of the Franciscan Order, Jerome of Ascoli (who became Pope Nicholas IV).

The remainder of Bacon's life is not well documented. There are accounts of his imprisonment because of his work, although this is disputed.

> **Mathematics is the gate and key of the sciences.**
>
> Opus Majus (1267)

Legacy, truth, consequence

- Debate continues on whether Bacon was truly an early proponent of modern scientific method, or whether his views were in fact shared by many of his contemporaries.
- His *Perspectiva* or *Perspective* (1267) led to the introduction of the study of optics at new universities in the Latin West.
- His studies of optics, and the refraction of light through lenses, probably contributed to the development of the telescope.

Key dates

c.1214	Born near Ilchester in South West England.
1250	Joins the Franciscan Order.
1266	Pope Clement IV orders Bacon to send him his work.
1267	Sends his *Opus Majus* and then *Opus Minus* to the pope.
1268	Sends his *Opus Tertium* to the pope, who dies later the same year.
c.1292	Dies in Oxford, England.

Essential science

Bacon is thought to have produced 80 published works but there is much controversy over who was the true author of many of them. His most important publications were his *Opus Majus*, *Opus Minus*, and *Opus Tertium*. Much of the *Opus Majus* deals with philosophy and theology subjects, but a large portion is devoted to science. He discussed **optics**, perspective, the experimental sciences, mathematics, and physics. Little survives of the *Opus Minus*, which was intended as an aid to understanding the *Opus Majus*. In the *Opus Tertium*, Bacon gave detailed explanations of subjects covered in the previous two works; unfortunately half of this third work has been lost.

In the *Opus Majus* Bacon wrote strong criticisms of his peers' methods: their neglect of important fields of science, and the errors and misconceptions in their interpretations of the subjects they studied, because of their ignorance of the original languages in which the works were written.

His works described the reflection and **refraction** of light, mirages, and burning mirrors (said to have been invented by **Archimedes**, that focused the sun's rays, producing an intense heat to burn invading ships). Other descriptions included eclipses, the light of the stars, and the diameters of and distances between the celestial bodies. Bacon provided proof that the **Julian calendar** was wrong, spoke of mathematics as the fundamental science underlying the other sciences, and reasoned on matters in geography and astronomy.

His predictions included the possibility of steam vessels, aerostats (hot air balloons and airships), microscopes, telescopes, and other inventions that only appeared several centuries later.

Jacob ben Machir ibn Tibbon

The medieval Jewish scholar Jacob ben Machir ibn Tibbon was a translator, mathematician, doctor, and primarily an astronomer who invented the useful instrument known as Jacob's Quadrant. His work formed part of the corpus of knowledge that filtered through to Western Europe and contributed to the later development of science there.

Jacob ben Machir (Makir) ibn Tibbon or Jacob ben Tibbon was known in most Christian countries by his Latinized name, Prophatius or Profatius Judaeus, but in Provence, where he came from, he was known as Don Profiat.

From a long line of distinguished Jewish doctors and translators who had been settled in the south of France since Jews were expelled from Muslim Spain in 1148, he continued the family academic tradition, studying in Lunel and then specializing in medicine at the University of Montpellier. He only branched out into mathematics and **astronomy** well after he had already established himself in his first careers as a doctor and as a translator.

Among the many books he translated into Hebrew were original works by Arabic thinkers such as **Ibn al-Haytham**, **al-Zarqali**, and al-Ghazali (1058–1111), as well as Arabic editions of Greek mathematical and astronomical works, including **Euclid**'s *Elements*, *Data*, and *Optics*, and **Ptolemy**'s *Almagest*.

Ibn Tibbon supported the ideas laid down by the great Jewish philosopher Maimonides (1135–1204), who encouraged science and rationalist thinking, and argued that the Torah (the first part of the Hebrew Bible) could be interpreted rationally. This approach brought him into conflict with some Jewish authorities in Montpellier who disapproved of Maimonides; just three years before ibn Tibbon was born, Maimonides' most famous philosophical book had been publicly burnt in Montpellier.

Legacy, truth, consequence

■ His astronomical tables were well known in southern Europe soon after his death, since the Italian playwright Dante mentioned them in his play *Divine Comedy*, written between 1308 and 1321. They were also referred to by the Italian astronomer Andalo di Negro in a 1323 study of almanacs.

■ Ibn Tibbon was one of the astronomers who **Nicolaus Copernicus** acknowledged as having provided him with useful background material.

■ Small, cheap, and handy, Jacob's Quadrant was popular with many star-gazers, including sailors.

And here is geometry, the basis for all mathematical sciences ...

Marginal comment by ibn Tibbon in his copy of Euclid's *Elements*.

Key dates

c.1236 Born in Marseilles, Provence, in France.
c.1266 Lives in Gerona, northern Spain for a short while.
c.1305 Dies in Montpellier, France.

Essential science

Jacob's Quadrant

In his book translated as *Jacob's Quadrant*, ibn Tibbon describes how to make a new improved quadrant, or instrument for measuring the altitude of heavenly bodies, that he had invented. This came to be called the *quadrans novus* (new quadrant) as opposed to the traditional old quadrant or *quadrans vetus*. It was effectively a quarter circle with all edges marked off with various arcs and scales to show angles of planets above the horizon or the sun's position in the zodiac, and to help the user solve **trigonometry** problems. A weighted thread would be used to align positions on the quadrant's face.

Jacob's Quadrant was basically a part of a simplified **astrolabe**, and since it was flat rather than three-dimensional, could be made cheaply out of wood or even card.

Astronomical tables

Ibn Tibbon also wrote a book of astronomical tables, or "perpetual almanac" as they were then called, simply titled *Luhot* (*Tables*). His version was written specifically for French users, and he pointed out the rising times of the planets over Paris. He probably made use of other tables, such as al-Zarqali's eleventh-century *Toledan Tables* or the thirteenth-century *Alfonsine Tables*, to produce his version.

Abraham Zacuto

An astronomer, mathematician, historian, and rabbi, Abraham ben Samuel Zacuto provided a great deal of the astronomical and navigational equipment and knowledge that helped the European explorers sail across the oceans and reach the Americas and the East Indies by sea.

Zacuto lived through the main tragedies of the Jewish people in the late fifteenth century: the expulsion of Jews from Spain and then from Portugal. The long-established Jewish communities in the Iberian Peninsula had benefited from contact with Arab cultures, and produced many great scholars. Zacuto (sometimes spelt Zacut, Cacuto, or Cacoto) was one of them, a **Renaissance** man with wide-ranging interests.

As well as a thorough education in traditional Jewish law, the young Zacuto also learnt secular sciences. He studied astronomy at the University of Salamanca, one of the leading institutions of its time, and later taught **astronomy** and mathematics there. The head of the university, Gonzalo de Vivero, bishop of Salamanca, admired Zacuto's work and encouraged him to write down his ideas; Zacuto's first astronomical work, *Ha-Hibbur ha Gadol*, is dedicated to the bishop. Along with his later books, this provided a set of astronomical tables showing the declination of the sun, which helped sailors to calculate their latitude at any time.

Zacuto went on to teach astronomy at the universities of Zaragoza and Cartagena, and, at the same time, he became a rabbi for his religious community. His writings were mostly in Hebrew, and were later translated into Spanish and Latin.

Zacuto was friends with the explorer Christopher Columbus, and encouraged him to persevere with his dream of sailing to Asia.

When, in 1492, the monarchs Ferdinand and Isabella demanded that Jews either convert to Christianity or leave Spain, Zacuto was besieged with offers from Christian centers of learning to join them if he converted. However, he was an observant Jew all his life, and he left Spain for Portugal.

Settling in Lisbon, he soon gained a position as royal astronomer and historian. Consulted by King Manuel and by the sailor Vasco da Gama, Zacuto agreed that a voyage of exploration to the east would be feasible. It is not known when he completed his improved astrolabe, an astronomical instrument, but it must have been before 1497, when he gave a set to da Gama.

That same year Manuel also issued an ultimatum to Portuguese Jews to convert or leave. Zacuto and his son Samuel were among the few to escape in time, but on their journey to sanctuary in North Africa they were twice captured by pirates and held to ransom. This was a common risk for Jewish travelers at the time, since bandits knew that Jewish communities would always raise the funds to save members of their religion.

Zacuto eventually landed in Tunis, but the ever-present fear of Spanish invasion forced him to move on, and he wandered around North Africa before settling in Turkey. There he finished an important book on the chronology and genealogy of the Jewish people, although he died before he could see it printed.

Essential science

The astrolabe

When Zacuto was born most European sailors hugged the coastline, following well-known routes, and had rudimentary navigational skills. Portuguese sailors were more advanced, partly because they had already adopted the use of an astrolabe. This scientific instrument, that calculated the position of the sun or a planet in relation to the horizon, had previously only been used for astronomical observation and to calculate the time. Zacuto was one of the fifteenth-century scientists who promoted its use for navigation.

The mariner's astrolabe was a circular disc with a moveable sight and with a scale in degrees engraved around the edge of the disc. The zero mark was aligned with the horizon, and the sight bar was aimed either at the sun during the day or at the Pole Star during the night. The altitude of the star was indicated by where

the sight bar aligned on the scale. Zacuto's versions were the first to be made from metal, so they were more stable than earlier wooden ones. His were also far more accurate than any of the previous clumsy contraptions; with his new design the astrolabe for the first time became an instrument of precision. Together with his astronomical tables, his astrolabe allowed for safer navigation at sea.

The solar tables

Almanacs of tables of the position of the sun and planets were in common use for astronomy and astrology, but the tables in Zacuto's *Ha-Hibbur ha-Gadol* and *Almanach Perpetuum* were among the first that were used for navigation. His books provided more than 300 pages of tables showing the sun's

Christopher Columbus taking possession of the new country; Columbus relied heavily on Zacuto's books.

Legacy, truth, consequence

- Zacuto's solar tables were astonishingly accurate, and were widely copied and reprinted.
- The Portuguese explorer Bartholomeu Dias used Zacuto's tables when he led the first European expedition to sail round the Cape of Good Hope in South Africa in 1487–8.
- Christopher Columbus relied heavily on Zacuto's books, which once directly saved Columbus' life. During his third voyage to the New World (1498–1500), he and his crew were in danger of

being killed by a group of natives. Columbus knew from Zacuto's tables that a lunar eclipse was due, so he frightened the group off by claiming he had caused the eclipse and would take away sunlight permanently if he was threatened again.
- Zacuto's astrolabes, charts, and navigational tables helped Vasco da Gama on his voyage of discovery in 1497–9, when he became the first European to reach India by sea.
- The Zagut crater on the moon was named after him.

> ## My astronomical charts circulate throughout all the Christian and even Muslim lands.
>
> *Sefer ha-Yuhasin* (c.1515)

declination, or angular height above the horizon, for every day for several years in advance, as well as giving practical instructions: the astrolabe measurement of the sun's highest point should be deducted from 90 degrees, then added to the figure in the table for that day to indicate latitude.

His books also provided tables for the moon and planets, and contained other valuable astronomical information that would help sailors venturing into unknown seas, including Vasco da Gama (c.1460–1524) and Pedro Álvares Cabral (c.1467–c.1520) in their voyages to India and Brazil respectively.

Other works

Zacuto wrote several treatises on astrology, a Hebrew–Aramaic dictionary, and other works on astronomy.

Key dates

c.1450	Born in Salamanca, Spain, to a wealthy Jewish family.
1473–8	Writes his great astronomical work, *Ha-Hibbur ha-Gadol* (*The Great Composition*) at the request of the bishop of Salamanca.
1486	Meets the sailor and explorer Christopher Columbus in Salamanca, and encourages him to plan a naval expedition to Asia.
1492	Takes refuge in Lisbon, Portugal, after the expulsion of Jews from Spain.
1496	Publishes a new astronomical book in Latin, *Almanach Perpetuum* (*Perpetual Almanac*) containing part of his previous work and some new material.
1496	Teaches the explorer Vasco de Gama and his sailors how to use his improved astrolabe, tables, and maritime charts.
1497	Goes to Tunis, North Africa (modern Tunisia), when Jews are expelled from Portugal.
1504	Works on his *Sefer ha-Yuhasin* (*The Book of Genealogy*), the first proper chronicle of the Jewish people.
c.1510	Dies in Turkey, probably in Constantinople (modern Istanbul).

Leonardo da Vinci

Scientist, engineer, painter, architect, anatomist, mathematician, and much more, Leonardo da Vinci was a visionary genius like no other. He enlightened a superstitious world with his dedication to reason, truth, and learning. His contributions to the fields of anatomy, optics, hydrodynamics, and civil engineering were unconstrained by the technology and the perceived thinking of his world. Many of his ideas were not improved upon, or even understood, for several hundred years.

Leonardo di Ser Piero was born illegitimately near Vinci, Italy, to a young notary and a peasant girl. While growing up he had access to scholarly texts in his father's house, which helped to satisfy his great curiosity. Vinci, and the nearby city of Florence, were steeped in painting tradition and it was in that discipline that Leonardo first excelled. As a teenager he was apprenticed to the renowned Florentine workshop of Andrea del Verrochio. He was fascinated by drawing real things and representing them as accurately as possible on paper, with all their movement, light, and shade. He also wanted to know how these things worked – what, for example, made up the human eye? How did birds fly?

In 1482 he moved to Milan to enter into service for the Duke of the city. Over the next 17 years Leonardo reached new heights of scientific and artistic achievement alongside his professional work. He subsequently traveled throughout Italy working as a military architect and engineer for the Pope's son, Cesare Borgia. Leonardo was a peaceful man at heart, but he had to earn a living and the work enabled him not only to satisfy his appetite for knowledge and invention, but also to put some of his ideas into practice. It also made this charming, erudite, and handsome man a much sought-after presence in both the military arena and the royal court.

As he grew older, Leonardo became jaded by his lifelong association with unfinished projects (due to himself and to others). He moved to France to spend his last few years working as "first painter, architect and engineer" for King Francis I, though he was in effect an honored guest whom the young king was thrilled to have in his extended household.

Essential science

Mechanics

Leonardo's lifelong fascination with **mechanics** led to the invention or initial development of a vast number of machines and devices. Integral to his work was his dedication to precision – his drawings are laced with a myriad of levers, various types of gears, hydraulic jacks, screws, and swiveling appliances that between them accounted for every single tiny part of the potential working machine.

His studies in the world of nature led him to conclude that the universe was beholden to a basic mechanical energy that determined its structure and function. His study of bird flight, for example, led to his most famous visionary idea of developing a flying machine. Furthermore, he believed that the physical appearance of all organic and inorganic forms in the natural world was caused by the motion and the force within each one.

Painting as a science

At the core of Leonardo's belief that painting was a scientific art was his assertion that "*the eye deludes itself less than any other senses*". It followed, then, that the importance of the eye in painting gave the art form a genuine, scientific quality, based on a definite visual experience, rather than, for example, poetry, which was more open to the boundless, creative muses of the writer. He argued that painting gives, "*immediate satisfaction to human beings in no other way than the things produced by nature itself*".

He further linked science and painting by explaining that the ten optical functions of the eye, "*darkness, light, body and color, shape and location, distance and closeness, motion and rest*" were all vital constituents of painting.

Human anatomy

Leonardo originally studied human anatomy as part of his artistic training. Ultimately his unique skill was to represent the three-dimensional mechanics of the human body in a clear and complete two-dimensional way on paper. He drew the body in see-through layers, employing innovative devices such as using dotted lines to show hidden parts and drawing muscles as an elongated cluster of strings.

Astronomical theories

During the **Renaissance**, scientific thinkers such as Leonardo da Vinci, **Nicolaus Copernicus**, **Galileo Galilei**, and **Johannes Kepler** attempted to refine earlier thought on astronomy. Although Leonardo is less well known for his astronomy studies,

Legacy, truth, consequence

- Although Leonardo did not publish his anatomical findings, because it was not a field of his work in which he was professional, the techniques discovered in the drawings published after his death shaped the foundation of contemporary scientific illustration.
- In all Leonardo left 13,000 pages of drawings and technical notes, illustrating and describing a truly visionary range of creations and conceptions such as parachutes, swimming fins, machine guns, water turbines, submarines, apparatuses for breathing underwater, giant crossbows, cranes, pulley systems, compasses, armored cars, street lighting, and contact lenses.
- In 1502 he proposed a bridge across the Gulf of Istanbul. It was deemed to be an unfeasible project and was never built. In 2001 a smaller version of his bridge opened to foot and bicycle traffic in Norway, and a full-scale Leonardo bridge is currently being planned in Turkey.
- Leonardo's writing, which read from right to left and with the aid of a mirror, has for many years been a source of discussion. One practical reason for it could be that, being left-handed, and thus having an advantage in mastering the style, his writing hand stayed ahead of the words and did not smudge the ink.

Many will think they may reasonably blame me by alleging that my proofs are opposed to the authority of certain men held in the highest reverence by their inexperienced judgments; not considering that my works are the issue of pure and simple experience, who is the one true mistress. These rules are sufficient to enable you to know the true from the false ...

Quoted from *The Notebooks of Leonardo da Vinci* (collected and translated by Jean-Paul Richter, 1888)

and his thoughts in this area are often contradictory, he did propose the hitherto unlikely theory that the earth rotated around the sun, which Copernicus would later develop:

"*The earth is not in the center of the Sun's orbit nor at the center of the universe, but in the center of its companion elements, and united with them. And any one standing on the moon, when it and the sun are both beneath us, would see this our earth and the element of water upon it just as we see the moon, and the earth would light it as it lights us.*"

The Notebooks of Leonardo da Vinci

The moon itself, he proposed, was lit by reflected sunlight:

"*The moon has no light of itself but so much of it as the sun sees, it illuminates. Of this illuminated part we see as much as faces us.*"

The Notebooks of Leonardo da Vinci

Key dates

1452	Born in Anchiano, near Vinci (in present-day Italy), from where he takes his name.
1460	Moves with his father and stepmother to nearby Florence.
c.1466–69	Works at Verrocchio's workshop.
1472	He is registered as a painter working on his own for the first time.
1482	Moves to Milan where he offers the court of Ludovico il Moro a wide range of services, including architect, engineer, and organizer of court feasts.
1490	He works on projects to irrigate the countryside.
1495	Begins work on the painting, *The Last Supper*.
1499	Leaves Milan after it falls to the French.
1500	Returns to Florence.
1502	Begins work as a military engineer.
1503	Works on a way of connecting Florence by water to the sea. A canal along his route is built several centuries later.
c.1503–6	Paints the *Mona Lisa*.
c.1509	Engineers a project to channel the course of the Adda river.
1513	Moves to Rome and spends more time working on hydraulic engineering and geometry.
1516	Moves to the Castle of Cloux (in present-day France). Designs the royal residence at Romarantin.
1519	Dies at Cloux and is buried in nearby Amboise, Indre-et-Loire, in present-day France. He bequeaths his artistic and scientific drawings, manuscripts, and instruments to his pupil, the painter Francesco Melzi.

The Vetruvian Man, the drawing by Leonardo da Vinci, illustrates the mathematical proportions found in human anatomy as described in the Golden Section – a proportion used by many artists and architects since the Renaissance.

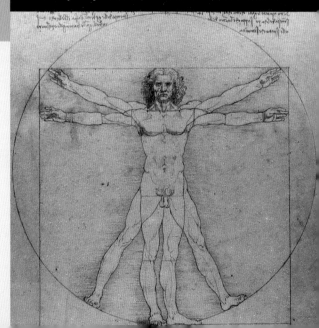

Nicolaus Copernicus

Catholic cleric, mathematician, physician, military leader, and economist: Copernicus was all of these, but it was in astronomy that he made his mark on the world. His heliocentric theory of the planets, demonstrating that the sun, and not the earth, is at the center of what is now called the solar system first caused an outcry. It challenged Ptolemy's geocentric model of the universe, endorsed by the Christian Church, and signalled the beginning of the Scientific Revolution.

Born to a wealthy Polish family, Nicolaus Copernicus was raised by his cleric uncle after his father's death. He enrolled at the University of Cracow, in Poland, where he studied **astronomy** among other subjects. An astronomy course at the time would have introduced him to **Aristotle** and **Ptolemy**'s view of the universe in order to calculate calendar dates or navigate the sea, and to astrological practices of calculating horoscopes of people. Copernicus then studied at Bologna, renting rooms from the famous astronomer Domenico Maria Novara da Ferrara and assisting him in observations.

Through his uncle's influence Copernicus was appointed a canon in the cathedral of Frauenburg in 1501. He continued his studies for several years with the permission of the cathedral chapter, this time in Padua, where he studied "astrological medicine" – in medieval Europe, physicians made use of **astrology**, in the belief that the position and aspect of the stars influenced the course of human affairs.

Copernicus' clerical position gave him access to the highest circles of power, but it also allowed time for him to spend in solitary study. He is said to have made his celestial observations from a turret in the fortifications of the town – there were no instruments to help him as the telescope would not be invented for another hundred years. He lived during the turbulent period of the Reformation, but managed to continue his observations despite the unrest.

Around the year 1514, Copernicus circulated among a few friends an anonymous treatise, subsequently titled *Commentariolus* (*Little Commentary*), an early stage in the development of his theory that attributed to earth a daily rotation around its own axis and a yearly revolution around a stationary sun. *De Revolutionibus Orbium Coelestium* (*On the Revolutions of the Heavenly Orbs*), a detailed exposition of his **heliocentric** system, now regarded as the starting point of modern astronomy, was printed on his deathbed. A Lutheran minister oversaw the printing and inserted an anonymous preface presenting the theory, contrary to Copernicus' opinion, as a practical, mathematical device for charting the movements of the planets, and not as a truth about the world. It wasn't until the early seventeenth century, when **Galileo Galilei** attempted to convert the establishment to the Copernican system, that the Church was provoked to object.

Essential science

The Ptolemaic (geocentric) system

During Copernicus' time, the prevailing theory in Europe was still based on the **cosmology** of Aristotle, which had been developed by Ptolemy around 150 CE. This viewed the earth as a stationary center of the universe ("cosmos"), with "heavenly bodies" (other planets and stars) orbiting around the earth. In Ptolemy's theory, the moon was the closest heavenly body to the earth, followed by the planets, and then the stars, which were fixed points of light in a rotating sphere. He also explained why a planet appears to move closer to or further away from the earth during its orbit, and sometimes backwards (retrograde), and faster or slower, rather than uniformly. His system was used by sailors to navigate the seas. It was given additional credibility by certain passages of Scripture, and it appealed to human nature: it seemed to accord with the appearance of the skies to any casual onlooker; and it placed man at the center of things.

The Copernican system

Early traces of a heliocentric model are found in several texts composed in ancient India before the seventh century BCE, and an ancient Greek astronomer, Aristarchus of Samos (third century BCE), developed a scientific model along the lines of Copernicus' but it was rejected in favor of the **geocentric** scheme. Copernicus' contribution was to create an integrated astronomical system without the complex sets of geometrical techniques for each of the planets characteristic of Ptolemaic astronomy.

> *In the center rests the sun. For who would place this lamp of a very beautiful temple in another or better place than this where from it can illuminate everything at the same time.*
>
> De Revolutionibus Orbium Coelestium (1543)

Legacy, truth, consequence

■ The publication of his *De Revolutionibus Orbium Coelestium* in 1543 brought the first serious challenge to the geocentric system. It ushered in a new way of thinking that gave man his place among all things existing around him, replacing the idea of man at the center of the cosmos. As Goethe said:

> "Of all discoveries and opinions, none may have exerted a greater effect on the human spirit than the doctrine of Copernicus. The world had scarcely become known as round and complete in itself when it was asked to waive the tremendous privilege of being the center of the universe. Never,

perhaps, was a greater demand made on mankind – for by this admission so many things vanished in mist and smoke! What became of our Eden, our world of innocence, piety and poetry; the testimony of the senses; the conviction of a poetic – religious faith? No wonder his contemporaries did not wish to let all this go and offered every possible resistance to a doctrine which in its converts authorized and demanded a freedom of view and greatness of thought so far unknown, indeed not even dreamed of."

■ In 1609, **Johannes Kepler** modified Copernicus' system, leading to a significant improvement in the prediction of planetary positions.

■ In December 1610, **Galileo Galilei** used a telescope to make observations that were basically incompatible with the Ptolemaic system. He felt that he had to answer the objection that the new science contradicted certain passages of Scripture, and his campaign led to the Catholic Church placing Copernicus' book on the Index of Prohibited Books in 1616. However, the ban and Galileo's subsequent trial for heresy had little effect on the increasing acceptance of heliocentrism: the "Copernican Revolution".

■ Copernicus was correct to place the sun at the center of the solar system, but wrong to think that the planets execute uniform circular motion. As was discovered later, by Kepler, the orbits of the planets are "flattened circles" or ellipses.

An early printed rendition of a geocentric cosmological model.

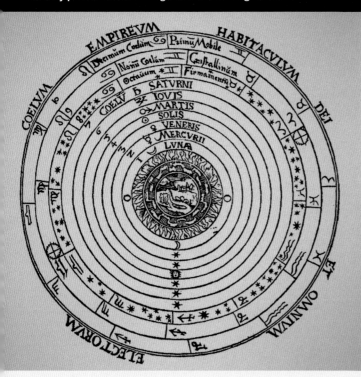

Thus indeed, as though seated on a royal throne, the sun governs the family of planets revolving around it.

De Revolutionibus Orbium Coelestium (1543)

The basic elements of Copernicus' heliocentric model are: the sun is close to the center of the universe, and the earth and all other planets revolve around it. Around the sun, in order, are Mercury, Venus, the earth and the moon, Mars, Jupiter, Saturn, and the fixed stars. The earth rotates around its axis daily (this explains why the stars appear to an observer to rotate around the earth), and the moon revolves around the earth once a month. The earth revolves around the sun annually (explaining why the sun appears to rotate around the earth), and tilts on its axis. Most importantly, the motions of the earth in Copernicus' system eliminate the need for the idea of a retrograde motion of the outer planets as a physical reality, as in Ptolemy's theory.

The Copernican model did not do away with the need for epicycles, as in the Ptolemaic system (see pages 18–19), but needed fewer **epicycles** as a result of moving the sun to the center. The great advantage of the model was its simplicity.

Key dates

1473	Born in Torun, northern Poland.
1491	Enters the University of Cracow.
c.1496	Studies canon law, University of Bologna, Italy.
1447	Notification of his appointment as a canon at Frauenburg (now Frombork) Cathedral, Poland.
1501	Obtains permission from the cathedral chapter to study astrological medicine in Padua.
1504	Begins collecting observations and ideas pertinent to his heliocentric theory.
c.1514	Makes available to friends his *Commentariolus* (*Little Commentary*), a short handwritten text.
1533	Johann Albrecht Widmannstetter delivers a series of lectures in Rome on Copernicus' theory.
1536	Archbishop of Capua Nicholas Schönberg urges Copernicus to publish his findings in astronomy.
1543	His major theory, *De Revolutionibus Orbium Coelestium* (*On the Revolutions of the Celestial Spheres*), is published, dedicated to Pope Paul III. Copernicus dies the same year, in Frauenburg.

Georgius Agricola

Agricola has been called the "father of mineralogy" and the "father of metallurgy". A classicist, historian, physician, philosopher, philologist, and outstanding chemist, he produced the first classification of minerals using scientific techniques to explain their physical properties and relationships. His study of geology and mining laid the scientific foundations of the industry as it is known today.

Little is known about Agricola's family or childhood. Born Georg Bauer in Saxony, eastern Germany (Bauer means peasant or husbandman), his name was Latinized, which was customary at the time, to Georgius Agricola. In Europe the **Renaissance** was well established. Inventions, such as the printing press, brought about an expansion in literacy and thirst for knowledge at the secular level that had never happened before. This growth of literature, arts, advances in science and technology, and the new ways of thinking, was an inspiration to Agricola. After excelling at school, he attended the University of Leipzig to study the classics, philosophy, and philology. He was awarded his degree in 1517 – the year the Protestant Reformation against the Roman Catholic Church is thought to have begun in Wittenberg by Martin Luther.

Agricola returned to his old school in Zwickau where he taught Greek and Latin, becoming the principal within a year. However, he was anxious to pursue his studies in chemistry, medicine, and physics, and went back to the University of Leipzig as a lecturer. His friend and supporter at the university with whom he had been in correspondence, Petrus Mosellanus (Peter Schade, 1493–1524), was a professor of classics and a German humanist well known for his works on rhetoric. Agricola was a loyal Catholic, and remained so until his death, but the influence of

Mosellanus' humanist philosophy, which included the promotion of human dignity and fulfilment by means of reason and scientific method, seems apparent in Agricola's rigorous approach to scientific study, and in his tolerant religious attitude.

On the death of Mosellanus in 1524, Agricola went to Italy to continue his studies in the natural sciences, philosophy, and medicine at the universities of Bologna, Padua, and Venice. He became a friend of the Christian humanist scholar, Desiderius Erasmus (1466–1536), whose ideas had influenced Protestant Reformers. After taking his doctorate, Agricola returned to Germany. He was appointed physician at the town of Jaochimsthal, in the center of one of the major mining and smelting works in Europe, which enabled him to continue studying medicine, while observing the various mining techniques and the treatment of ores. After three years he left to tour and study the mines around Germany. He published *Bermannus, Sive De Re Metallica Dialogus* around this time, which brought him much public attention. He then settled in Chemnitz, a prominent mining town, and over the years published several works on mining and mineralogy, as well as other subjects. During this time, most of Germany had converted to the Protestant cause. Chemnitz had adopted the Lutheran creed and Agricola, who remained a faithful Catholic, had to resign his office

Essential science

Mining and geological science

Before Agricola, the main source of information on metallurgy and mining techniques in the Western world was Pliny the Elder's *Historia Naturalis* (*Natural History*) from c.77 CE. Knowledge of chemistry was very basic; chemical analysis as recognized today was unknown, the exception being the use of fire in the crude analysis of ores. It is thought that Agricola may have been one of the first to describe simple and compound substances.

De re metallica

Agricola's definitive work *De Re Metallica* (*On the Nature of Metals/Minerals*), which is believed to have taken around 20 years to complete, is a collection of 12 books covering detailed aspects of the processes

and problems involved in mining operations, and other related areas, including: administration, assaying, construction, miners' diseases, geology, marketing, prospecting, refining, smelting, surveying, use of timbers, ventilation, and water pumping. Water removal from mines was a major problem at the time, and Agricola's examination of this subsequently led to the discovery of air pressure. Also included is an appendix with German translations of Latin technical terms, and many wood-cut technical illustrations based on actual engineering practices of the time, which had never been published before.

The first textbook on mineralogy

De Natura Fossilium (or *De Veteribus et Novis Metallis*), Agricola's second most important work, offers the first classification of

Fire-setting, a mining method described by Agricola.

■ Agricola was a brilliant chemist and pioneer who was among the first to apply systematic observation techniques instead of speculation to found a natural science. *De Natura Fossilium* was the first scientific classification of minerals and their properties, relationships, and occurrence.

■ Agricola's famous work, *De Re Metallica*, remained the classic, standard publication on metallurgy and mining for around two centuries. Considered the most important technological treatise of the sixteenth century, it undoubtedly contributed to the developing science of geology. Expensive to produce and with distribution limited, some copies were chained in churches where the priest translated the Latin text on request. By 1700, more than a dozen editions had been published in German, Italian, and Latin. The first English translation was made in 1912 by a mining engineer and future president of the United States, Herbert Hoover (1929–33), and his wife, Lou Hoover, a geologist and Latin scholar.

■ Agricola frequently referred to the work of Pliny. Analysing the type of black rock found at the Schlossberg at Stolpen (in Saxony), he judged it was the same as Pliny's basalt and gave it the same name – the **petrological** term that is used in geology today.

During 180 years it was not superseded as the text–book and guide to miners and metallurgists ... That it passed through some ten editions in three languages ... is in itself sufficient evidence of the importance in which it was held ...

Herbert C. Hoover and Lou H. Hoover, preface, English edition of *De Re Metallica* (1912)

of town burgher because of protests. It is thought that he died while having a heated argument with a Protestant. Despite the good Agricola had done for Chemnitz, hostile religious feelings prevented his burial there. Demonstrations were held as his body was taken to Zeitz, some 30 miles (50 km) away.

minerals by geometric form. This was important in the process of identifying minerals, as form can reflect the structure of the chemical composition and environmental origins.

Other works

Bermannus, Sive De Re Metallica Dialogus is the first work to put forward the scientific study and order of minerals based on practical knowledge. *De Ortu et Causis Subterraneorum* is a pioneering study which laid the foundations of the science of physical **geology**, and criticized the ancient Greek theories. of **Aristotle**. It describes wind and water as geological forces, and explains the causes of earthquakes and volcanic eruptions.

Key dates

1494	Born in Glauchau, Saxony, in eastern Germany.
1517	Awarded his degree at the University of Leipzig.
1524	Goes to Italy for two years and takes his doctor's degree.
1527	Appointed physician at Jaochimsthal, Bohemia, a major area of mining and smelting.
1530	Publishes *Bermannus, Sive De Re Metallica Dialogus* with a foreword by Erasmus. He is appointed historiographer by Prince Maurice of Saxony and goes to Chemnitz to observe mining methods.
1533	Appointed town physician of Chemnitz by its citizens.
1543	Marries a widow from Chemnitz (with whom he is believed to have had five children).
1544	Publishes *De Ortu et Causis Subterraneorum*.
1546	Publishes *De Natura Fossilium* (also known as *De Veteribus et Novis Metallis*). He is appointed a burgher of Chemnitz.
1555	Dies in Chemnitz.
1556	*De Re Metallica*, his celebrated and famous treatise on mining and metallurgy, is published; first English translation by H. Hoover and L. Hoover, is published in *Mining Magazine* (London), 1912.

Garcia de Orta

A pioneer of tropical medicine, Garcia de Orta was also a naturalist and a renowned physician of the Renaissance. He was the first European physician to study the acute infectious disease of cholera and to describe diseases specific to the tropics.

Garcia de Orta was born in Castelo de Vide, Portugal, around 1501. The de Orta family were Jews of Spanish origin, who had relocated to Portugal as "New Christians" to seek refuge from religious persecution in Spain. Also known as *conversos*, the New Christians professed to convert to Christianity but many continued to practice Judaism in secret.

De Orta returned to Spain to study medicine, philosophy, and art at the universities of Alcalá de Henares and Salamanca, graduating in 1523. He then went back to Portugal to practice medicine, before being appointed professor of logic at Coimbra University in 1530. In 1534 he left for India as chief physician aboard the fleet of Viceroy Martim Afonso de Sousa. He set up a medical practice in Goa, married Brianda de Solis, and fathered two daughters. It has been intimated that his marriage was an unhappy one; professionally, however, de Orta was very successful, leading to the publication of his experiences with tropical diseases and his invaluable description of Asian cholera.

In the Portuguese colonies de Orta was appointed as physician to many prominent politicians and religious figures, and was thus able to avoid the persecution of Jews taking place in Portugal. He managed to help his mother and sisters escape to live with him in Goa, but the persecution of non-Catholics eventually spread to the Far East, and several members of de Orta's family suffered torture and painful deaths. Despite his fame, de Orta was posthumously condemned as a Jew and his remains were removed from his grave and burned.

Legacy, truth, consequence

- As well as describing the graphic details of the sufferings of cholera victims, de Orta wrote about many tropical diseases that were previously unknown in Europe. His book became hugely popular, and was used in many European universities and medical schools.
- A prominent Portugese physician of the time, Dr Ruano, accused him of going against the authority of classical medicine handed down from the Greeks, Romans, and Arabs. De Orta defended his methods and theories as being truthful and not biased by tradition.

> *I say that one gets to know more in one day now ... [in India] than what knowledge was gathered in one hundred years by the Romans.*
>
> Garcia de Orta in a response to Dr Ruano (c.1563)

Key dates

c.1501	Born in Castelo de Vide, Portugal.
1523	Graduates from the universities of Alcalá de Henares and Salamanca.
1530	Appointed professor of logic at Coimbra University.
1534	Sails to India as the chief physician aboard the fleet of Viceroy Martim Afonso de Sousa.
c.1543	Marries Brianda de Solis.
1563	Publishes *Colloquies of the Simple, and Drugs and Medicinal Applications in India*.
1568	Dies in Goa.
1580	Posthumously convicted of being Jewish.

Essential science

Tropical medicine

De Orta's only known work, *Os Colóquios dos Simples e Drogas e Coisas Medicinais na India (Colloquies of the Simple, and Drugs and Medicinal Applications in India)*, was published in 1563 and printed on the first Asian printing press, installed in Goa in 1545. In this respect, "simple" means herbal medecine or remedy. The book reveals his knowledge of Eastern spices, gained from his dealings with merchants in Goa, and describes his experiences with tropical diseases, particularly cholera (also called Asiatic cholera) – the acute infectious disease of the small intestine. For his research he performed an autopsy on a cholera victim, the first recorded autopsy in India. He was the first European to give a detailed account of the signs, symptoms, and prognosis of Asian cholera.

Xu Guangqi

A scientist, translator, and government official, Xu Guangqi introduced Western mathematics and science to China at a time when the once-magnificent Chinese scientific tradition had begun to decline. He helped translate several significant European books into Chinese, was himself the author of an important book on scientific methods in agriculture, and was one of the first Chinese converts to Christianity.

As a young man Xu Guangqi (Hsu Kuang-ch'i) excelled in the official examinations for Chinese scholar-officials, and he entered Imperial service in the last decades of the Ming dynasty. Xu was well aware that Chinese science and mathematics were in decline, and when he came into contact with the recent European missionaries, with their new advanced scholarship, he became highly critical of the Chinese society that had allowed knowledge to be lost.

Xu worked with the Italian Jesuit missionaries Matteo Ricci (1552–1610) and Sabatino de Ursis (1575–1620) to translate Western texts on mathematics, geography, and hydraulics into Chinese, including **Euclid**'s comprehensive mathematical work *Elements of Geometry* dating from around 285 BCE. He also translated some of the key Confucian works into Latin, making ancient Chinese thinking available to the West. When he converted to Christianity in 1603, he took the Christian name of Paul Xu.

Eventually Xu rose high in Imperial service and was given titles such as "grand secretary" and even "second man after the Emperor".

> *Four things in this book are impossible. It is impossible to remove any particular passage, to refute it, to shorten it, or to place it before that which precedes it, or vice versa.*
>
> From Xu Guangqi's foreword, translation of Euclid's *Elements of Geometry*.

Essential science

Calendar reform

In 1629 the government held a competition to decide who would be awarded the job of reforming the calendar, by asking for predictions of the next solar eclipse. Using his new European astronomical knowledge (one of the Jesuit priests with whom he worked had studied with **Galileo Galilei**), Xu won. As well as improving calculations for dates, Xu also announced 23 new constellations derived from Western star catalogs.

Mathematics

In translating Western books on mathematics, Xu introduced to China several new concepts, particularly in **geometry** – such as parallel line, acute angle, obtuse angle – and had to invent words for the new terms.

Legacy, truth, consequence

- According to some scholars, Xu kick-started the "Chinese Enlightenment". Although at first there was some opposition to the new ideas he introduced, Xu eventually persuaded others that the new mathematics, particularly geometry, could be applied to practical problems.

- Xu persuaded the Chinese government to adopt Western firearms and military methods to push back the Manchu people from the north, gaining a valuable breathing space, but eventually the Manchu also gained Western weapons and conquered China, forming the last Imperial dynasty, the Qing.

- Through Xu the Roman Catholic Church gained a certain amount of influence in China, which it lost after his death. He was the most influential Christian convert before the twentieth century.

Key dates

1562	Born in Shanghai, China.
1600	Meets and becomes friends with the Jesuit missionary Matteo Ricci.
1603	Converts to Roman Catholicism.
1607	His translation of Euclid's *Elements of Geometry* is printed in China, introducing new mathematical concepts.
1625–8	Writes a complete compendium of agricultural knowledge.
1629	Wins a government competition to reform the calendar.
1633	Dies in Shanghai.

Agriculture

Xu had a particular interest in improving the rural economy, so he included in his massive tome on agriculture all the ideas on irrigation that he had received from his European friends. His book was a valuable practical guide to all aspects of land management, animal husbandry, and crops, and even included advice on disaster relief after a famine.

Galileo Galilei

A mathematician, physicist, and astronomer, Galileo was one of the pioneers who transformed physics into a mathematically-based discipline using experiments to establish facts. He built the first telescope powerful enough to observe the solar system in some detail, thereby revolutionizing astronomy. His support for Copernicus' heliocentric model led to a trial by the Inquisition for heresy, at which he was forced to recant his views, becoming a symbol of the scientific struggle to find the truth at all costs.

Galileo was born in Pisa, Italy, to a noble but impoverished family. His father hoped he would become a well-paid doctor, and sent him to university. But, bored with everything except mathematical problems and **natural philosophy**, Galileo left university without a degree.

Although he gained a reputation as a mathematician, as his father had feared, he became desperately poor. He turned to inventing and created a thermometer, a pump, a hydrostatic balance, and, in 1597, a compass which finally brought him success and some money.

Galileo's fortunes improved dramatically with his creation of a telescope in 1609, modeled on a Dutch invention he had never seen. He refined the instrument, eventually producing one which enabled him to make amazing astronomical discoveries, including evidence that the earth and planets revolve around the sun. His new fame also brought him a lucrative role as court mathematician to Cosimo de Medici, the Grand Duke of Tuscany.

As well as **astronomy**, Galileo was fascinated by the study of motion and other fields of physics. He produced a mathematical paradox and designed technical instruments such as a microscope. He was an excellent self-publicist, and did not hesitate to insult his colleagues in his writings. His support for **Copernicus'** **heliocentric** system, that put the sun at the center of the cosmos, clashed with Church doctrine. In 1600 the Papal Inquisition had burnt the philosopher and cosmologist Giordano Bruno at the stake, so perhaps with this example in mind, Galileo recanted.

Galileo did not marry, but he did have three children with his mistress Marina Gamba, who later married another man.

> *With regard to ... the movement of the sun and earth, the inspired Scriptures must obviously adapt themselves to the understanding of the people.*
>
> Letter to the Grand Duchess Christina (1615)

Essential science

Empirical methods

Galileo championed the use of systematic experiments and the application of mathematics to scientific problems. In particular, he initiated a new approach by using logic and experience to break a problem down to simple terms to make it easier to analyze.

Astronomical discoveries

Galileo's improvements to the design of the telescope meant that he was the first person to turn an effective magnifying instrument upon the heavens, and was the first person to report seeing the craters and mountains of the moon. This particular observation disproved **Aristotle's** theory that the heavenly bodies would be perfectly smooth spheres.

He made other original observations: of the four largest moons orbiting Jupiter, showing that at least some heavenly bodies did not go around the earth; the phases of Venus, suggesting that it orbited the sun; a vast number of stars indicating that the universe was far larger than previously believed; and – contemporaneous with other astronomers – the dark blotches known as sunspots, which he correctly concluded were part of the sun's surface, although contemporaries argued they were satellites passing around the sun. All in all, Galileo concluded that the Church was wrong to hold that the sun and other planets orbited around the earth.

Study of motion

When only 20 and watching a chandelier swinging in Pisa Cathedral, Galileo made an observation no one had previously recorded. Using his pulse to time it, he realized that a pendulum's swing always takes the same amount of time, regardless of the size of the arc of the swing. In actual fact, modern instruments can detect a tiny difference, but his discovery was an adequate basis for his later work on pendulum clocks.

Law of falling bodies

In a famous experiment (possibly legendary), Galileo dropped balls of different masses from the top of the Leaning Tower of

Legacy, truth, consequence

■ Galileo has been called by some the "father of modern science" and the "father of modern physics". Certainly he had a major impact on the development of science in general, and the quantitative experimental method he pioneered has become the standard scientific approach.

■ As the scientist who backed down before the Inquisition, he symbolized the tension between religion and scientific knowledge, but his work helped eventually to separate science from religion and philosophy.

■ His astronomical observations, made possible by his own ground-breaking work on telescopes, provided new knowledge of the solar system that astounded scientists and lay-people alike.

■ His law of uniformly accelerated motion has stood the test of time, and his systematic, mathematical approach to the study of motion was the foundation for modern **mechanics**.

■ Galileo had so many ideas that some were bound to be wrong. His explanation of tides, when he completely discounted the effect of the moon, was one.

Galileo's telescopes, c.1610.

Pisa, and observed that they all hit the ground at the same time. He certainly carried out several other experiments by rolling balls down a slope. He therefore disproved Aristotle's argument that in free fall objects with a greater weight have a greater velocity, but he was unable to produce a theory to explain his discovery. He did, however, later describe a law to cover the phenomenon of uniform velocity: in free fall an object will travel a distance proportional to the square of the elapsed time.

Galileo's law led him to an understanding of projectile motion and the conclusion that the path of a projectile must be a parabola.

Theory of tides
Galileo rejected the proposal of several contemporary scientists that the moon affects tides, instead attributing them to the earth's own motion. He argued that as the earth spins on its axis and revolves around the sun it effectively "swishes" the seas around.

One cannot understand it [the universe] unless one first learns to understand the language and recognize the characters in which it is written. It is written in mathematical language ...

The Assayer (1623)

Johannes Kepler

A German astronomer and mathematical genius, Kepler formulated the three laws of planetary motion that are named after him. He drew up the first modern set of astronomical tables, and he was also the first astronomer to openly support Copernicus' theory of a heliocentric universe, helping to establish the sun-centered model of the universe.

The son of a mercenary soldier and an innkeeper's daughter, Kepler was a premature baby and a sickly boy. His father vanished when Kepler was five, probably dying in battle opposing the Dutch Revolt. His interest in astronomy can be dated to 1577, when, at the age of six, his mother took him to a hilltop to see a comet.

The local Protestant scholarship system enabled Kepler to study at a seminary and enter the University of Tübingen. A life-long Lutheran, he intended to become a minister, but as was customary he took courses in other subjects as well as theology, including philosophy, mathematics, and **astronomy**, learning about **Copernicus'** new **heliocentric** theory that challenged the long-standing **geocentric** model of the cosmos. He came to support heliocentricity as a true description of the universe, and began to gain a reputation as an astrologer. At the university's recommendation, in 1594, he became a teacher of mathematics and astronomy at the Protestant school in Graz (in modern Austria). While there he corresponded with the Danish astronomer Tycho Brahe (1546–1601), who was working for the Holy Roman Emperor Rudolph II, and visited Brahe at his new observatory near Prague.

In 1600 Europe's growing religious strife intervened – as it was to do several times in Kepler's life. Protestants were banished from Graz and Kepler and his family took refuge in Prague, where Brahe found him work helping prepare a new set of astronomical tables (lists of the positions of the sun, moon, and planets) based on Brahe's copious, accurate planetary observations. After Brahe's sudden death in 1601 Kepler was appointed his successor as imperial mathematician and entrusted to finish the tables.

Kepler experienced the full gamut of seventeenth-century religious pressure. In 1620 his mother Katharina was imprisoned, accused of witchcraft by a woman who was in a financial dispute with Kepler's brother. While awaiting trial Katharina was terrorized with descriptions of the torture she would undergo, but Kepler abandoned his work to help with her legal case, and she was released in 1621. He was also excommunicated from his own Lutheran church over a doctrinal disagreement.

When Prague turned against Protestants, Kepler had to move to Linz (in modern Austria). He lost his first wife and a son to illness, remarried happily, but had to move again in 1626 when Catholic forces besieged Linz as part of the Thirty Years' War. Kepler found temporary work while arranging for the astronomical *Rudolphine Tables* to be printed, but died on his journey to collect unpaid wages from the Empire.

Essential science

Religious belief

Kepler's drive to understand the universe came from his Christian belief that this would draw him closer to understanding the nature of the universe's creator. He firmly believed that God had used a **geometrical** plan to build the cosmos, and that plan could be understood by reason. He embraced the heliocentric model partly because he thought it mirrored the spiritual universe: God the Father was the large and powerful sun at the center of creation.

Platonic solids

Kepler's first model came from this quest to reveal why God had structured the universe as He did. In 1595 Kepler realized that he had the answer all along: having studied **Euclid** he knew that there were supposed to be just five perfect regular polygons or three-dimensional shapes, the **Platonic solids**, each of which

could be bound by one **inscribed** and one **circumscribed** spherical orb. These, he reasoned, must be God's chosen spaces between the six known planets, Mercury, Venus, Earth, Mars, Jupiter, Saturn. Nesting the solids, encased in spheres, within each other, Kepler found that his model matched the available astronomical data of each planet's path, assuming the planets circled the sun.

War against Mars

Brahe's extensive, precise astronomical observations were the basis for Kepler's extrapolations. He focused on the lengthy data for Mars, and since accepted theory was that the heavenly bodies moved in circles or in a combination of circles, he spent years struggling to find a circular system that would explain the planet's sometimes erratic orbit. He called it a "war against Mars", and eventually the battle was won when he realized that the underlying assumption was wrong –

Legacy, truth, consequence

- Kepler's idiosyncratic ideas were not widely accepted. However, astronomers eventually had to agree with his elliptical orbits.
- Kepler overthrew the belief in heavenly circles that had dominated cosmology for more than 2,000 years, and transformed astronomy from a geometrical study to one involving physics.
- **Isaac Newton** based some of his work on universal gravitation upon Kepler's laws, providing the mathematical explanation for Kepler's theories.
- Kepler's laws are the foundation for much of our modern understanding of the workings of the solar system. They are also crucial for calculating the orbits of artificial satellites (a word he coined) and spacecraft.

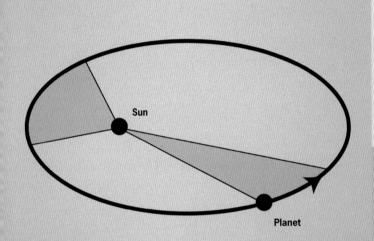

Kepler's Second Law: The line joining the planet to the sun sweeps out equal areas in equal intervals of time. This law can be used to determine the point at which a planet is located on its elliptical orbit at a given time.

[Geometry] supplied God with a model for the creation of the world. Geometry was implanted into human nature along with God's image and not through man's visual perception and experience.

Harmonices Mundi (1619)

"It seemed as if I woke from sleep and saw a new light break on me." The orbits could not be perfect circles, but were ellipses. It was from this work that he derived the rules that are now accepted as laws, although he himself never numbered these theories or singled them out from his other discoveries.

Kepler's laws of planetary motion

1. All planets move around the sun in an oval-shaped elliptical orbit with the sun at one focus.
2. An imaginary line from the center of the sun to the center of a planet traces out equal areas during equal lengths of time. This means that planets move faster when they are close to the sun, and slower when they are distant from the sun.
3. The squares of the sidereal periods (length of time of an orbit) of any two planets are directly proportional to the cubes of their mean distances from the sun. Thus by knowing the length of the orbital period, the distance from the sun can be calculated, and vice versa.

Other science

The Rudolphine Tables (1627), Kepler's new astronomical tables, were the most accurate yet produced and were widely adopted. But apart from astronomy, he had an astonishing range of important ideas. He explored the numerological links between music, universal structure, and the natural world, and correctly identified the function of the retina of the eye and explained how lenses work. He also invented a new two-lensed astronomical telescope, proved how **logarithms** work, invented a method of calculating that was ancestral to modern **calculus**, suggested an arrangement for packing spheres that became known as the Kepler conjecture, and discussed acceptable error in scientific observations.

Blaise Pascal

A French mathematician, physicist, and philosopher, Blaise Pascal is well known for building one of the first mechanical calculators. His mathematical brilliance enabled him to lay the foundations for probability theory and to formulate projective geometry theorums. He also contributed to a greater understanding of the concepts of atmospheric pressure and vacuum, and to the behavior of fluids.

Blaise Pascal's mother died when he was three years old, and two years later the Pascal family moved from Clermont (now Clermont-Ferrand) to Paris, France. His father, Étienne, decided to teach his children himself, and like his sister, Jacqueline, Pascal proved to be a child prodigy. While Jacqueline's focus was on literature, Pascal showed mathematical genius. In 1639, still a teenager, his father took him to a meeting of intellectuals, to present his paper, which contained several theorems of **projective geometry (a non-Euclidean geometry)**.

A year later, after the family had moved to Rouen, Pascal published his first paper, *Essay on Conic Sections*. The paper, based around work by the founder of projective geometry, Gérard Desargues (1591–1661), aroused attention and praise from established mathematicians. Over the next few years, financed by his father who was then a tax collector, Pascal worked on a long-term project – devising and modifying a calculating machine. Called the Pascaline, it took 50 modifications before it added and subtracted satisfactorily. The machine first went into production in 1642.

In 1646 Pascal took up the principles of Jansenism (a movement within the Roman Catholic Church, popular in the seventeenth century and considered by some to be heretical), which increasingly influenced his thoughts. However, in 1647 he turned his attention to completing a series of experiments on atmospheric pressure and vacuum, which followed the work of Italian scientist, Evangelista Torricelli (1608–47), who had died the same year. During this period, Pascal examined how barometers worked, and invented the syringe and a hydraulic mechanism based on his new pressure principle (now called "Pascal's law").

In spite of poor health due partly to overworking, Pascal pursued his research with renewed intensity from 1651 to 1654. During these years he examined the pressures exerted by liquids and gases, wrote a treatise on the arithmetical triangle ("Pascal's triangle"), and with Pierre de Fermat (1601–65) formulated probability theory (see below). His list of achievements came to an abrupt end in 1654 when he had an intense religious vision after being thrown from a carriage in an accident. He saw the experience as a sign that he should turn his back on the world and live a life of self-denial and prayer. His religious meditations, written during the last years of his life, were published posthumously as Pascal's *Pensées* (*Thoughts*).

Essential science

Pascal's theorum and projective geometry

Pascal's theorem of projective geometry, now known as "Pascal's theorem", established projective geometry as a formal discipline. It was discovered by Pascal when he was only 16 years old and involves the study of properties of figures that are unchanged when the figures are projected from a point to a line or plane. The theorem states that if a regular or irregular hexagon (six-sided figure) is drawn within any conic section (the curved shape obtained by slicing a flat surface through a cone, e.g. a circle or an eclipse), then the three pairs of the continuations of opposite sides meet on a straight line, called the Pascal line.

Mechanical calculator

Apart from a prototype built by Wilhelm Schickard in 1623, Pascal's was the first mechanical calculator to be built. In Pascal's calculator, rotors were used to dial in the numbers, and the answers were read through windows at the top of the lid. It went into production in 1642 but only fifty were built. It was too slow and expensive to be popular.

Pascal's law of pressure

Evangelista Torricelli had been the first to create a sustained vacuum in a laboratory and to suggest that air has weight. Pascal, unlike others such as René Descartes (1596–1650), believed there was a vacuum above the atmosphere. An experiment in which he carried a barometer up a hill in Auvergne in 1648 showed that the readings dropped with height. The experiment confirmed Torricelli's theory that the level of mercury in a barometer rises and falls according to the increase and decrease of surrounding atmospheric pressure. Pascal also confirmed that the pressure of the atmosphere could be likened to a weight, and that the pressure exerted by a vacuum was

Legacy, truth, consequence

■ Pascal's inventions that led to commonly used items today include the syringe and hydraulic press. His name has been given to the SI unit of pressure, the pascal (symbol: Pa), in recognition of his works on liquids and gases.

■ Pascal is well known for his arithmetical triangle. This arrangement of the binominal coefficients in a triangle has been studied by people around the world centuries before Pascal, but today is commonly called "Pascal's triangle". Work following Pascal's *Treatise on the Arithmetical Triangle* eventually led to Isaac Newton's discovery of the general binominal theorem of fractional and negative powers. This complex mathematical theory has gained a place in popular culture, for example in Mikhail Bulgakov's novel *The Master and Margarita*, the mysterious gentleman magician Woland says, "*But by Newton's binomial theorem, I predict that he will die in nine month's time…*" From this, "*It's hardly Newton's binomial theorem*" became a popular Russian expression.

■ Pascal's mechanical calculator looked similar to those of the 1950s, and, in working by counting **integers**, in some respect marked the beginning of the age of the digital calculator.

■ Following Pascal and Fermat's work, probability theory became a significant branch of mathematics. In physics, it became an important tool of **calculus** and is also a vital part of economics.

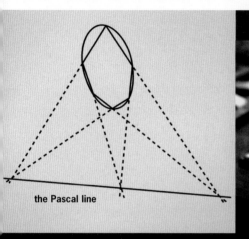

the Pascal line

Pascal's theorem of projective geometry.

Pascal's calculating machine, in spite of numerous modifications, was slow and cumbersome.

If God does not exist, one will lose nothing by believing in him, while if he does exist, one will lose everything by not believing.

Quotation known as "Pascal's wager", from *Pensées* (1670)

zero. This led on to Pascal's law of pressure (now called "Pascal's law" or "Pascal's principle"), which says that if pressure is applied to a non-flowing fluid in a container, then that pressure is transmitted equally in all directions within the container.

Probability theory

Pascal and Pierre de Fermat together studied **probability**. It is said that a wealthy Frenchman and gambler, Chevalier de Méré, asked for Pascal's help on probabilities in relation to a dice game. He wanted a method to work out how the stakes should be divided if a game of dice between two players was left unfinished. Pascal and Fermat found a mathematical method for how to divide the stakes in a game with two players, but were unable to go beyond this. Christiaan Huygens (1629–95) took the probability theory further after reading the letters of Pascal and Fermat.

Key dates

1623	Born in Claremont (now Clermont-Ferrand), Auvergne, France.
1640	Writes a paper on conic sections, based on studies of Desargues' writings on projective geometry.
1642	Produces a calculating machine to help his father calculate taxes.
1646	Meets and is influenced by two doctors tending his father, who are followers of Jansenism.
1647	Finds proof that a vacuum exists and writes *New Experiments Concerning Vacuums*.
1648	Takes a barometer up the Puy-de-Dôme in Auvergne and notes the changes in readings with height.
1653	Writes *Treatise on the Arithmetical Triangle*.
1654	Works on the probability problem with Fermat in the summer. Turns his back on the world after a religious vision in the winter.
1655	Goes on a retreat to a Port-Royal convent, and begins to write his first religious work, *Provincial Letters*.
1662	Dies of cancer at the age of 39, in Paris.
1670	Posthumous publication of Pascal's *Pensées*.

Robert Boyle

Robert Boyle was an Irish chemist and physicist who is best known for his investigations into the nature of gases, for formulating Boyle's law, and, more generally, as the first practitioner of systematic scientific experimentation. Although an alchemist, he was an important figure in the movement to transform chemistry from an occult enterprise into a science.

Born in Lismore Castle, in County Waterford, Ireland, in 1627, Boyle was the fourteenth of fifteen children. His father, Richard Boyle, was the first Earl of Cork and exerted wide social influence until the outbreak of the civil war.

As a child, Boyle displayed a prodigious memory and a talent for languages. For the most part he was privately educated although he did spend three years at Eton College, England, where he learnt mathematics and history. After leaving Eton he traveled in Europe, most notably being tutored on theology and philosophy in Geneva by the French Calvinist Issac Marcombes.

He returned to England in 1944, spent approximately ten years living at a family manor in Stalbridge, Dorset, where he read widely on philosophy, science, and theology, wrote on ethics and religion, and began to conduct a variety of scientific experiments, before moving to Oxford to join some of the leading scientists of the era, such as John Wallis (1616–1703), John Wilkins (1614–72), and Robert Hooke (1635–1703). Hooke became Boyle's research assistant and together they developed a new model of the air pump that was initially devised by Otto von Guericke (1602–86).

The majority of Boyle's significant scientific work was performed between the late 1650s and the early 1670s, and throughout this period he was extraordinarily prolific in his written output. Although much of this work derived from earlier experiments by scientists such at Guericke and his followers in Germany, Boyle's great contribution consisted in his ingenuity in constructing experiments and in his implementation of a strictly scientific methodology.

Leaving Oxford in 1668, he moved into the home of his sister Katherine, Lady Ranelagh, in Pall Mall, London, where, despite periods of ill health, he spent the rest of his life. In his later years he became particularly interested in diseases of all kinds and wrote a series of books on medical subjects.

Boyle died in 1691, at the age of 64, only a week after the death of his sister. He is buried at St Martin in the Fields, Trafalgar Square, London. The Irish inscription on his tombstone describes him as the "Father of Chemistry".

> **The Wisdome of God does confine the creatures to the establish'd Laws of Nature.**
>
> *Occasional Reflections* (1665)

Essential science

The air pump experiments

After hearing of Guericke's invention of an air pump, Boyle set out, in collaboration with his assistant Robert Hooke, to design an improved version in order to perform a series of experiments on the physical properties of air. Boyle performed experiments on the weight and expansion of air and he measured the effects of air on combustion and respiration, showing that in respiration air serves to rid the body of "noxious steams". Through these experiments he proved that sound is impossible in a vacuum, that air is necessary for life and for flame, and that air is permanently elastic. In 1660, he published *New Experiments Physico-Mechanical: Touching the Spring of the Air and their Effects*, which outlined the experiments and the results that had been gained using the air pump.

Throughout his life Boyle continued to perform experiments on the properties of air. Perhaps most notably, he showed that many fruit and vegetables contain air (actually carbon dioxide) which they give off during **fermentation**.

Boyle's law

Boyle's experiments with the air pump on the elasticity of air ultimately yielded the famous law that bears his name. In 1662, while writing an appendix to the second volume of *New Experiments*, which attempted to reply to some of the objections raised by his critics, Boyle managed to develop his findings into a quantitative relationship. Boyle's law, which is concerned with the nature of gases, states that the volume occupied by a gas is inversely proportional to the pressure of the gas. In its modern formulation, Boyle's law can be written as: $pV = K$ (where "p" is pressure, "V" is volume, and "K" is a constant). Strictly speaking, Boyle's law applies only to **ideal**

Legacy, truth, consequence

- Boyle wrote over 40 books on science, religion, and the nature of the relationship between them, which were read throughout Europe during his lifetime and after his death. Given his prolific output and his standards of experimentation, Boyle is considered one of the originators of science as a professional discipline.
- He was one of the creators of modern chemistry. His landmark publication *The Sceptical Chymist* was the first to make a distinction between chemistry and **alchemy** and is often credited as helping to make a science out of chemistry.
- In 1676 he wrote the first book dealing with the science of electricity, presenting his discovery that electrical attraction may occur in a vacuum and that an attracted body pulls the charged body as strongly as it itself is pulled.
- He endowed a series of lectures in his will – known as the Boyle lectures – which aim to prove the truth of Christianity. These lectures are still held annually.
- Towards the end of the twentieth century there was much renewed interest in Boyle scholarship. Birkbeck College, London, for example, run the Robert Boyle Project, which is devoted to his work. Many of these contemporary scholars have emphasized Boyle's intellectual debt to alchemy. Despite his emphasis on the separation of chemistry from alchemy, Boyle was still an alchemist and believed in the **transmutation** of metals.

> *Much may be said to Excuse the Chymists when they write Darkly and Aenigmatically, about the Preparation of their Elixir, and Some few other grand Arcana, the divulging of which they may upon Grounds Plausible enough esteem unfit; yet when they pretend to teach the General Principles of Natural Philosophers, this Equivocall Way of Writing is not to be endur'd.*
>
> The Sceptical Chymist (1661)

gases. Real gases follow Boyle's law only at low pressures and high temperatures.

Despite the fact that the law bears his name, Boyle was not the first to formulate it. Henry Power had originally discovered the same gas law in 1661. In addition, throughout continental Europe Boyle's law is sometimes known as Mariotte's law, after Edme Mariotte (1620–84) who, unaware of Boyle's work, published an essay that stated the same law, in 1676.

Key dates

1627	Born in County Waterford, Ireland.
1635–8	Studies at Eton College.
1645	Settles at a manor left to him by his father, in Stalbridge, Dorset.
1654–5	Moves to Oxford to learn from some of the leading scientists and philosophers of the day.
1657	Hears of Otto von Guericke's air pump.
1659	Finishes designing his own air pump.
1660	Publishes *New Experiments Physico-Mechanical: Touching the Spring of the Air and their Effects*.
1661	Publishes *The Sceptical Chymist*.
1662	Discovers Boyle's law. Publishes second edition of *New Experiments*. Appointed first governor of the Corporation for the Propagation of the Gospel in New England.
1663	Publishes *Experiments and Considerations Touching Colors*.
1665	Publishes *Occasional Reflections on Several Subjects*.
1666	Publishes *Origin of Forms and Qualities*.
1670	Suffers a serious stroke.
1676	Publishes *Experiments and Notes about the Mechanical Origine or Production of Particular Qualities*.
1680	Elected president of the **Royal Society of London**, although he declines the post.
1686	Publishes *A Free Enquiry into the Vulgarly Received Notion of Nature*.
1689	Falls ill. Resigns as governor of the Corporation for the Propagation of the Gospel in New England.
1691	Dies in London.

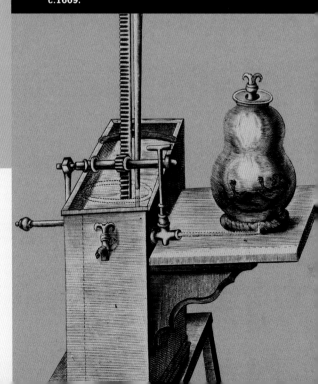

An air pump and glass receptacle, being used here to demonstrate that a small animal will die in a vacuum, c.1669.

Antony van Leeuwenhoek

Considered to be the first microbiologist, the Dutch lens-maker Antony van Leeuwenhoek opened up the microscopic world to eighteenth-century scientists. Making his own accurate microscopes, he discovered bacteria, single-celled organisms, spermatozoa, blood cells, the banded structure of muscle tissue, and microscopic creatures such as nematodes, as well as many other features of nature.

Christened Thonis Leeuwenhoek, he signed his name Antonij – Antony in English – and added the "van" to his name in 1686. In the seventeenth century most scientists (or natural philosophers) were gentlemen who had enjoyed a good university education, including classical and modern languages, but Leeuwenhoek came from a very different background. His family were tradesmen (his father was a basket-maker and his mother's relatives were brewers) and they were not particularly well-off; as a youngster he had only elementary schooling, and he never learnt any language other than Dutch.

When he was 16, Leeuwenhoek was apprenticed to a textile merchant in Amsterdam, where he spent six years learning the trade before returning to Delft to set up business for himself as a linen-draper. At that time magnifying glasses were used in the textile trade to examine the quality of cloth by counting the density of threads in material. The glasses, usually fixed on a stand, could magnify up to the power of three. Thus Leeuwenhoek would have early on encountered the principle of magnification, but it is thought that his interest in exploring the microscopic natural world was sparked off by the popular book *Micrographia*, written by the English scientist Robert Hooke (1635–1703) in 1665, which contained reports and vivid pictures of his observations of tiny objects and creatures such as fleas and lice.

Within three years Leeuwenhoek was grinding lenses to make his own microscopes and telling scientists in Delft about his discoveries. In 1673 a doctor, Regnier de Graaf, described Leeuwenhoek's reports to the secretary of the **Royal Society of London**, the world's oldest surviving society for the advancement of science, which had only recently been founded in 1660. The letter was published in the Society's *Philosophical Transactions*, and the secretary requested an introduction to Leeuwenhoek, beginning a correspondence that was to last until Leeuwenhoek's death at the age of 90. Written in Dutch and translated by the Society into Latin or English for its *Transactions*, he sent hundreds of illustrated letters and several of his specimens to the Society, and others to the Paris Academy of Science and to private individuals. In 1680 he was elected a full member of the Royal Society, although he never attended their meetings.

Leeuwenhoek must have had very good eyesight to be able to distinguish the details that he did. But he could not draw, so he hired an illustrator to provide the drawings that went with his written descriptions. His letters to the Royal Society were rambling and conversational, nothing at all like scientific papers, except that he described the meat of his studies factually, clearly, and accurately.

Essential science

Microscopes

Leeuwenhoek made more than 500 microscopes, although technically they were simply powerful magnifying glasses, not modern microscopes that have compound or multiple lenses. Early compound microscopes had been known since about 1595, but they could not magnify beyond 30 times natural size, whereas Leeuwenhoek achieved up to 300 times magnification. He inserted a single lens between two metal plates – brass, copper, or silver – that were then riveted together and fixed three or four inches above the base of the instrument. This usually had a spike for holding the specimen, and had screws to allow the focus to be adjusted by raising, lowering, or rotating the specimen. Some of his lenses were as small as a pin head.

Not only did he achieve greater magnification than other instrument makers, but his lenses were also notably clear and bright. He kept some of his techniques secret, and we still do not know exactly how he achieved his results. Perhaps he discovered a way of obliquely illuminating the specimens in order to enhance the lens, or, as other researchers think, perhaps he used the properties of spheres to improve his images, either encasing his specimens in spherical drops of fluid or using balls of glass for lenses instead of ground lenses.

Discoveries

Curious about everything that could be placed under a microscope, Leeuwenhoek examined plants and animal tissues, insects, fossils, and crystals. As a result he was the first person to describe many microscopic aspects of life such as living spermatozoa, from which he correctly concluded that an egg is

Legacy, truth, consequence

■ Leeuwenhoek is held to be the "father of **microbiology**" not just because he identified so many microscopic creatures, but also because of the way he confined himself to the pure facts in his descriptions.

■ His discoveries disproved the commonly held belief that lower forms of life could be spontaneously generated or born out of the corruption of natural material. For example, fleas were thought to be produced from sand or dust, and flour mites were thought to spring from rotten wheat. Leeuwenhoek showed that these tiny creatures actually had the same life-cycles as larger insects.

Leeuwenhoek is thought to be pictured in this painting by Johannes Vermeer, *The Astronomer*, c.1668.

... my work ... was not pursued in order to gain the praise I now enjoy, but chiefly from a craving after knowledge ... I have thought it my duty to put down my discovery on paper, so that all ingenious people might be informed thereof.

Letter (June 1716)

Key dates

1632	Born in Delft, the Netherlands.
1648	Begins apprenticeship to a textile merchant.
1654	Returns to Delft; sets up as a draper.
1660–99	Serves as chamberlain to the city sheriffs.
c.1665	Reads Robert Hooke's *Micrographia* and is inspired to use microscopes to look at the natural world.
1668	By now has learnt to grind lenses, make microscopes, and observe microscopic life.
1673	Begins correspondence with the Royal Society of London.
1676	Reports the discovery of single-celled organisms.
1676	Serves as executor of the estate of the artist Jan Vermeer.
1680	Elected a full member of the Royal Society.
1698	Demonstrates microscopes to Russian Tsar Peter the Great.
1723	Dies in Delft.

Many of his reports were published separately and widely circulated, making him famous as a man who had discovered the secrets of nature. Kings and queens were among the many curious people who visited him to look through his microscopes.

Since Leeuwenhoek was executor for the estate of the artist Johannes Vermeer, some art historians think that Leeuwenhoek is pictured in two Vermeer paintings: *The Astronomer* and *The Geographer*.

fertilized when the sperm penetrates it. However, perhaps because he lacked a scientific education and background, he seldom theorized about what he described. Among his most important discoveries are:

1. Unicellular organisms: On September 7, 1674, Leeuwenhoek wrote about pond water:

> *"I found floating therein divers earthy particles, and some green streaks, spirally wound ... The whole circumference of each of these streaks was about the thickness of a hair of one's head ... all consisted of very small green globules joined together: and there were very many small green globules as well."*

He had here discovered the single-cell structure of spirogyra algae. Until Leeuwenhoek reported his observations of pond life, single-celled or unicellular organisms were unheard of. The Royal Society thought this report was so unlikely that it sent a special mission comprising a vicar, medical doctors, and lawyers to examine his studies, and, in 1680, Leeuwenhoek's observations were fully confirmed.

2. Bacteria: Among some of the earliest accounts of bacteria are his studies of plaque on teeth, such as this letter of September 17, 1683: *"a little white matter, which is as thick as if 'twere batter."* He coined the world animalcule (little animal) for many of the tiny creatures he saw, as described in this extract:

> *"... there were many very little living animalcules, very prettily a-moving. The biggest sort ... had a very strong and swift motion, and shot through the water (or spittle) like a pike does through the water. The second sort ... oft-times spun round like a top ... and these were far more in number."*

Isaac Newton

A mathematician and physicist, Isaac Newton is considered to be one of the greatest scientific intellects of all time and the major figure of the seventeenth-century Scientific Revolution, which laid the foundations for modern experimental and investigative science. He is best known for his laws of motion and theory of gravity, which provided the first scientific explanation of how the universe was physically held together, but he made other significant discoveries in physics and mathematics.

Isaac Newton was born on 25 December 1642. Britain was still using the old Julian calendar rather than the Gregorian calendar which moved dates ten days ahead, but historians adopting retrospective Gregorian dates record his birth as 1643, not 1642.

His family were gentlemen farmers, but early on he realized that the farming life was not for him. He went to Cambridge University, but formulated many of his greatest ideas while the university was closed because of plague. During the two years of 1665–6, while working at home, he made the most intense scientific advances ever known in such a short period, although he did not publish his ideas for some years. He experimented in **optics** and **mechanics**, and observed (or was hit on the head by) a falling apple, leading him to conclude that it was the same force – **gravity** – which acted on the apple and on the moon's orbit.

> *If I have seen a little further it is by standing on ye shoulders of giants.*
>
> Letter to Robert Hooke (1676)

He was morbidly sensitive to criticism, and after an initial setback did not offer his great ideas for debate. As a result, his theories were finally published almost by chance. In 1684 the astronomer Sir Edmund Halley (1656–1742) consulted Newton about planetary orbits. He was astounded to find that Newton had a complete scientific theory: gravity as a universal force holding together the structure of the universe. Halley persuaded Newton to publish, and arranged and paid for the work.

The result was *Philosophiae Naturalis Principia Mathematica* (*Mathematical Principles of Natural Philosophy*), commonly known as *Principia*, published in 1687. This is considered to be the cornerstone of modern science, and has been described as the greatest single work in the history of science. Covering gravity and the laws of motion, the treatise became the accepted scientific view of the universe.

Newton wrote extensively on alchemy, ancient history, and Bible studies. He also held unorthodox Christian views, rejecting the Trinity. Partly because of these interests, he is associated with the organization known as the Priory of Sion, which, as popularized in Dan Brown's *The Da Vinci Code*, is supposed to hold secret religious knowledge.

Essential science

Scientific approach

He considered himself to be a natural philosopher, speculating on the nature of reality. This speculation led him to the possibility of universal forces underlying the mechanistic nature of the universe. He held that understanding can be reached by experiment, and that mathematical reasoning is the best way to describe the physical world.

Theory of universal gravitation

Newton showed that the same force – gravity – acts both at close quarters and at vast distances: pulling an apple to earth and holding the planets in orbit around the sun. An object with more matter, or mass, exerts the greater force of attraction. His Inverse Square Law of Attraction explains the force of gravity as related to the inverse square of the distance between the two objects: $F = Gm_1m_2/r^2$ where F is the gravitational force, G is the universal gravitational constant, r is the separation between the objects, and m_1 and m_2 are the two objects.

Calculus

An essential tool for advanced mathematical analysis, this enables calculation of the area bounded by a curve and the slope of a point along a curve, among many other applications. Newton discovered these problems are inversely related, and resolved them by using "fluxions", algebraic expressions to calculate the magnitude of "flow" of the curve. Although his theorem is valid, **Gottfried Leibniz**'s terms of differentiation and integration are now used instead of "fluxions".

Optics

Before Newton's experiments, it was thought that white light was homogeneous, and that color belonged to the colored object, not to a property of light. He was not the first to observe that light passed through a prism separates into the full **spectrum** of colors,

Legacy, truth, consequence

■ Newton's explanation in mathematical terms for how a large part of the physical universe works – both on earth and in the skies – was a revolution: before him scientists had only observed, not explained, planetary orbits and other mechanical phenomena.

■ His work inspired a new generation of experimental scientists using his investigative and analytical methods to explain other aspects of the natural world.

■ Newton's theories of universal gravitation and motion underpin modern disciplines such as rocket science, and remain valid except for within the comparatively new fields to do with near-light speed relativity and **quantum mechanics** (see **Max Planck**, pages 132–3).

■ His theory of the mathematical properties of light and color established the modern discipline of optics.

> *Plato is my friend, Aristotle is my friend, but my best friend is truth.*
>
> Certain Philosophical Questions (c.1664)

but he continuously refined his experiments until he could prove that white light is composed of colors, each with different properties.

First law of motion, or law of inertia
An object at rest will stay at rest and an object in motion will remain in motion until acted upon by an external force.

Second law of motion, or law of acceleration
A force acting upon an object will change the object's velocity in the direction of the force, directly proportional to the force applied and inversely proportional to the mass of the object, as often reduced to the equation $F = ma$ (force = mass X acceleration).

Third law of motion
For every action there is an equal and opposite reaction.

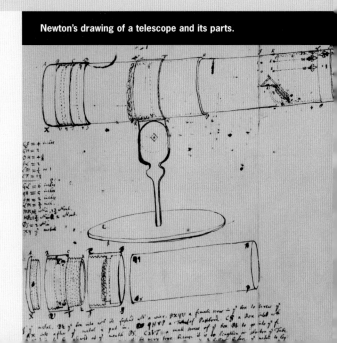

Newton's drawing of a telescope and its parts.

Benjamin Franklin

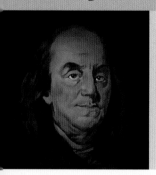

Printer, inventor, scientist, statesman, revolutionary, philanthropist, and much more, Benjamin Franklin was a omnipresent figure throughout the development of eighteenth-century America. Such was his importance that had he been only a scientist, or only a philanthropist, or only a politician, he would still have achieved enough in any one of those careers to be considered a major luminary.

Benjamin Franklin is recognized as a great American, but he was born to English parents, the fifteenth of seventeen children, in Boston, Massachusetts. He was only schooled for two years, and after a stint learning the trade of his chandler father he became an apprentice printer for his brother James, who had just set up Boston's first original newspaper, *The New England Courant*. One of the most successful articles was a series of recurring advisory letters written by a mysterious woman called Silence Dogood. It turned out that they'd been penned by Benjamin, and the subsequent fallout led to his search for further employment.

Printing was a progressive and increasingly significant trade in eighteenth-century America and Franklin decided to stick with it. He settled in Philadelphia where, interrupted by an ill-starred couple of years in London, he set up his own printing business, took over the running of a newspaper and published annually his *Poor Richard's Almamac*. His reputation was such that, still only in his twenties, he began to receive contracts for government work.

So busy did Franklin become in so many aspects of professional life that his interest in science almost seems like something he did to pass the time. In the mid-1800s, with some successful inventions, such as the Franklin stove, already to his name, he begun studies on electricity. His famous, and rather dangerous, experiment of flying a kite in a thunderstorm proved the connection between lightning and electricity and led to his invention of the lightning conductor.

Franklin's later years were spent in politics. He regularly worked with the English. His diplomatic trips backwards and forwards between the two countries culminated in his conclusion that American independence was the way forward. He was consequently a significant contributor to the resultant Declaration of Independence.

His son, William, became the Royal governor of New Jersey. William remained loyal to England, a decision that caused a permanent rift between him and his father.

Essential science

Electricity

In the Europe of the 1740s, the idea that electricity could have more application than just being a novel form of entertainment was beginning to take hold. An electrical machine was dispatched from England to Franklin's Library Company and he began experimenting in 1746. The fact that he had little idea of what the Europeans had discovered possibly helped him maintain a open mind in his observations.

One of his initial conclusions was that electricity was a form of energy that could spread through many substances and that, once touched by an electric current, the natural order always reverted to the norm. It was traditionally thought that electricity was made up of two kinds of fluid. Franklin showed there was only one fluid that had a positive charge and a negative charge occurring in equal amounts. This is called the law of conservation. He also experimented with a capacitor – a simple box made of charged glass plates that could store electrical charges – which he called a battery. The word is derived from its alternative definition of a physical attack, supposedly because that is what the effect of an electric shock from the device felt like.

Lightning rods

The eighteenth century's many wooden buildings were always susceptible to fire by lightning strikes. Franklin's discovery that lightning and electricity are identical prompted him to develop the lightning rod. In his experiment he ran a length of cable down the side of the house. The bottom end was buried several feet into the ground. A long rod pointing up to the sky was attached to the top end. Lightning struck the rod and ran its charge down the cable into the ground, bypassing the house altogether and significantly cutting down the risk of a fire.

Bifocal glasses

This was an invention borne out of personal necessity. As he grew older, Franklin found it hard to see both close-up and far away, and had to constantly swap between two pairs of glasses. It was a tiresome task so he put his mind to developing a lens with two focal points. The final product actually entailed fitting two lenses into the frame of the spectacles, one for distance at the top and one for seeing close-up at the bottom, where it would be better suited to aid reading.

Legacy, truth, consequence

- A look at the long list of institutions Franklin helped to found, many of which still thrive today, gives an idea of the range of his influence – The Philadelphia Union Fire Company, the American Philosophical Society, and the Pennsylvania Hospital to name but three. Franklin himself considered his work in public service more consequential than his scientific contributions.

- His will stipulated that a significant amount of money be used to support future enterprises. Many social and technical advances that benefit his country today, especially around Pennsylvania, are a legacy of his foresight.

- Franklin was the first to use electrical science words such as battery, electrify, charge, conductor.

- He was a passionate advocate for the abolition of slavery. The year before he died he wrote an anti-slavery treatise, and after his death, his will decreed that his son-in-law should "*set free his negro man Bob*".

> ## *Our new Constitution is now established, and has an appearance that promises permanency; but in this world nothing can be said to be certain, except death and taxes.*
>
> Letter to Jean-Baptiste Leroy (1789)

Heat-efficient stove

Franklin redesigned the inefficient household fireplace of the day. By adding a canopy-like structure on the front of the stove, a ventilator at the back and a new arrangement of flues within, he created a fire that produced twice as much heat and burned one quarter the amount of fuel – an eight-fold improvement in performance.

He declined to take out a patent, because the stove's purpose was for the benefit of society.

> ## *Does thou love life? Then do not squander time, for that's the stuff life is made off.*
>
> Poor Richard's Almanack (1741)

Key dates

1706	Born in Boston, Massachusetts, US.
1718	Becomes an apprentice printer to his brother, James.
1724	Moves to London for two years.
1729	Takes over publication of the *Pennsylvania Gazette*.
1730	Marries Deborah Read.
1731	Founds the first public library.
1732	Prints the first *Poor Richard's Almanack*. Continues publication until 1758.
1740	Invents the Franklin stove.
1748	Retires from printing.
1751	*Experiments and Observations on Electricity* is published in London.
1752	Flies a kite in a thunderstorm.
1753	Becomes postmaster general of the British Colonies in America.
1757	Travels to England as colonial representative.
1771	Begins his autobiography.
1776	Helps to draw up the Declaration of Independence.
1778	Negotiates the Treaty of Alliance with France. The two countries agree to help each other in the event that either is attacked by the British.
1783	Signs the Treaty of Paris with Britain.
1784	Invents bifocal glasses.
1790	Dies in Philadelphia; 20,000 people attend his funeral.

Franklin's reception at the court of France, 1778, where he negotiated the Treaty of Alliance with France.

Carolus Linnaeus

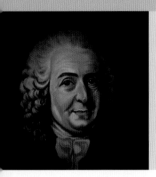

Carolus Linnaeus was a botanist who developed the first complete system of classification for organisms and who first used binominal nomenclature – two names (genus and species) for each living thing. Linnaeus' classification and naming methods were welcome at a time when explorers and botanists had begun to discover the vast range of the world's plant and animal kingdoms.

Born in Sweden, Carl Linnaeus (also called Carolus Linnaeus and Carl von Linné) developed an interest in plants as a boy. Inspired by the family's large garden and his father's interest in plants, he was known as "the little botanist" at eight years old.

Rather than follow his father into the ministry, Linnaeus chose to study medicine, largely because it included the study of **botany**. But at the University of Lund in 1727, and in 1728 at the University of Uppsala, a lack of lectures on botany meant that Linnaeus had to pursue his interest alone. Then a meeting in 1729 with eminent botanist Olof Celsius (1670–1756) changed his life. Impressed by Linnaeus' knowledge and collection of over 600 pressed native plants, Celsius provided him with a lodging and his library in which to continue his work on identifying plants and developing a new system of plant classification.

The following year Linnaeus became lecturer in botany at Uppsala. In 1732, botanical investigations on a trip to Lapland resulted in Linnaeus' work *Flora Lapponica*, in which he used his own system of classifying plants by their stamens and pistils, and rejected the older Tournefort system of describing them according to their flowers and fruit.

In order to complete his medical degree, Linnaeus went to Holland in 1735. By the time he left Holland for Sweden in 1738, Linnaeus had published 14 books on botany. As well as setting out his new method of classification, he had also established rules and principles on how plants should be named. By then he had acquired a reputation as a well known and respected botanist. Because his works were written in Latin, he became known as Carolus Linnaeus.

In 1742, after a brief period as a practicing and successful physician in Stockholm, he took the chair of professor of medicine and botany at Uppsala University. Teaching, research, and writing filled his later years. In 1753, *Species Plantarum* was published. It described 7,300 species of plant using his binominal system, for the first time omitting the old, lengthy names. It was revolutionary in simplifying the naming of plant species. In 1762 he was knighted, and took the name of von Linné.

Essential science

Taxonomy

Taxonomy, the classification of living things, had been attempted by botanists and naturalists before Linnaeus. Plants had been classed according to their location, time of flowering, and medicinal use. Swiss naturalist Konrad von Gesner (1516–65) had used the more scientific method of grouping them by their fruits, and Andrea Cesalpino (1519–1603) developed a precise method based on the parts of fructification (the seed-bearing structure). Cesalpino is sometimes considered to be the first modern botanist because he was the first to study the parts of plants. Many systems, such as French botanist Joseph Pitton de Tournefort's (1656–1708), classified plants according to their flowers. Tournefort took this further by grouping plants into genus, then species. The genus was based on the flower form and the species on the position of the fruit.

Linnaeus was the first to realize the importance of a complete system. Once the classification groups had been agreed, scientists could then use them as a universal system that all could refer to and understand. Linnaeus' system classified plants according to the stamens and pistils of flowers and ranked them into classes, orders, genera, and species.

Increased knowledge has meant that many more levels of the hierarchy of order have been added in both botany and **zoology**. For example, in botany phylum, class, subclass, etc, lie above genus, and subspecies lie below species.

Binominal nomenclature

Before Linnaeus' system was taken up, plants were often given a group name followed by a long description. The names often differed according to each botanist's classification and system, causing confusion for readers and students. To avoid this and for ease of reference Linnaeus suggested that certain rules should be followed. In *Critica Botanica* (published 1737) he explained that all plants of one genus should always be given the same generic name. The Latinized name should be easy and short, and if possible reflect a major characteristic of the plant, such as *Helianthus* (which means "flower of the sun") for a flower that is "*modeled on the sun's shape*".

Legacy, truth, consequence

■ Linnaeus simplified the method of naming plants and brought order to their classification. By the early nineteenth century the Linnaean system was the most accepted by English-speaking scholars and provided a foundation for today's taxonomy. However, his system was limited in being based solely on the number and arrangement of the reproductive organs (stamens and pistils).

■ Systems today use Linnaeus' binominal nomenclature, but classify and name plants using many more techniques beyond Linnaeus' focus on plant stamens and pistils. Darwin's evolutionary concepts, for example, are applied to botanical systems today, as well as research into **plant genetics** and **biochemistry**.

Key dates

1707   Born in Rashult, Sweden.

1729   Meets scholar Olaf Celsius and takes a room in his house.

1730   Gives his first lecture on botany to the public, in Uppsala botanical garden.

1732   Explores Lapland for the Uppsala Academy of Sciences.

1735   *Systema Naturae* is published, a 12-page booklet outlining his system for classifying the world's animals, plants, and minerals.

1737   First edition of *Genera Plantarum* is published, incorporating genera of every plant known and a revised classification system. It established Linnaeus as an important botanist.

1753   *Species Plantarum* is published, in which he describes and names (by genus and species) 7,300 species of plants.

1762   Knighted and takes the name of Carl von Linné.

1767   Twelfth edition of *Systema Naturae* is published, now incorporating 2,300 pages, 15,000 plant species, and 4,300 animal species.

1778   Dies in Uppsala, Sweden, after a stroke and long period of ill health.

It is a distinguishing mark of a very good name that the plant should offer its hand to the name and the name should grasp the plant by the hand ...

Preface to *Critica Botanica* (1737)

Linnaeus in Lapp costume, after his exploration in Lapland.

An example of Linnaeus' simplification of names is the wild briar rose. Previously botanists had named it as *Rosa sylvestris inodora seu canina* or *Rosa sylvestris alba cum rubore, folio glabro*. He simplified it to *Rosa canina*. Today even those names he published long ago in *Species Plantarum*, in 1753, are still used.

Herbaria

Linnaeus had his own herbarium, which consisted of dried plant specimens, pressed and mounted on sheets of paper and labeled. He studied his ever-growing collection while trying various methods of classification, and described his collection of plants "*without doubt the greatest collection ever seen*". Collections such as these are still important in the study of plant **taxonomy** and provide a record of the change in plant growth and species over time.

Exploration

Linnaeus was an important figure in the study of plants from around the world. He sent many students on exploration voyages, including Daniel Solander (1733–82), the Swedish naturalist who went on James Cook's first world voyage, between 1768 and 1771, and who returned with the first specimens from Australia and the South Pacific.

James Hutton

James Hutton was a Scottish scientist who is credited as the founder of modern geology. Employing extensive field research on various rock formations, he hypothesized that the surface of the earth is formed by ongoing geological cycles of erosion and deposition. This led to his greatest achievement: his proof that the earth was far older than had previously been believed.

James Hutton was born in Edinburgh, Scotland, in 1726. He was the son of an Edinburgh merchant who held the office of City Treasurer but died when Hutton was very young.

Interested in both chemistry and mathematics, Hutton attended first the High School and then (after a brief spell as a lawyer's apprentice) Edinburgh University as a medical student, at the age of fourteen – the standard age to begin university at the time. After a period of research in Paris, he gained a Doctor of Medicine degree at Leyden (in the Netherlands) with a thesis on blood circulation, before returning to Scotland in 1750.

At this point Hutton abandoned medicine and moved to the family-owned farm in Berwickshire, where he studied agriculture and used the farm as a base from which he traveled around England, France, Belgium, and Holland studying **geology**.

In the late 1760s he left the farm and made Edinburgh his permanent home. He became an influential figure in the **Scottish Enlightenment**, moving in social circles that included economist Adam Smith (1723–90), philosopher David Hume (1711–76), chemist Joseph Black (1728–99), and Hutton's future biographer, the scientist John Playfair (1748–1819). Together Hutton, Smith, and Black founded the Oyster Club, an organization that conducted weekly meetings devoted to their varied intellectual pursuits, although often focusing on scientific matters. Hutton, Smith, Black, and Playfair were also founding members of the Royal Society of Edinburgh in 1783. During this period, inspired by the striking physical phenomena he witnessed in and around Edinburgh and also on his travels outside the city, Hutton began work on his most famous book, *Theory of the Earth*.

Hutton died in Edinburgh in 1797. Surviving him was an illegitimate son; none of Hutton's friends knew about him until after Hutton's death.

Essential science

Theory of the Earth

Hutton's theory of the earth grew in his mind after his return to Edinburgh. During the 1780s he presented his ideas to the Royal Society of Edinburgh and published papers on the theory in the Society's *Transactions*. He finally published his great work *Theory of the Earth* in two volumes in 1795, two years before he died. Prior to this time there had been some interest in earth sciences, but geology as an individual branch of science was barely recognized. Hutton's theory was therefore the first major geological contribution.

Hutton claimed that the earth was experiencing a continual pattern of self-restoration. He proposed a geological cycle, where the erosion of the land was followed by the deposition of the eroded matter on the sea floor. These deposited particles would consolidate into **sedimentary rock**, which would then rise to form new land, then to be eroded again, with the process repeating itself over and over. Hutton observed evidence of this process in the sedimentary rock at Inchbonny, Jedburgh, in 1787, in what is now known as "Hutton's unconformity". The following year he saw similar evidence, also in the Scottish Borders, at Siccar Point, Berwickshire.

From extensive field research on various rock formations, especially the wave-cut benches found on many coastlines, Hutton concluded that the geological cycle was an extremely slow process, since it must have been repeated an indeterminate number of times in the past. Contrary to the age of the earth specified in the Bible, Hutton claimed that the earth was in fact far more ancient. Equally, he could find no evidence to suggest that this cyclic process would ever cease and he assumed it would continue indefinitely. Thus, he conceived of his geological cycle as a process without beginning or end.

At the basis of Hutton's theory was the assumption, supported by his observations, that geological evidence, such as rock formations, provides both a key to the past and an indication of the course of future events. After Hutton, this principle came to be known as Uniformitarianism.

Although a lively and interesting man in conversation, Hutton was not known for the clarity of his written exposition. Due to the prolix style of his prose, many readers first came to appreciate Hutton's ideas through reading John Playfair's *Illustrations of the Huttonian Theory* (1802), which endeavored to present Hutton's theory in a more readable form.

Legacy, truth, consequence

- Hutton's greatest achievement is that he saw the earth was much older than the 6,000 years claimed by biblical scholars. However, he was unable to provide an *exact* age for the earth because this required knowledge of the rate of decay of naturally occurring **radioactive** elements and radioactivity was unknown at this time.

- He foresaw the possibility of **evolution** in both the physical world and living creatures before Charles Darwin was born. In fact, Hutton's view that the world is much older than had previously been believed was highly influential on the formation of Darwin's theory.

- Hutton was a man of wide scientific and intellectual interests. He published a three-volume treatise on metaphysics and moral philosophy and made other contributions to the physical sciences, in particular in chemistry, physics, and **meteorology**. In one notable case, Hutton and his friend Jamie Davie extracted sal ammoniac, a rare mineral composed of ammonium chloride, from chimney soot and conducted a joint investigation into its properties and production.

We find no vestige of a beginning, no prospect of an end.

Hutton's description of the geological cycle given in his "Theory of the Earth" lecture to the Royal Society of Edinburgh (1788)

Key dates

1726	Born in Edinburgh, Scotland.
1744–7	Attends Edinburgh University.
1749	Gains a Doctor of Medicine degree at Leyden (the Netherlands) with a thesis on blood circulation.
1750	Returns to Edinburgh. Conducts chemical experiments.
1752	Visits Norfolk to study innovative farming.
1754	Returns to Scotland.
1764	Goes on a geological tour of north Scotland.
1767–8	Returns to Edinburgh.
1770	His house is built, St John's Hill, Edinburgh.
1767–74	Involved with work on the Forth and Clyde Canal.
1777	Publishes *Considerations on the Nature, Quality and Distinctions of Coal and Culm*.
1783	Royal Society of Edinburgh is founded.
1785	Lectures on his "Theory of the Earth" to the Royal Society of Edinburgh.
1787	He notices the "Hutton unconformity" at Ichbonny, Jedburgh (Scottish Borders).
1788	Printed versions of his "Theory of the Earth" lectures to the Royal Society of Edinburgh are circulated.
1795	Publishes two-volume *Theory of the Earth*.
1797	Dies in Edinburgh.

Siccar Point, on the east coast of Scotland, where Hutton observed evidence of the angular unconformity now known as "Hutton unconformity".

The past history of our globe must be explained by what can be seen to be happening now. No powers are to be employed that are not natural to the globe, no action to be admitted except those of which we know the principle.

"Theory of the Earth" Lecture to the Royal Society of Edinburgh (1785)

Antoine Lavoisier

A French chemist who launched the "Chemical Revolution" at the end of the eighteenth century, Lavoisier dismantled the old phlogiston theory of combustion, introduced his experimentally-based theory of the nature of oxygen, and coauthored the basis of the modern system of chemical naming. His scientific work was cut short by the guillotine during the Age of Terror that followed the French Revolution.

The son of a wealthy Parisian family, Antoine-Laurent de Lavoisier followed in the footsteps of his father and maternal grandfather by studying law. But he also explored natural sciences, writing papers that won him election to France's foremost natural philosophy society, the Academy of Sciences, in 1768. Apart from his books, most of his research was published through the Academy.

At about the same time Lavoisier invested in the General Farm, a finance company that lent money to the government every year, and for reimbursement was allowed to collect some sales and excise taxes. Investors, known as tax farmers, usually made a great deal of money and were hugely unpopular.

In 1771 he married 13-year-old Marie Anne Paulze, the daughter of a fellow tax farmer. She has been called the "mother of chemistry" for her role experimenting with Lavoisier, translating English papers, and illustrating his works. As well as his scientific experiments, Lavoisier worked on several government committees proposing social reforms and played a major part in the adoption of the metric system to rationalize weights and measures. He also ran the French Gunpowder Administration for several years, making sure that the future revolutionary armies were self-sufficient in gunpowder.

Like many other philosophers and scientists of his time, Lavoisier approved of the rational policies of the new republic after the French Revolution in 1789. Even when the Terror began and everyone who had benefited from wealth or privilege under the old regime was threatened, he believed that loyal, useful scientists would never be killed. He was wrong. As a former tax farmer he was doomed, especially since he had made a personal enemy of the radical Jean-Paul Marat after once scorning his scientific ideas. Marat in fact was stabbed to death in his bath before Lavoisier faced trial, but along with his fellow tax farmers, Lavoisier was sent to the guillotine in 1794.

Essential science

Methods
Lavoisier wanted to give chemistry the same standards of investigation that were emerging in physics, so he was one of the first chemists to insist upon carefully recorded measurements in experiments and the use of a standardized chemical language. As a wealthy man he had an impressive laboratory and could afford all necessary equipment for his own grand experiments.

In 1787 Lavoisier and three prominent colleagues published the *Méthode de Nomenclature Chimique* (*Method of Chemical Nomenclature*), doing away with the old fanciful names of substances (such as "flower of zinc") and calling for standard, logical naming principles: substances should have one fixed name, preferably from Latin or Greek roots, and the name should reflect the known composition.

Two years later he published *Traité Élémentaire de Chimie* (English title: *Elements of Chemistry*) that set out proper methods for investigation and report, and included his list of 33 known elements, which he defined as substances that he could not break down any further. Up until then some scientists still believed the ancient idea that all matter was made up of the four elements of earth, air, fire, and water.

Phlogiston: an outdated theory
The prevailing view of combustion was that a colorless, odorless, fiery substance called phlogiston – the very element of fire itself – was released from flammable materials during combustion, and was also released in respiration and during the **calcination** of metals.

The theory held that some materials were phlogiston-rich, such as charcoal, which is why there was so little residue when it was burned. A metal calx, or powder, when mixed with burning charcoal took the charcoal's released phlogiston to return the metal to a solid form. Other substances, like air, could absorb phlogiston but were never able to contain much of it. So, fires in an enclosed space burnt out because the air quickly reached saturation point of phlogiston, and a creature locked in an airtight space died because air again reached saturation point and was unable to absorb any more phlogiston from the body, in what was thought to be the role of air during respiration.

Oxygen theory of combustion
One of the problems with the phlogiston theory was that to explain some results of burning it was variously credited with some weight, no weight, or even a negative weight. It was from

Legacy, truth, consequence

- Lavoisier is considered to be the founder of modern chemistry and the inspiration of the **Chemical Revolution**. His appeal to fellow investigators for a standard methodology and description helped move chemistry from a disorganized muddle to a modern science.
- By disproving the phlogiston theory he also helped shift chemistry away from semi-mystical hypotheses dating back to medieval or even older **alchemy**.
- His naming system of chemistry is fundamentally the one used today.
- Lavoisier carried out many careful experiments, but was a theorist at heart. His experiments were mainly developments of other people's work (which he did not always acknowledge) in that he took their original ideas but tried to establish the correct explanation.

> *We must trust to nothing but facts. These are presented to us by Nature, and cannot deceive. We ought, in every instance, to submit our reasoning to the test of experiment, and never to search for truth but by the natural road of experiment and observation.*
>
> *Elements of Chemistry* (1789)

A plate in the 1789 edition of Lavoisier's *Elements of Chemistry*.

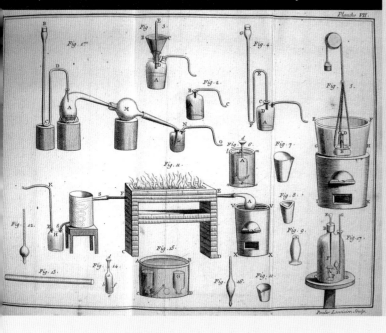

Key dates

this point of view that Lavoisier first looked at combustion in the early 1770s, to explain why during calcination the calx or powder weighs more than the original metal, and if it is returned to a solid form, it loses weight. Rather than fire, Lavoisier wondered if it was air that caused the weight changes.

He did his experiments in closed containers, weighed everything involved in the process of combustion, and showed that every time a substance gained weight, air was absorbed, and every time a substance lost weight, air was released.

In 1774 the British chemist Joseph Priestley (1733–1804) isolated "dephlogisticated air" and Lavoisier immediately realized that this might be the element involved in burning.

More experiments along Priestley's lines involved burning a candle in an airtight container, then showing how a small animal could not breath the air that was left over. Lavoisier showed that the same part of air that was burnt up – the "dephlogisticated air" – was the same part that metal residues absorbed from air making them heavier after heating. Lavoisier named this gas "oxygen" (meaning "acid generator" in Greek) because he at first thought

that it was also responsible for the acids made when chemicals such as sulfur are mixed with water.

Lavoisier gave the name "azote" (Greek for "without life") to the unbreathable gas left behind by burning. It is now called nitrogen.

Law of conservation of mass

Lavoisier's careful weighing of the substances before and after chemical reactions led to an early statement of the law of conservation of mass, that the total weight of substances remains the same, but the weight may be redistributed as the substances are rearranged. In other words, matter cannot be created or destroyed, but it may be rearranged.

Other science

Working with Pierre-Simon Laplace, Lavoisier proved that water is a compound of oxygen and hydrogen (a name he chose meaning "water-maker"). He also showed that respiration produces carbon dioxide and water, and that organic compounds contain carbon, oxygen, and hydrogen.

Joseph Banks

Sir Joseph Banks was an English naturalist, botanist, and patron of the sciences. Well known for his exploratory voyages to far away lands and the vast quantities of rare plant specimens that he collected on his travels, Banks did a great deal to promote science throughout the world.

Joseph Banks' birth in 1743 was deemed so significant by his wealthy and important London family that it was announced in *The Gentleman's Magazine*. Interested in natural history through his early years, the young Banks was disappointed when he entered Christ Church, Oxford, at the university's emphasis on teaching the classics, and he sought private lessons from botanist Israel Lyons (1739–75) in order to further his scientific interests in **botany** and **entomology**.

Shortly after graduation he inherited a large fortune from his father, who had died three years earlier. It was not long before he set off on the first of a trilogy of exploratory expeditions from which he made his name. On the 1766 voyage to Labrador and Newfoundland, Canada, Banks was commissioned as an onboard naturalist for a fishery-protection ship, and throughout the trip he accumulated numerous plant specimens.

His election to the **Royal Society of London** was followed, in 1768, by a second voyage, again as resident naturalist, but this time to the southern hemisphere on the *Endeavour*, commanded by Captain James Cook (1728–79). This expedition was later known as Captain Cook's first great voyage: around Cape Horn, then westwards across the Pacific to Tahiti. With botanist Daniel Solander (1736–82) as his assistant, Banks' main objective on the trip was to oversee the astronomical observation of the transit of Venus, which he achieved a year after the ship set sail.

On his return, the large collection of plant specimens he brought back with him attracted wide public interest and made him a celebrity. Notably, King George III summoned Banks to Windsor Castle to describe of the expedition and this meeting marked the beginning of a lifelong friendship between the two men.

Banks' final expedition, undertaken a year after he returned to London, was to Iceland, in order to conduct an investigation into the geysers (hot springs).

Back in England in 1778, Banks became president of the Royal Society amid some opposition, and he remained so until his death more than 40 years later. The position enabled him to use his connections to bring wealthy patrons into the Society and promote its international reputation, and to reestablish the connections between the organization and King George III, which had become strained following a disagreement over the best shape for the ends of lightning conductors.

He married Dorothea Hugesson in 1779 and they settled in a large house in Soho Square, London. Their home was frequently a hive of scientific activity; a popular meeting place for scientists and eminent intellectuals.

In later life he suffered repeatedly from gout and after 1805 was wheelchair-bound, although he continued to conduct meetings and even pursued a new interest in archeology. He died in Isleworth, near London, in 1820.

Essential science

Kew Gardens – London
Banks was the honorary director of the Royal Botanic Gardens at Kew, London, and, with the assistance of the king, made the institution an important center for botanical research. In order to gain as many new plant species for the garden as possible, Banks oversaw the distribution of expert plant collectors to countries all over the world. Another aspect to his position was to promote a functional role for the botanical center. In this regard, he initiated several important projects. He exported merino sheep from Spain to the UK and Australia; tea plants from China to India; and, despite an initial setback brought about by the mutiny on the *Bounty*, the ship being used for transportation, he exported the breadfruit tree from Tahiti to the West Indies.

Botany Bay colony
During his voyage on the *Endeavour*, Banks explored the coasts of New Zealand and Australia and developed a lifelong affection for the area. In 1788 he played a key role in establishing the first British colony at Botany Bay, Australia, and subsequently oversaw its growth and stability.

The African Association
Banks was at the forefront of African exploration, when he led the African Association in 1788. The aim of the organization was to fund various expeditions to Africa to chart its geography, in particular to discover the origin of the Niger River and the location of Timbuktu, "the lost city of gold".

Legacy, truth, consequence

- Some 75 species of plants bear Banks' name. He introduced eucalyptus, acacia, mimosa, and the genus Banksia to the Western world.
- Numerous places around the world also bear his name, including Banks in Canberra, Bankstown in Sydney, the Banks Peninsula on South Island in New Zealand, Bank Island in Canada, and the Banks Islands in Vanuatu.
- When the *Endeavour* arrived in Australia, near modern-day Sydney, Banks' extensive plant collecting gave the area Botany Bay its name. From this trip alone he brought back over 800 previously unknown species of plant.
- His methods for mounting and storing specimens at his home in Soho Square, London, subsequently became the standard practice for museums worldwide. His natural history library and world-renowned collection of plant specimens are now kept in the British Museum, London.

This 1795 caricature shows "the metamorphosis of Sir Joseph Banks from a caterpillar to a butterfly upon his investiture with the Order of the Bath (a British order of chivalry) as a result of his South Sea expedition".

Scientific patronage

Banks used his wealth and position to positive effect by bestowing financial assistance on talented young scientists. Robert Brown (1773–1858), the eminent botanist and discoverer of **Brownian motion**, was one such recipient.

Key dates

1743	Born in London, England.
1752–60	Educated at private schools: Harrow and Eton.
1760–3	Attends Christ Church College, Oxford.
1764	Inherits a large fortune.
1766	Voyage to Labrador and Newfoundland. Elected fellow of the Royal Society.
1768	Voyage to the southern hemisphere.
1771	Returns to England.
1772	Voyage to Iceland.
1778	Appointed president of the Royal Society. The first colony established at Botany Bay.
1779	Marries Dorothea Hugesson.
1781	Made a baronet, gaining the title Sir Joseph Banks, 1st Baronet.
1788	Leads the African Association.
1797	Becomes a member of the Privy Council.
1808	Honorary founding member of the Wernerian Natural History Society of Edinburgh.
1820	Dies in Isleworth, near London.

Of plants here are many species and those truly the most extraordinary I can imagine ... probably no botanist has ever enjoyed more pleasure in the contemplation of his favorite pursuit than Dr Solander and myself among these plants; we have not yet examined many of them, but what we have turned out in general so entirely different from any before described that we are never tired with wondering at the infinite variety of Creation, and admiring the infinite care with which providence has multiplied his productions, suiting them no doubt to the various climates for which they were designed.

Journal entry, on the *Endeavour* voyage (January 20, 1769)

Alessandro Volta

An Italian physicist, Alessandro Volta was a pioneer in the early history of electricity. He is most famous for his research on animal electricity and as the inventor of the first electrochemical cell, known as the voltaic pile, the prototype for the modern electric battery.

Alessandro Giuseppe Antonio Anastasio Volta was born in Como, Lombardy, Italy, then under Austrian rule, into a family of aristocratic origins, who at that time were enduring financial struggles.

Although he did not speak until he was four years old, Volta subsequently made great progress with his academic studies. A diligent student at both the public school and then the Royal Seminary in Como, he was very inquisitive and eager to learn, especially about natural phenomena.

Entering the academic profession, he secured his first position, as professor of physics in the Liceo, Como, in 1774. Throughout this time Volta was interested in a wide range of intellectual pursuits. He traveled to Switzerland in 1777 and met with great thinkers such as Horace-Bénédict de Saussure (1740–99) and Voltaire (1694–1778).

Two years later he was appointed to the chair of physics at Pavia. In the following years he continued to travel, first to Bologna and Florence, and then on to Germany, Holland, England, and France, where he exchanged ideas with such intellectual luminaries as Georg Christoph Lichtenberg (1742–99), Martinus Van Marum (1750–1837), Joseph Priestley (1733–1804), **Antoine Lavoisier**, and Pierre-Simon de Laplace (1749–1827).

In 1794, at the age of 49, he married Donna Teresa Perigrini Ludovico, the youngest of eight daughters of the "royal delegate" in Como, and over the next few years began to establish his scientific reputation. His invention of the voltaic pile was made public on March 20, 1800, when Volta addressed a letter on the subject to the **Royal Society of London**. The following year he lectured on **galvanism** at the French Institute in Paris and decomposed water with the electric current from his cells. Napoléon (1769–1821) saw the importance of this feat and awarded him a gold medal, the cross of the Legion of Honor, and 6,000 francs. Later he made Volta a senator of the Kingdom of Italy.

Having retired from his university post in 1815, Volta died 12 years later in Como, at the age of 82, after suffering from a fever.

Essential science

Animal electricity and the voltaic pile

In 1786 physicist Luigi Galvani (1737–98) observed a striking phenomenon regarding a recently dissected frog in his laboratory. When the frog's legs were hung on copper hooks so that they touched an iron railing and thus completed a circuit, they contracted even though no electrical machine was in operation in the vicinity. It appeared as if an electric discharge from the frogs' legs had taken place.

Galvani's results, published in 1791, stimulated much thought. He attributed the muscle contraction in the frogs to a kind of animal electricity, believing that animals' bodies were storing electricity within them.

The following year Volta publicly disagreed with Galvani. He claimed the source of electricity is not in the animals themselves but rather in the junction of the metals. Repeating Galvani's experiments with a dead frog, and also using other animals, as well as his own tongue, Volta showed how violent convulsions could occur whenever a connection between the nerves and the muscles was made by a metallic circuit. Volta concluded that since the biological tissue was a fluid layer between two metals then, by the same principle, putting a damp cloth between two metal plates could generate a continuous current. This invention paved the way for the modern battery.

His "voltaic pile", as it was known, was a cylindrical stack that consisted of a number of discs of zinc and copper separated by paper or leather and pieces of wet cloth, soaked in salt solution or dilute acid, and arranged in a vertical column. By testing it on a variety of sources, Volta observed that this apparatus produced an electric current.

Almost immediately, chemist William Nicholson (1753–1815) and surgeon Anthony Carlisle (1768–1840) used this battery to decompose water by electrolysis. Volta himself did little more work on the pile and left it to other scientists to prove that a chemical reaction produced the electricity and that the electricity from a voltaic pile was identical to that produced by electrical machines and electric eels.

The electrophorus

Using his extensive research on the nature and quantity of electrostatic charge (i.e. charges that are not in motion) produced

Among the many things which indeed give me great pleasure, I do not delight in believing I am more than what I am; and to a life upset by vainglory, I prefer the peace and sweetness of domestic life.

Letter to his wife (1801)

Legacy, truth, consequence

■ Before Volta very little was understood about the science of electricity. However, his invention of the electric battery, which he announced in 1800, in a letter to **Joseph Banks**, president of the Royal Society of London, transformed the discipline because it enabled scientists to employ a ready and reliable source of electric current. His work in this area made possible the research on **electrolysis** later conducted by Humphrey Davy (1778–1829) and **Michael Faraday**.

■ The electrical measurement known as a "volt" was named in his honor in 1881. A volt is the unit of electromotive force or electric potential and is symbolized by "V". The domestic electricity supply is 110V in the US and 230V in the UK.

■ In September 1927, there was a seven-day meeting of 61 scientists from around the world at Lake Como for the centenary celebrations marking Volta's death. The group included many famous names from the world of science, such as **Niels Bohr**, James Franck (1882–1964), **Max Planck**, Max Born (1882–1970), **Werner Heisenberg**, and **Enrico Fermi**.

[Metals] should be regarded no longer as simple conductors, but as true motors of electricity, for with their mere contact, they disrupt the equilibrium of the electrical fluid, remove it from its quiescent, inactive state, shift it, and carry it around.

Transunto sull'elettricità animale ed alcune nuove proprietà del fluido elettrico (1792),
quoted in Marcello Pera, *The Ambiguous Frog* (1993)

by various materials, Volta developed a device for the production of electric charges. His electrophorus, developed in 1775, consisted of a disc made of turpentine, resin, and wax, which was rubbed to generate a negative charge. A tinfoil-covered plate was then lowered by an insulated handle onto the disc, which produced a positive charge on the lower side of the foil. Volta then repeated the process to build up greater charges. From these experiments on electrostatic **induction**, he observed that the quantity of charge produced is proportional to the product of its tension and the capacity of the conductor.

Key dates

The "voltaic pile".

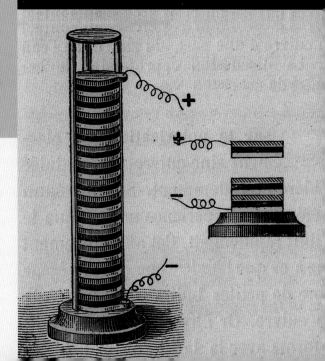

Edward Jenner

Surgeon and nature enthusiast Edward Jenner was the pioneer of vaccine research. Through his unique observations and hypothesis that the swinepox, cowpox, and smallpox vaccines are linked, he developed the idea of inoculating patients with a strain of the virus in order to induce immunity. When his vaccination succeeded, Jenner continued his work and received international renown for his achievements in vaccine research, earning him the title of "father of immunology".

Born in 1749 in the country town of Berkeley, Gloucestershire, in western England, Edward Jenner was apprenticed to a local surgeon at the age of 14. He continued his medical studies in London and upon completing his training at St George's Hospital returned to his hometown, where he was to spend the rest of his life practicing medicine as the local surgeon.

A nature enthusiast from a young age, Edward Jenner counted bird-watching, wildlife observation, and the study of fossils among his hobbies. It was through his interests in nature and agriculture that he was to gain recognition from the **Royal Society of London** in his early career, in particular his experimental observations of the nature of the cuckoo. Later on, nature and medicine would coincide again to help him develop the concept of immunization, for which he is revered. It is thought that after the success of his ground-breaking experiment on the cowpox vaccine, Jenner was so confident, despite the hesitation from the Royal Society, that he even vaccinated his own 18-month-old son.

Essential science

The nesting habits of the cuckoo

Early in his career, and as a result of his bird-watching hobby, Edward Jenner conducted an experimental study on cuckoos in the nest, which was published by the Royal Society in 1787. In this study, using observation, experiment, and the first recorded dissection of a cuckoo, he clarified many previous misconceptions about the bird. Recognized for their parasitic nesting habits, it was not known how the baby cuckoos survived in the adopted nest alongside foreign birds once left there by their parents, nor was it known how exactly the cuckoo parent removes the original eggs or chicks from the adopted nest to make room for its own chicks. Through his observations, Jenner showed that it is the baby cuckoos that evict the nest's rightful inhabitants from their home and not the cuckoo parent. Using dissection, he also identified that baby cuckoos have a special cavity in their backs for the purpose of cupping the eggs or other chicks and pushing them away; the cavity disappears before the cuckoo is 12 days old. It was these extraordinary findings that earned Jenner his fellowship with the Royal Society in 1789.

Developing the smallpox vaccine

Smallpox was one of the most feared diseases of Jenner's time. The disease particularly affected infants and small children, and had a very high death toll; the few survivors were often horribly disfigured by the disease. Blessed with an inquisitive mind and a natural instinct, Jenner was convinced that somehow the **viruses** affecting animals could be linked to the human form of smallpox, and that perhaps this link could somehow lead to disease prevention or cure. Jenner had heard folk stories of local farmers' children and milkmaids who had suffered from cowpox and then appeared to be immune to smallpox. In 1796 he used this theory on eight-year-old James Phipps, by inoculating him with pus from the wounds of Sarah Nelmes, a dairymaid, who had caught cowpox from a cow named Blossom. Jenner transferred the diseased pus from the wound in Sarah's arm using a stick and placed it directly into a suture on James. Apart from initial symptoms of fever and general discomfort, James Phipps did not contract smallpox. Jenner tested the success of the vaccine further by inoculating James with variolous (smallpox) material, and found similar results, proving that immunization was successful.

James Phipps was one of several patients that Jenner vaccinated using this method, and all were included in his first research paper, which he presented to the Royal Society. Ever cautious, the Royal Society did not publish the work, stating that there was not enough evidence to support such a revolutionary finding, and Jenner's efforts were met with not only hesitation, but outrage by the general public. It was not until several years later, in 1798, that the results of his study were finally published. Jenner continued to meet with disapproval, but carried on vaccinating patients and improving upon his theory and technique. It was not long before the reality of his work and the results of his vaccine won over his critics. The full implications were thus appreciated – and a new chapter in medicine was born, that of **immunology**.

Legacy, truth, consequence

■ Although farmer Benjamin Jesty (1736–1836) had successfully inoculated his family with cowpox in 1774, some 20 years before Edward Jenner, it is thought that Jenner arrived at the same findings independently and added value to them by experimenting and helping the process to be better understood.

■ As a consequence of Jenner's groundbreaking work in immunology, smallpox was declared an eradicated disease by the World Health Organization in 1980.

■ Less well known are Jenner's studies of the hibernation and migration of birds. In these he checked the body temperature of hibernating animals to see if, when placing food in their stomachs, the digestion system still worked at normal levels. It was assumed up until Jenner's time that birds hibernated in river mud when they disappeared for the winter. Jenner was one of the first to notice that birds returning in the spring were neither muddy nor starving.

I hope that some day the practice of producing cowpox in human beings will spread over the world – when that day comes, there will be no more smallpox.

Attributed to Edward Jenner

Key dates

1749	Born in Berkeley, Gloucestershire, western England.
1770	Moves to London to train at St George's Hospital.
1772	Graduates and returns to Berkeley.
1787	His study on the life of the cuckoo is published in the Philosophical Transactions of the Royal Society.
1789	Elected Fellow of the Royal Society.
1796	Successfully vaccinates eight-year-old James Phipps against smallpox.
1798	Results of Jenner's research on inoculation are published by the Royal Society.
1803	The Jennerian Institution, a society promoting the eradication of smallpox through vaccination research, is formed.
1805	Joins the Medical and Chirurgical Society the year it forms, now known as the Royal Society of Medicine.
1806	Petitions the king and parliament to support his continuing vaccine research and is granted 20,000 pounds sterling (about 40,000 United States dollars).
1808	The Jennerian Institution, on receiving the government aid gained by Jenner and his supporters, becomes the National Vaccine Establishment.
1821	Appointed Physician Extraordinary to King George IV, and also mayor of his hometown, Berkeley.
1823	In what was to be his final year, he presents his paper "Observations on the Migration of Birds" to the Royal Society, before suffering a stroke from which he is unable to recover.
1840	The British Government bans the process of variolation, and provides vaccination free of charge.
1980	The World Health Organization declares smallpox an eradicated disease.

In this 1802 caricature of a vaccination scene, Dr. Jenner is vaccinating a frightened young woman, and cows are emerging from different parts of people's bodies.

Future nations will know by history only that the loathsome smallpox has existed and by you has been extirpated.

Thomas Jefferson, 1806, in a letter to Edward Jenner

John Dalton

An English chemist, physicist, and meteorologist, John Dalton was a pioneer in the development of atomic theory. He also published the first paper on the subject of color-blindness and was instrumental in turning the largely amateur study of meteorology into a serious scientific field.

John Dalton, the youngest of three surviving children, was born in Cumberland, England. His father, a weaver, and his mother were both from Quaker families. When he was just 12 he became a teaching assistant to his older brother. During his own schooling, in Kendal, he was greatly influenced by two men: the scientific Elihu Robinson (1734–1807) and the classic scholar and mathematician, John Gough (1757–1825). They not only tutored Dalton in their specialized subjects, but also introduced him to a subject in which they both had an amateur interest – **meteorology**. Living in the wild and hilly Lake District, Dalton was ideally placed to experience, observe, and understand meteorological happenings. He began to enter weather records into a journal, a practice he continued throughout his life.

In his mid-twenties he moved to Manchester to teach mathematics and natural philosophy at the New College, an academy set up for Dissenters (members of a non-established church) who could not afford other universities. In the same year he published his paper on *Meteorological Observations and Essays*,

which, though it was little noticed at the time, contained the seeds of the ideas that would elevate him from schoolteacher to chemical pioneer.

When, in 1799, New College relocated to York, Dalton stayed behind in Manchester and began to teach as a freelancer. By this time he had already been noticed for his studies on color-blindness, a condition that he and his brother both suffered from, and in the next few years he became renowned for the further papers he presented to the Manchester Literary and Philosophical Society (of which he became secretary).

Dalton's great skill in identifying and interpreting patterns in his experimental data eventually turned a meteorological problem into a groundbreaking scientific idea: his work, initially on the constitution of mixed gases, led him to secure his position as an architect in the development of **atomic** theory.

In the last third of his life, Dalton made few significant scientific contributions, but he died an affluent and much-bestowed man. He stuck rigidly to an austere Quaker lifestyle throughout.

Essential science

Atomic theory

The workings that led Dalton to develop his atomic theory are somewhat vague – even he had only a partial recall in the matter. Certainly he first proposed the idea in his paper "On the Absorption of Gases by Water and Other Liquids" (1803). Within a gas, he reasoned, it was only the alike atoms that reacted against each other – the unalike atoms were unmoved by each others' presence. Ultimately he stated that all elements were made up of atoms and that the atoms within a given element were identical. The atoms in each element were unique to that element alone and were distinguishable by their respective weights. Chemical compounds were formed when the atoms from two different elements were combined, and chemical reactions occurred when the atoms were rearranged. Finally he stated that atoms could not be manufactured or destroyed.

Some aspects of his propositions have subsequently been shown to be wrong. However, he was right to challenge the erroneous and long-held belief that atoms in many elements were similar, and

it was by finally laying to rest this theory that he paved the way for others to build on his views.

Atomic weights

The seeds that led to Dalton's pioneering work on atomic theory are in his 1793 paper "Meteorological Observations and Essays". In this he noted that each gas exists and reacts independently and physically. During later meteorological studies, he deduced that water, after evaporating, remains in the air as an independent gas. Addressing the puzzle of how air and water are able to simultaneously occupy the same space, he proposed that, assuming air and water are composed of separate particles, evaporation could be the merging of the two. After experimenting by combining a variety of mixed gases with individual gases and observing the effects on their properties, he came to the conclusion that the sizes of the different particles must be different. And so:

"it became an object to determine the relative sizes and weights … a train of investigation was laid for determining the number and weight

> ## We might as well attempt to introduce a new planet into the solar system, or to annihilate one already in existence, as to create or destroy a particle of hydrogen.
>
> *A New System of Chemical Philosophy (1808)*

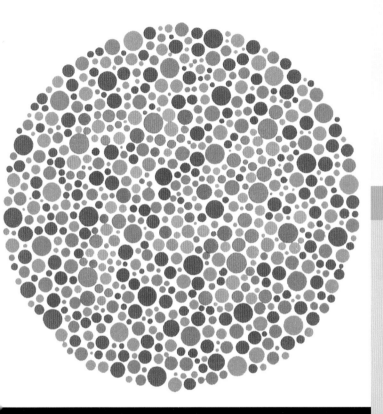

Example of a color-blindness (or Daltonism) test.

of all chemical elementary particles which enter into any sort of combination one with another".

(Lecture delivered at the **Royal Institution** in London, 1810)

Setting hydrogen at 1, he determined the weight relative to this of oxygen, carbon, nitrogen, and other elements depending on how they combined with the set masses of the other gases.

Color-blindness

Dalton and his brother both suffered from the same sight defect. He explained their problem in a paper on the subject:

> "*That part of the image which others call red appears to me little more than a shade or defect of light. After that the orange, yellow and green seem one color which descends pretty uniformly from an intense to a rare yellow, making what I should call different shades of yellow*".
>
> (*Extraordinary Facts Relating to the Vision of Colors*, 1794)

His theory for the defect was that the thin watery fluid that fills the space between the cornea and the iris, known as the aqueous humor, had become bluish and in this condition was filtering out

Legacy, truth, consequence

- Although some details of Dalton's atomic theory were not correct, its central concepts that atoms have peculiar properties, and that chemical reactions are caused by the merging and separating of these atoms, form the cornerstone of modern physical science.
- Dalton received widespread recognition of his work in his lifetime. Contemporaries lauded him not only as the "father of meteorology", but also the "father of chemistry". While he may have had competition for those titles in either individual field, it was unique to be lauded in both.
- In the eighteenth century the patterns of the weather were often still explained by old mythological theories rather than being regulated by scientific phenomena. Dalton was largely responsible for changing this attitude and determining meteorology as a serious scientific activity.

Key dates

1766	Born in Eaglesfield, Cumberland (now part of Cumbria), England.
1778/9	Becomes a teaching assistant at his brother Jonathan's village school.
1781	Moves to teach and study in Kendal (in Cumbria).
1793	Moves to Manchester (in Lancashire, England) to teach at the New College. Publishes *Meteorological Observations and Essays*.
1794	Publishes the essay "Extraordinary Facts Relating to the Vision of Colors".
1802	Presents his paper on atmospheric gases, "Experimental Enquiry into the Proportion of the Several Gases or Elastic Fluids, Constituting the Atmosphere".
1803	Proposes atomic theory.
1808	Publishes first volume (of three) of *A New System of Chemical Philosophy*.
1817	Elected president of the Literary and Philosophical Society.
1825	Awarded first annual **Royal Society** prize.
1844	Too ill to be president, he is appointed vice-president of the British Association for the Advancement of Science.
1844	Dies of a stroke in Manchester. He is accorded a state-like funeral that attracts 400,000 people.

colors. The word "Daltonism" was subsequently used to describe color-blindness in general, although it's likely that Dalton suffered from an unusual form of color-blindness called deuteroanopia.

Amedeo Avogadro

Amedeo Avogadro was an Italian mathematical physicist and teacher whose hypothesis on the molecular composition of gases was belatedly recognized as fundamental in clarifying the foundations of molecular physics and chemistry.

Lorenzo Romano Amedeo Carlo Avogadro was born into a noble family in Turin, in the kingdom of Sardinia and Piedmont (now in Italy). His father was an eminent lawyer and senator. The family had been steeped in a tradition of legal work for many generations, and initially Avogadro also ventured into this world, receiving his doctorate in ecclesiastical law when he was 20.

Although he showed great promise in this fledgling legal career, his vocation lay in the sciences. He undertook private studies in mathematics and physics, and after flirting with research in the field of electricity, he joined the Academy of Sciences of Turin, becoming a demonstrator when he was 30.

Apart from an 11-year period teaching **natural philosophy** in nearby Vercelli, Avogadro stayed in Turin all his life. It was while he was in Vercelli, however, that he published an article called, not misleadingly, "Essay on Determining the Relative Masses of the Elementary **Molecules** of Bodies and the Proportions by which they enter these Combinations" (1811). This contained the hypothesis describing what is now known as "Avogadro's law", and which led to the calculation of "Avogadro's number".

Unfortunately for Avogadro's career, the Turin of the early 1800s was not a hotbed of progressive scientific activity and he was somewhat isolated from other prominent scientists of the time. It was as much an intellectual isolation as a geographical one, and many of his hypotheses that did gain the attention of his contemporaries either created confusion or were ignored altogether. His habit of referencing his own work to prove a point further isolated him from his peers.

A hard-working and humble man, Avogadro was devoted to his home, his wife, Felicita, and their six children. There is also some evidence that he was devoted to certain revolutionary causes that were taking place in Sardinia following the downfall of Napoléon Bonaparte of France in 1815. By the time of his death, the situation had settled and Turin was five years away from becoming part of the new Kingdom of Italy.

Essential science

Avogadro's law
In 1811, with the aim of improving the scientific world's comprehension of molecular composition, Avogadro took as his starting point Joseph Louis Gay-Lussac's (1778–1850) law of combining volumes, which states that the volume of gases involved in a chemical reaction show simple whole number ratios to one another when those volumes are measured at the same temperature and pressure. From this Avogadro formed the following principle, now known as Avogadro's law:

"Equal volumes of gases, at the same temperature and pressure, contain the same number of molecules".

(Note sent to *Journal de Physique, de Chimie et d'Histoire Naturelle*, 1811)

Although Avogadro never attempted to calculate the number of particles in a liter of any gas, in the 1860s the Austrian chemist Josef Loschmidt (1821–95) gave an approximate value, and at the beginning of the twentieth century a definitive value representing the number of molecules in a gram molecule (the molecular weight of a substance expressed in grams) was given as 6.022×10^{23}. This is called Avogadro's number and is identical for all gases, from the lightest (hydrogen) to a heavy gas such as bromine. It is used to help chemists calculate the amounts of substances produced in a particular reaction.

Molecular formulas
Avogadro's ideas provided a way of determining molecular formulas of compounds independently of mass measurements. For example, Gay-Lussac had stated that water is formed by the combination of two parts hydrogen with one part oxygen. Based on Avogadro's law, therefore, the molecular formula for water was H_2O.

Molecules and atoms
Avogadro did not use the word "**atom**". Instead he spoke of three types of molecule, with the *molecule élémentaire* being equivalent to an atom. The other two were the molecule of a compound and the molecule of an element – *molecule intégrante* and *molecule constituante* respectively. He argued that simple gases were compound molecules composed of more than one atom (*molecule élémentaire*). This notion sat uncomfortably with many chemists, however, because it meant that elementary gases contained polyatomic molecules (molecules with more than two atoms), an idea that was inconsistent with chemical philosophy of the time.

Legacy, truth, consequence

- Avogadro's number is now recognized as an integral constant of physical science. It allows the calculation of quantities of chemical elements or compounds involved in a chemical reaction. It is also used to determine how much heavier a molecule of one gas is than a molecule of another. By comparing the weights of equal volumes of gases it can be used to determine the relative molecular weights of gases.

- Although Avogadro did not use the precise terms "atom" and "molecule", he succeeded in distinguishing the different levels of an element's structure, and clarifying that gases are composed of molecules, and molecules in turn are composed of atoms.

- Avogadro claimed that his theory for determining the relative molecular masses of gases was broadly acknowledged. In reality, however, there was still much confusion over the whole concept, in part because of his isolation from the core of prominent scientists. His work was not taken seriously until a few years after his death, when the studies of the young Italian chemist Stanislao Cannizzaro (1826–1910) began to instill the view that Avogadro's hypothesis may in fact have been correct. In 1909 the number was named in his honor, when the French physicist Jean-Baptiste Perrin (1870–1942) wrote:

> *this invariant number N is a universal constant, which may, with justification, be called Avogadro's constant*.
>
> (*Mouvement Brownien et Réalité Moléculaire*, 1909)

- In 1911, one hundred years after the publication of his famous essay, the anniversary was celebrated in Turin in the presence of King Victor Emmanuel III.

Key dates

1776	Born in Turin (now in Italy).
1787	Inherits the title Count of Quaregna.
1792	Graduates as a bachelor of jurisprudence.
1801	Becomes secretary of the prefecture of the department of Eridano.
1803	Carries out his first scientific research, with his brother Felice, on electricity.
1804	Arrives at the Academy of Sciences of Turin as a corresponding member.
1806	Becomes a demonstrator.
1809	Becomes professor of natural philosophy at the Royal College of Vercelli.
1811	Publishes his hypothesis that first introduces the science world to Avogadro's law. Stipulates the correct molecular formulas for water, carbon monoxide, nitric and nitrous oxides, hydrogen chloride, and ammonia.
1814	Provides formulas for sulfer dioxide, carbon dioxide, carbon disulfide, and hydrogen sulfide.
1820	Accepts position of chair of mathematical philosophy at the University of Turin.
1821	Works out the correct formulas for ether and alcohol.
1822	Civil unrest, and possibly his involvement with it, causes Avogadro to lose his position at the university.
1834	Re-appointed chair at the University of Turin.
1843–50	Writes four memoirs detailing his work on atomic volumes.
1850	Retires from academic life.
1856	Dies in Turin.

It is the task of experiment to confirm or correct ... theoretical estimates.

"Essay on Determining the Relative Masses of the Elementary Molecules of Bodies and the Proportions by which they enter these Combinations" (1811)

"Avogadro", an open source molecular editor software – for use in computational chemistry, molecular modeling and related areas – has been named in honor of the physicist.

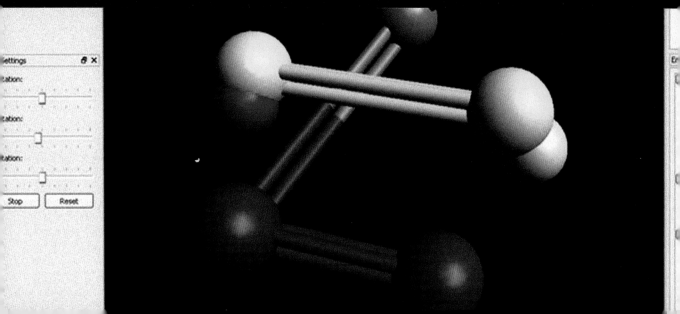

Carl Friedrich Gauss

Considered to be one of the world's greatest mathematicians, the German polymath Carl Friedrich Gauss was interested in a wide range of scientific disciplines from geometry, probability theory, and number theory to astronomy, cartography, electromagnetism, and surveying. In every field he studied, he left important discoveries or inventions behind him before moving on to another subject.

Born Johann Carl Friedrich Gauss, the only child of poor parents, Gauss showed so much potential at his elementary school, including the ability to carry out complex calculations in his head, that when he was 14 his mother and teachers asked for support from the Duke of Brunswick. The Duke provided a stipend that allowed Gauss to remain at school and go to university in Göttingen, earning him the lasting gratitude and loyalty of the child prodigy.

Still a teenager, Gauss independently "discovered" several mathematical laws and came to grips with advanced **number theory**. Mathematical ideas came to him fast and furiously, and when he published several of them in 1801, in *Disquisitiones Arithmeticae*, he earned immediate acclaim from other mathematicians. By then he had also established his basic scientific and mathematical approach: intense **empirical** investigations, followed by reflection, then the construction of a theory.

In 1808 Gauss moved back to Göttingen from Hanover to organize and run a new observatory. He made several contributions to theoretical astronomy in those years, as well as working on further mathematical theory, and developed his unique mixture of practical and theoretical skills.

Gauss learnt from everything he studied. In 1851 he began to make money from investments after he learnt about finance from sorting out the Göttingen University's widows' fund.

Never close to his father (although he supported his mother for 22 years), not making friends easily, and distraught by the death of his first wife, Gauss was usually a distant and aloof personality. He did not like change, which is why he hardly ever left Göttingen once he had settled there when he was 30, although he could probably have taken a senior academic position in a major city such as Berlin, where he might have found like-minded scholars.

Essential science

Gauss studied an amazing range of subjects. His early passion was arithmetic and number theory, but he branched out into other areas of mathematics and into both theoretical and empirical research in **astronomy**, **geodesy**, surveying, **geomagnetism**, **electromagnetism**, **optics**, and just about anything that was brought to his attention. Throughout his life he continued to explore different approaches to mathematics, often coming up with varied ways of proving or solving problems.

Geometry

Before he was 20, Gauss achieved one of the greatest advances in **geometry** since the ancient Greeks, when he proved that a regular 17-sided polygon could be constructed using a ruler and compasses. In 1796 he went further and proved that only regular polygons with a certain number of sides can be constructed this way.

Number theory

In his 1801 mathematical masterpiece *Disquisitiones Arithmeticae* (*Arithmetical Disquisitions*) Gauss wrote the first systematic textbook on **algebraic** number theory or "higher arithmetic". He summarized the somewhat scattered writings on the subject, included many of his own theories on outstanding problems, and laid down the definitive analysis of the concepts and research areas.

Astronomy

Gauss also became famous in 1801 when he was the first person to correctly predict the reappearance of the asteroid Ceres. This question had occupied scientists ever since the Italian astronomer Giuseppe Piazzi (1746–1826) had briefly spotted it the previous year. Gauss used a technique of dealing with observational error known today as the method of least squares or **regression analysis**, and he later developed his interest in astronomy into an important work on the numerical calculation of orbits.

Surveying

In the course of directing a modern survey of the territory of Hanover, Gauss invented the heliotrope and came up with a mathematical theory of map-making. The heliotrope used mirrors and a telescope to focus the sun's rays into a reflected beam that could be seen over long distances; it made surveying more accurate. His work on maps proved that the curved surface of the earth's sphere cannot be unfolded onto the flat surface of

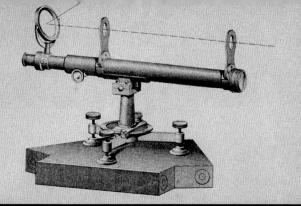

Gauss invented the heliotrope. This model is from c.1866.

The enchanting charms of this sublime science reveal themselves in all their beauty only to those who have the courage to go deeply into it.

Letter to Sophie Germain (1807)

So he had few close colleagues, although he corresponded widely. A complex man, he refrained from publishing some of his scientific and mathematical theories, perhaps because he was more interested in moving on to a new area of study, or because he was a perfectionist and would only publish when he thought that everything was complete and exactly right. His published texts and letters were only a small part of the scientific legacy he left behind.

a map without some form of distortion. He experimented with preserving angles but at the cost of distorting distances and areas. This work won him a prize from the Danish Academy of Sciences in 1823.

Magnetism

In 1832 Gauss took part in the attempt by the well-known scientist Alexander von Humboldt (1769–1859) to map the earth's magnetic field. He invented a magnetometer, consisting of a bar magnet suspended from a gold fiber, which he used to measure the strength and direction of the magnetic field in a particular place. With his colleague Wilhelm Weber (1804–91), he built the first **electromagnetic** telegraph, transmitting messages over 5,000 feet (1.5 km). More importantly for him, he arrived at several mathematical principles from studying magnetism, and by 1840 wrote three major papers. Some of his conclusions have implications today for the study of forces such as **gravitation** and electromagnetism. He also supplied an empirical definition of the earth's magnetic force, produced an absolute measure for it, showed why there can only be two magnetic poles, and proved a theorem relating horizontal magnetic intensity to the angle of inclination.

Legacy, truth, consequence

- Gauss' mixture of theoretical ideas, empirical research, and practical inventions in a wide range of fields shows that he was one of the greatest scientific minds ever. Many of his ideas are applied today in different fields, such as **Gaussian curvatures** and **Gaussian probability distribution**. However, although other theoreticians recognized the importance of his work, Gauss never received much public recognition. Of course, he did not embark on public lecture tours that would have brought him to the notice of non-scientists.

- Gauss produced 178 publications, but left behind 226 unpublished papers, along with notes, commentaries, and memoirs. These covered many new ideas as well as proving that he had worked on some problems – such as the application of **differential equations** to infinite series – before other people had published their conclusions on them. In keeping many ideas to himself Gauss hindered the development of mathematics in some areas.

- The unit for magnetic field strength is called a "gauss" in his honor.

Key dates

1777	Born in Brunswick (now in Germany).
1781	Receives a stipend or scholarship from the Duke of Brunswick.
1795	Studies at Göttingen University, Hanover (in Germany).
1799	Returns to Brunswick; gains a doctorate.
1801	Establishes his scientific and mathematical reputation with the publication of *Disquisitiones Arithmeticae* (*Arithmetical Disquisitions*) and by predicting the rediscovery of the asteroid Ceres.
1807	Becomes director of Göttingen Observatory.
1818–32	Directs a new geodesic survey of Hanover.
1823	Wins Danish Academy of Sciences prize for his studies of map projection.
1855	Dies in Göttingen.

Non-Euclidean geometry

In letters and comments from the early 1800s onwards, Gauss revealed an interest in the possibility of a non-**Euclidean geometry**, but he also feared that his reputation would suffer if he openly admitted a belief in it, and never published his theories. When Gauss' old friend Farkas Bolyai (1775–1856) sent his son Janos' work on non-Euclidean geometry to Gauss in 1831, Gauss replied: "*To praise it would mean to praise myself.*" Later research into non-Euclidean geometries would eventually lead to **Albert Einstein**'s theory of **general relativity**.

Michael Faraday

Michael Faraday was an English physicist and chemist who is often referred to as the father of electricity. His innovations in the fields of electricity and magnetism, in particular his model for the first electric generator, provided the foundations for the modern electricity supply industry.

Born in 1791, in Newington (now part of London), Faraday had a difficult childhood. His father was a village blacksmith who was unable to work full-time because of ill health and, as a consequence, the family suffered economic difficulties and Faraday received little formal education.

Despite this situation, Faraday showed great determination for success. While employed as a bookbinder, he read many books on chemistry and electricity, and in his spare time conducted chemical experiments. In 1810 he began attending the City Philosophical Society, where he heard lectures on a variety of scientific topics. Three years later he gained his first job in science, as a laboratory assistant at the **Royal Institution**, London, working with surgeon and chemist Humphry Davy (1778–1829).

In the same year Davy took Faraday on an 18-month tour of Europe, during which Faraday learnt French and Italian and met many of Europe's leading scientists. After marrying Sarah Barnard in 1821, Faraday focused his research on chemistry, in particular on the manufacture of steel and glass. This was interspersed with experiments on electricity. In 1831 he abandoned his work on steel

and glass to devote more time to his research on electricity. Success came almost immediately: in the autumn, he discovered both **electromagnetic** and **magneto–electric** induction, and the following year he showed that there is only one kind of electricity.

From 1838 onwards Faraday suffered various debilitating ailments. However, he continued to work intermittently and, in 1845, through his experiments with light and magnetism, he discovered what became known as the "Faraday effect". In later life, as a public figure, the government and the monarchy sought his advice on scientific matters. In 1846, with geologist Charles Lyell, Faraday headed an investigation into the explosion at Haswell Colliery, County Durham, which killed ninety-five miners, and in 1850 he was on the committee for preserving paintings at the National Gallery, London.

Despite his achievements he remained very down to earth in his principles. He refused a knighthood and twice refused the presidency of the Royal Society. In 1858 Queen Victoria offered Faraday a Grace and Favor Residence at Hampton Court for the rest of his life. He died there in 1867, and is buried in Highgate Cemetery, London.

Essential science

Electromagnetic rotation

For some time prior to Faraday scientists had speculated about the exact relationship between electricity and magnetism. Faraday built two devices designed to prove that "electromagnetic rotation" characterized this relationship. In the first, Faraday showed that an electrically-charged, hanging wire will rotate around a securely fixed magnetic bar, and in the second he showed that a magnetic bar fixed at only one end will rotate around a fixed, electrically-charged wire.

Electromagnetic induction

Faraday's famous "induction ring" consisted of two coils of wire wound around an iron ring. Coil A, on the left side of the ring, was connected to a battery, while coil B, on the right side of the ring, was connected to a piece of wire which passed over a compass three feet (approximately one meter) away. By passing an electric current through coil A, Faraday observed an effect in the needle of the

I must remain plain Michael Faraday to the last.

Faraday's response when he declined the presidency of the Royal Society (1857)

compass under coil B. He had discovered "electromagnetic induction": the **induction** of electricity by means of magnetism.

Magneto-electric induction

Faraday sought to produce electricity from magnetism. He took an iron bar with a coil of wire wrapped around it. This he placed inbetween two magnetic bars so that the top of the iron bar touched the top magnet, the bottom of the iron bar touched the bottom magnet and the two magnets themselves were touching each other at the side, thus creating a kind of triangular magnetic circuit. The coil on the iron bar was connected to a **galvanometer**. When the iron bar was removed, thus breaking

- Faraday's discoveries of electromagnetic and magneto-electric induction form the basis of modern electrical technology. His induction ring was the first **electrical transformer**, an essential instrument in the modern electricity supply industry.

- Modern **electric generators** employ the ideas used in his magneto-electric induction device, using rotary motion between the magnet and the coil. The French instrument maker Hippolyte Pixii (1808–35) constructed the first generator after reading of Faraday's discoveries.

- Through his pioneering research in **electrochemistry**, he introduced terms, such as **electrolyte**, **electrodes**, **anode**, and **cathode**, which have become the standard nomenclature.

- He was a keen student of the art of lecturing. His *Advice to a Lecturer*, published in 1960 after his death, is considered to be an excellent handbook for teachers and lecturers. Faraday also did much to emphasize the importance of teaching science in schools to the Public Schools Commission.

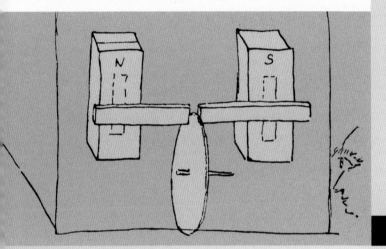

1791	Born in Newington (now part of south London), in England.
1810–11	Attends the City Philosophical Society.
1813	Becomes a laboratory assistant to Humphry Davy at the Royal Institution.
1813–15	Tours Europe with Humphry Davy.
1815	Resumes work at the Royal Institution.
1816	Publishes *Analysis of Native Caustic Lime of Tuscany*. Gives a series of lectures to the City Philosophical Society.
1818–23	Conducts research on steel.
1821	Invited to contribute an article on the history of electromagnetism to the scientific journal *The Annals of Philosophy*. Discovers electromagnetic rotation. Marries Sarah Barnard.
1825	Appointed director of the Laboratory at the Royal Institution.
1831	Discovers electromagnetic and magneto-electric induction.
1832	Seeks to prove that electricity is always the same regardless of its source.
1833	Appointed Fullerian Professor of Chemistry at the Royal Institution.
1834	Publishes his findings in electrochemistry.
1845	Discovers the "Faraday effect".
1846	Conducts mining disaster inquiry with Charles Lyell.
1857	Refuses presidency of the Royal Society.
1858	Receives a "grace and favor" house at Hampton Court.
1864	Declines the presidency of the Royal Institution.
1867	Dies at Hampton Court.

Sketch by Michael Faraday of an electric generator.

the magnetic circuit, the galvanometer recorded a brief electric current. When the iron bar was replaced, thus remaking the circuit, the galvanometer showed the electric current flowing in the opposite direction. Faraday had discovered "magneto-electric induction": magnetism being converted into electricity. He later produced a second device to prove the same results. This time he took a wire coil and wrapped it around a hollow paper cylinder. When he rapidly inserted and removed a magnetic bar the galvanometer recorded an electric current.

One kind of electricity

Faraday sought to prove that **magneto-electricity**, **volta-electricity**, and **static electricity** were not different kinds of electricity and that, therefore, there is only one kind of electricity, regardless of how it is produced. He conducted a series of experiments designed to show that they are all characterized by criteria such as their heating, magnetic, chemical, and spark-producing effects.

Electrochemistry and the laws of electrolysis

Through experiments designed to measure the chemical effects of electricity, Faraday discovered his First Law of Electrolysis: that the chemical effect produced by an electric current on a substance – whether that chemical effect is a gas or a solid – is always proportional to the amount of electricity that flowed. On the basis of this law, he made an instrument for measuring electricity, which he called a "volta electrometer" (or "voltameter") and used it to prove the Second Law of Electrolysis: that the electrochemical equivalent of a substance is proportional to its ordinary chemical equivalent.

The "Faraday effect"

By experimenting with magnetism, light, and glass, Faraday found that the plane of **polarization** of polarized light could be rotated with a magnetic field aligned in the direction the light is moving. This effect is also known as "Faraday rotation". In his notebook he wrote that he had succeeded in "*magnetizing a ray of light*".

Roderick Impey Murchison

Sir Roderick Impey Murchison was a Scottish geologist who conducted numerous geological field expeditions, including intrepid trips to Russia, and, most famously, discovered the Silurian system or geological period while examining rocks in south Wales in Great Britain. As frequent collaborators, he and his wife Charlotte are considered to be the most prolific British geologists of the nineteenth century.

Roderick Impey Murchison was born into an ancient Scottish Highland landowning family in 1792. After the death of his father, when he was only four years old, Murchison's family relocated to England, where the young Murchison was sent to a military college at Great Marlow. He excelled at horsemanship and foxhunting, and in 1808 he served briefly in the Peninsular War. After resigning his commission in 1814, he married Charlotte Hugonin, daughter of General Francis. Charlotte proved to be a source of great encouragement and inspiration throughout his career.

Following a winter in Hampshire, receiving instruction in natural history from Charlotte's family, Murchison and Charlotte traveled in Europe for two years, studying the arts and natural sciences. On their return, Murchison sold his family estate and the couple moved to Country Durham, where Murchison occupied himself with foxhunting for several years while Charlotte devoted herself to the study of **mineralogy** and **conchology**.

In 1824 Murchison exchanged this lifestyle for more intellectual pursuits after a chance meeting with scientist Humphrey Davy (1778–1829). Inspired by Davy's passion for the sciences, Murchison sold his horses to supplement his private income, and he and Charlotte moved to London, where he began attending lectures at the **Royal Institution**. Influenced by William

Buckland (1784–1856), and by Charlotte's interests, Murchison focused on **geology**. He was elected fellow of both the Geological Society of London and the **Royal Society**, where he mixed with many famous scientists of the era, including **Charles Darwin**, Charles Lyell (1797–1875), and Adam Sedgwick (1785–1873).

Almost every summer for the next twenty years, Murchison embarked upon the geological expeditions for which he received worldwide recognition. Lyell or Sedgwick often accompanied him on these trips through Britain, France, and the Alps. Charlotte too was ever-present, acting as his fossil hunter and geological artist. He published *The Silurian System* in 1839, based on research in south Wales. This trip was followed soon after by a famous geological expedition to Russia.

With the benefit of a large inheritance from Charlotte's mother, the couple moved into a house in affluent Belgravia, London, where for a period they established themselves at the center of an intellectual social scene, which included well-known scientists, politicians, and literary figures of the time. Knighted in 1846, Murchison is said to have become increasingly argumentative, dogmatic, and enamored by his own sense of importance in later life. By the time he died in 1871, he had fallen out with many of his contemporaries.

Essential science

The Silurian system

In 1839 Murchison produced his major work, *The Silurian System*, detailing his research on the "greywackes", or old slate rocks, in south Wales, underlying the Old Red Sandstone and dating from the lower **Palaeozoic** era. Despite the fact that most geologists believed that these slate rocks contained very few fossils, Murchison held that they contained the possibility of discovering the earth's earliest life forms. He named the strata "Silurian" after the Silures, a tribe that had lived in the region.

The Silurian was a major system of strata, and the beginning of the period has been dated to around 444 million years ago. It has a distinctive fauna, with many invertebrates but very little in the way of vertebrates or land plants. Murchison concluded that it marked a major period in the history of life on earth.

The Devonian system

In collaboration with Sedgwick, Murchison established the Devonian system in Devon, southwest England, and in the Rhineland, Germany. Sometimes called the "Old Red Age" after the Old Red Sandstone associated with this period (also dating from the Paleozoic era), the Devonian fossils indicate that this is when the first fish evolved legs and started to walk on land, and the land became covered in forests. The (older) Silurian period led up to the Devonian period, which began around 416 million years ago.

The discovery followed a controversy with Henry de la Beche (1796–1855), in which Murchison argued that there couldn't be coal below the Silurian system, as the strata beneath must be older than Silurian, and coal was associated with younger rocks. Findings by Murchison and Sedgwick showed that de la Beche's strata were

Legacy, truth, consequence

- For his achievements, Murchison gained many prestigious positions and received a number of awards. He was knighted, became director of the Geological Survey, he was a founding member of the British Association, a recipient of the Copley Medal from the Royal Society, and he was president of the Royal Geographical Society.

- As an imperialist, he was a supporter of the missionary activities of explorer David Livingstone (1813–73) in Africa. The Murchison Falls in Uganda are named after him. There are also landmarks bearing his name in Canada and Australia.

- His wife Charlotte was a significant collaborator who provided a major contribution to Murchison's success. Not only did she provide excellent resources, such as her collection of fossils and her sketches and drawings, but she was also a valuable source of scientific advice. In addition to her achievements with Murchison, Charlotte, along with her friend Mary Somerville (1780–1872), successfully campaigned for women to be allowed to attend geology lectures at Kings College, London. Although they subsequently experienced setbacks in their endeavors, this was perceived as the first step towards the admission of female students to British universities.

A fossil trilobite from the Silurian age. The Silurian system was first identified by Murchison.

Key dates

1792	Born in Tarradale, Ross and Cromarty, Scotland.
1808	Serves in the Peninsular War.
1814	Resigns his commission.
1815	Marries Charlotte Hugonin.
1816–18	Travels in Europe.
1824	Meets Humphrey Davy.
1825	Elected fellow of the Geological Society in London and presents his first paper there.
1826	Elected fellow of the Royal Society.
1827	Secretary of the Geological Society.
1831	Elected president of the Geological Society.
1834	Publishes *The Geology of Cheltenham*.
1839	Publishes *The Silurian System*. Moves to Belgravia in London.
1840–1	Expeditions to Russia.
1841	Elected president of the Geological Society for a second time. Publishes *On the Geological Structure of the Northern and Central Regions of Russia in Europe*.
1845	Publishes *The Geology of Russia in Europe and the Ural Mountains*.
1846	Knighted.
1849	Awarded the Royal Society's Copley Medal.
1854	Publishes *Siluria*.
1855	Succeeds Henry de la Beche as director of the Geological Survey.
1856	Appointed to a royal commission to report on coal reserves.
1864	Awarded the Wollaston Medal.
1871	Dies in London.

not pre-Silurian but were the lateral equivalents of the Old Red Sandstone (Devonian).

Although their definition of a Devonian system was initially controversial due to its reliance on purely paleontological criteria, Murchison and Sedgwick went on to provide further **empirical** evidence to support their original evaluation.

Despite their achievements together, Murchison and Sedgwick later fell out, each holding increasingly divergent views with regard to the base of the Silurian system. Murchison's book, *Siluria* (1854), a popularized version of his overall position, was an attempt to deliver a "knockdown blow" to opponents like Sedgwick and de la Beche.

The Permian system

After his expedition to Russia in 1841, accompanied by Edouard de Verneuil (1805–73) and Count Alexander von Keyserling (1815–91), Murchison defined the Permian system, named after the strata of the Perm region, which has now been dated to between 250 and 290 million years ago.

Jean-Baptiste Dumas

A chemist, teacher, and politician, Jean-Baptiste Dumas was a pioneer in the study and analysis of organic chemistry. He is best known for his theory of substitution which contradicted and eventually superseded the accepted molecular structure theory of his time. His revolutionary and inspirational teaching methods helped form the bedrock of modern chemistry education and research.

Jean-Baptiste Dumas was born in Alais, France, where his father was the town clerk. After an education in the classics he planned to join the Navy, but thought better of it after Napoleon's final defeat, becoming instead an apothecary's apprentice. In 1816 he moved to Geneva and studied chemistry, pharmacy, and **botany**. His lively work ethic soon attracted the attention and the encouragement of his professors.

Dumas returned to France in 1823 and, after a brief time as a lecture assistant to the French chemist Louis-Jacques Thénard (1777–1857) at the École Polytechnique in Paris, became professor of chemistry at the academic institute, the Athenaeum, the first of several academic positions that he would hold simultaneously. A good deal of his working day was spent traveling from school to school.

Dumas succeeded Thénard as professor of chemistry at the École Polythechnique in 1835, but by then his belief in the importance of scientific laboratories had already led him to establish a self-financed school of chemistry, the École Centrale des Arts et Manufactures.

Dumas' students were witness not only to his motivational teaching, but also to his passion for experimentation and especially his willingness to whole-heartedly throw himself into the center of the various controversies that plagued **organic chemistry** in the nineteenth century. His resultant discoveries and theories won him many plaudits, but they also won him enemies.

Dumas was initially a feeble public speaker, but by the late 1840s his elocutionary skills had matured to such an extent that, with the majority of his significant scientific work completed, he turned to politics. Posts under the Emperor Napoléon III led to a stint as the president of the municipal council of Paris, where his restructuring of the city's drainage system prompted the emperor to call him "the poet of hygiene". He ended his political career as permanent secretary of the Academy of Sciences before, in 1870, returning to scientific studies.

He was a devout Catholic for all of his life.

Essential science

Early discoveries

Working on his own and with colleagues, Dumas identified the molecular make-up of many important compounds, including cymene, urethane, and, most significantly, methanol. He and the French chemist, Eugène Melchior Péligot (1811–90), achieved this by distilling wood and prepared derivatives and it was this isolation of methyl alcohol that led to Dumas' search for hydrocarbon radicals – **molecules** with at least one unpaired **electron**.

The law of substitution

Dumas' attempts to formulize organic chemistry along dualistic, **inorganic** lines led to his "law of substitution". He challenged the early nineteenth-century dualistic theory of the great Swedish chemist, Jöns Jacob Berzelius (1779–1848), that all compounds were either positive or negative and chemical combinations resulted from the attraction of the opposing parts. He found that burning candles bleached with chlorine produced fumes of hydrogen chloride. His conclusion was that *"during the bleaching, the hydrogen in the*

hydrocarbon oil of turpentine became replaced by chlorine." Dumas had demonstrated that, in certain instances, it was possible to substitute the electropositive hydrogen atoms with the electronegative oxygen **atoms** without there being any dramatic structural alteration.

Others, such as the German chemist, **Justus von Liebig**, and not surprisingly Berzelius himself, disputed Dumas' findings to such an extent that Dumas felt compelled to retreat in the face of the attacks. Worse was to come. Auguste Laurent (1807–53), a former Dumas student, was his co-worker when researching the "law of substitution". When Dumas later expanded on their previous ideas, the credit for the original theory became a matter of dispute and the two fell out irreconcilably. Dumas did, however, patch things up with Liebig.

Vapor density and atomic weights

Early in his career at the École Polytechnique, Dumas worked on improving methods of measuring vapor densities. His method, which involved finding the mass, volume, temperature, and pressure of a substance in its vapor phase, led directly to the revised

■ Organic chemistry is arguably the most important field of chemistry in the world today. Dumas' influence on the subject prompted his former student, Charles-Adolphe Wurtz (1817–84) to describe him as "the founder of organic chemistry".

■ He was one of the first teachers of chemistry to realize and act upon the importance of scientific laboratories. His research students went on to form their own laboratory school comparable to Liebig's better known school in Germany. Among them were many important

French chemists, including Wurtz, Henri Etienne Sainte-Claire Deville (1818–81), and **Louis Pasteur**.

■ Much of Dumas' work in his initial paper on the atomic theory is visionary. He wrote about ideas which are often assumed to belong to a subsequent period.

■ Dumas was not above using his eminent position in the Academy of Sciences to hinder the careers of younger chemists whom he saw as threat to his reputation.

The surrender of Napoléon III to the Prussian army at the Battle of Sedan in September 1870 marked the end of the French Second Empire. Dumas withdrew from political life.

In all elastic fluids observed under the same conditions, the molecules are placed at equal distances.

Paper on the atomic theory (1826)

Key dates

atomic weights of 30 elements, including carbon, revised to 12.02, and hydrogen, valued as 1. This represented half the total number of known elements at this time and pre-dated the periodic table of Dmitri Mendeleev (1834–1907) by thirty years.

Estimating the amount of nitrogen in an organic compound
Dumas created this method by taking a nitrogen sample of known weight and, in his words, "*The sample is then heated with copper oxide and oxidized completely in a stream of carbon dioxide; the gaseous products of the combustion are passed over a heated copper spiral and the nitrogen collected in a gas burette over concentrated potassium hydroxide solution*". The resultant sample was then "*estimated by direst measurement*". It is a procedure that still forms the basis for modern analytical methods.

The function of kidneys
It was long thought that kidneys produced urea. By showing that urea was present in the blood of animals from which the kidneys had been removed, Dumas proved that one of the functions of kidneys was actually to remove urea from the blood.

Justus von Liebig

Liebig's work arguably straddled more fields of chemistry than any other nineteenth-century chemist. He was a particular authority on agricultural and biological chemistry. His involvement in the rise of scientific agriculture was key to the increase of food production at a time of great population growth. He was also a charismatic and innovative teacher who commanded great devotion among students from many countries.

Justus von Liebig was born in Darmstadt, Germany. His father was a chemical manufacturer who owned a shop and, more importantly, a laboratory in which his son was permitted to play with chemicals. Legend has it that the young Liebig's experiments with the more dangerous substances led not only to an explosion at his local school, but also to severe structural damage to his own room. If true his parents' subsequent decision to apprentice him to an apothecary in Heppenheim may have been as much to keep their house in one piece as to enhance their son's career.

By 1820 he was studying chemistry at the Prussian University of Bonn and, after receiving his doctorate, he continued his studies in Paris. It was here that Liebig's fascination with explosives, namely silver fulminate, bore fruit and led to the concept of isomerism (see below). Then after a brief stay back in Darmstadt, during which he found time to get married, he became a professor at the University of Giessen at the age of just 21.

Liebig was a prolific writer. In the 1830s alone he published 300 papers. As editor and founder of the monthly periodical *Annalen der Pharmacie und Chemie* (*Annals of Pharmacy and Chemistry*) he was not afraid to express in words the fiery and often critical views that many fellow chemists had witnessed in person. At the same time he favorably promoted his own work and that of his pupils.

Liebig initially worked in pure **organic chemistry**, but by the 1840s he had turned his attention to applied chemistry, especially concentrating on the application of chemical science to food, nutrition, and agriculture.

In later years he concentrated on his writing and popular lecturing. Through his *Familiar Letters on Chemistry* (1843) he kept his forthright views to the fore, commenting on issues as diverse as science methodology and the dangers of failing to recycle waste, the latter prompted by a visit to England where he was horrified to watch sewage being sent out to sea.

Essential science

Isomerism

The concept of isomerism was conceived concurrently by Liebig and fellow German chemist Friedrich Wöhler (1800–82), studying fulminic acid and cyanic acid respectively. The acids were identically composed, but possessed different chemical properties. That is to say their molecular formulas were the same, but their molecular structures and hence their properties, were different. This coincidental joint discovery instigated a lifelong friendship between the two men and a fruitful professional collaboration.

Their discovery of the benzoyl **radical** showed for the first time that in organic substances there are groups of **atoms** which hold together and, in reactions, act like **elements**. They also devised a way to analyze the amounts of carbon and hydrogen present in organic compounds.

Organic analysis

In 1831 Liebig perfected a procedure that was a significant factor in the improvement of organic analysis and in his own success. As a way of determining the carbon content of organic compounds, he burned copper oxide with an organic compound and identified the oxidation products, carbon dioxide and water vapor, by weighing them, immediately after absorption, in a tube of calcium chloride and in a unique five-bulb apparatus containing caustic potash. Before Liebig's breakthrough, a procedure such as this could take a day. His simple method could be performed six or seven times a day and with more precise results.

Chemistry in relation to agriculture

Liebig was adamant that the chemist alone could help the farmer. The chemist could tell the farmer the best way of feeding crops, the nature of different soils and how particular manures reacted with particular soils. Liebig's analysis of soils led to his debunking of the "humus theory" – he proved that a plant's carbon content came not from leaf mould, but from atmospheric **photosynthesis**.

He subsequently developed his own chemical manures. They were not, at first, a success. He erroneously believed that plant nitrogen was more directly sourced from atmospheric ammonia and soil nitrates than from manures. The fertilizers were too

Legacy, truth, consequence

■ When Liebig died, a former student, A. W. Hofmann, wrote, "*No other man of learning, in his passage through the centuries, has ever left a more valuable legacy to mankind*".

■ Liebig was a radical teacher. It was he who insisted that chemistry should be taught not as a part of general pharmaceutical studies, but as an independent subject. No doubt remembering his youthful experiments, he instigated the practice of laboratory experimentation to such an extent that it became a standard part of training.

■ Several key schools founded in the mid-1800s, including Harvard University's Lawrence Scientific School and London's Royal College of Chemistry, based their educational methods on Liebig's model.

■ Some of the conclusions that Liebig came to have since been shown to be incorrect. For example, he would not acknowledge that yeast was a living substance. However, these errors, and the subsequent research that they inspired, have led to further important discoveries and his influence on scientific agriculture is beyond question.

> *The production of all organic substances no longer belongs just to the organism. It must be viewed as not only probable but as certain that we shall produce them in our laboratories. Sugar, salicin [aspirin] and morphine will be artificially produced.*
>
> Liebig and Wöhler, article in *Annalen der Pharmacie* (1838)

German and Italian advertising cards for the "Liebig Extract of Meat", c.1900.

insoluble for the plants to absorb. This was corrected and he produced more encouraging results. His efforts prompted further investigation by others that led to the discovery of super-phosphates, which were developed into fertilizers.

Animal chemistry

Liebig's successful analytical work on blood, urine, and bile prompted his interest in the metabolism of animals. He proposed that the ability of muscles to work came from energy provided by **oxidation** of foodstuffs such as carbohydrates and fats.

Liebig later experimented with ways of cooking meat that preserved its nutritional qualities. He evaporated soup from lean meat at low pressure and produced a glutinous liquid he called an "extract of meat". He believed that this highly nutritional and cheap alternative to real meat could be used as part of a restorative diet for the sick and under-nourished, particularly in places such as South America where obtaining extract from the countless slaughtered cattle would be an extremely economic operation. Within 20 years his extract of meat was being marketed as a nutritional food for "invalids and the laboring classes".

Charles Darwin

Charles Darwin was an English naturalist and is the most famous exponent of the theory of evolution. His doctrine of natural selection revolutionized biology and forever altered our idea of life on earth.

Born in Shrewsbury, England, in 1809, into an upper-middle class family, Darwin was a late developer academically, although he was always very interested in natural history. After attending Christ's College, Cambridge, he gained an opportunity that would prove to be one of the defining moments of his life. In 1831 Captain Robert Fitzroy required a naturalist for his scientific expedition aboard his ship HMS *Beagle*, which provided Darwin with the opportunity to sail around the world. He later described this voyage as *"by far the most important event in my life"*, one that *"determined my whole career"*. On board, Darwin studied *The Principles of Geology* by Charles Lyell (1797–1875) and was deeply influenced by Lyell's discussion of **James Hutton**'s view that the earth is much older than biblical scholars had claimed. Throughout the many excursions on the trip, Darwin was fascinated by the earth's varied species of animals and plants.

He conceived his theory of natural selection in 1838 and devoted the next 20 years to exploring this new idea of evolution.

Published in 1859, *On the Origin of Species by means of Natural Selection*, usually shortened to *The Origin of Species* was the result. His theory of **evolution** through the process of natural selection attempted to show that the evolution of life could be explained without postulating a supernatural being such as God. Predictably this drew an angry reponse from members of the Christian Church. On one occasion, Samuel Wilberforce, Bishop of Oxford, subjected *The Origin of Species* to a range of objections at the British Association, a learned society formed to promote science, although supporters of Darwin were present to defend the Darwinian view.

In 1839 Darwin married his cousin Emma Wedgwood. Together they had ten children and lived in Down, Kent. As the author of many books on the natural world, he became very famous and in later life was regarded as something of an elder statesman of the scientific community. After several years troubled by illness, he died at home in 1882, aged 73. He is buried in Westminster Abbey, London.

Essential science

Theory of evolution

While on his voyage with the HMS *Beagle*, Darwin visited the Galápagos Islands, an archipelago in the Pacific Ocean inhabited by numerous endemic species. He discovered that the tortoises exhibited slight physical differences on each island. It occurred to Darwin that the tortoises had not in fact been created differently but were developing differences – or evolving – as they responded to the differing environmental conditions of each island. While this seemed to be true for one species, Darwin came away believing that the theory of evolution as a general thesis warranted scientific study. Darwin's research on other species proved that the process of evolution had in fact occurred: that rather than life on earth being the product of a creator, it developed from simple to more complex organisms in accordance with their reaction to their surrounding environment.

Natural selection

In 1838, after his return from his trip on HMS *Beagle*, Darwin discovered "natural selection", the mechanism with which to explain the theory of evolution. While evolution was conceived of as a *process*, natural selection was identified as its *cause*. Darwin described *natural* selection by analogy with *artificial* selection: the idea that a breeder can play a modifying role in breeding domestic plants and animals by artificially selecting mates. With natural selection, however, there is no breeder. Instead environmental conditions, such as reproductive competition and an aptitude for survival, serve to shape the futures of individual species by the natural process of selecting the fittest organisms and eliminating the unfit ones. The general idea of competition was also employed in Darwin's overall theory in order to explain phenomena such as extinction and diversification through time.

Alfred Russel Wallace

Darwin hurried to complete *The Origin of Species* in 1859 because he was aware that other scientists were developing similar theories. In particular, Alfred Russel Wallace (1823–1913) developed a theory of evolution independently of Darwin, starting it as early as 1855. The two men became friends and corresponded regularly from 1858. Many contemporary scholars credit Wallace as the co-discoverer of evolution.

Legacy, truth, consequence

■ Darwin's *Origin of Species* was an immediate bestseller, appearing in six editions in Darwin's lifetime. Evoking a storm of controversy on its initial publication, particularly in relation to the prevalent **creationism** of the time, it continues to stimulate thought to the present day. It is undoubtedly one of the most important scientific books in history.

■ Darwin's theory was a direct challenge to orthodox religion. For many, it constitutes a challenge to the existence of God, in particular with its connection to the **argument from design**. Darwin's theory continues to be frequently cited by proponents of atheism, such as **Richard Dawkins**. Equally, Darwinism has proven controversial in recent years, particularly in the US, where Christian groups have questioned its cogency as a theory and argued that **intelligent design** should be taught in schools as a genuine alternative.

■ Some ideas only loosely related to Darwin's theory developed that Darwin would not have endorsed himself, for example, eugenics, which applied the concepts of Darwinism to human society, in particular the inheritance of certain characteristics. Eugenics in the twentieth century became stigmatized after it was taken up in the rhetoric of the Nazi Germans in their drive for genetic "purity". Social Darwinists also attempted to apply concepts of "survival of the fittest" and evolution to society and economic systems in the late nineteenth and early twentieth centuries. Darwin himself dismissed the idea of policies for social change being instigated by governments based on the theory of evolution and natural selection.

Key dates

1809	Born in Shrewsbury in the West Midlands of England.
1818	Boards at Shrewsbury School.
1825–7	Attends Edinburgh University to study medicine but leaves without completing his degree.
1828	Attends Christ's College, Cambridge.
1831	Gains BA (without honors).
1831–6	Sails the world on HMS *Beagle*.
1839	Marries Emma Wedgwood. Publishes *The Voyage of the Beagle*.
1842	Moves to Down, Kent, with his wife and family.
1859	Publishes *On the Origin of Species By Means of Natural Selection*.
1860	Samuel Wilberforce, Bishop of Oxford, criticizes *The Origin of Species* at the British Association for the Advancement of Science at Oxford.
1871	Publishes *The Descent of Man and Selection in Relation to Sex*.
1872	Publishes *The Expression of the Emotions in Man and Animals*.
1881	Publishes *The Formation of Vegetable Mold through the Action of Worms*.
1882	Dies at home.

Detail of the Tree of Life, used by Darwin as a model for the theory of evolution, from the 1859 edition of *The Origin of Species*.

The old argument of design in nature ... fails, now that the law of natural selection has been discovered. We can no longer argue that, for instance, the beautiful hinge of a bivalve shell must have been made by an intelligent being, like the hinge of a door by man. There seems to be no more design in the variability of organic beings and in the action of natural selection, than in the course which the wind blows. Everything in nature is the result of fixed laws.

Autobiographies (1887 & 1958)

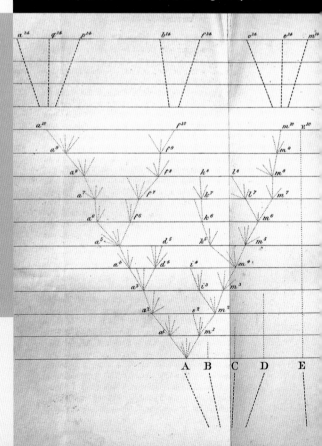

Other work

In addition to his work on evolution, Darwin published technical books on **geology**, a treatise on barnacles, and a widely celebrated account of his voyage on the HMS *Beagle*.

Robert Wilhelm Bunsen

Robert Wilhelm Bunsen was a German chemist who made a wide range of discoveries and inventions in a variety of fields. He is most famous for inventing the gas burner that bears his name.

Born in Göttingen, Germany, in 1811, Robert Wilhelm Eberhard Bunsen was the youngest of four sons. His father, Christian Bunsen, was librarian and professor of modern languages at the University of Göttingen.

After beginning his schooling in Göttingen, Bunsen transferred to the high school in Holzminden, where he graduated in 1928. He then entered the University of Göttingen and received his doctorate in physics two years later.

With a grant from the Hanoverian government, Bunsen traveled around Europe from 1830 until 1833, visiting factories, laboratories, and other places of scientific interest and meeting prominent European scientists of the time. He was a man of wide scientific interests; his main concerns lay in chemistry, physics, **geology**, and the application of experimental science to industrial problems.

On his return to Germany, Bunsen gained his first job at the University of Göttingen. He worked at the Polytechnic School, Kassel, in 1836 and two years later was appointed professor at the University of Marburg. In 1852 he succeeded German chemist Leopold Gmelin (1788–1853) at the University of Heidelberg and remained there for the rest of his career, despite having been offered the chance to succeed Eihard Mitscherlich (1794–1863) at the University of Berlin in 1863.

Bunsen was a very devoted teacher who always set aside a great deal of time to assist his students. He never married, partly because his teaching, research, and his love of travel occupied almost all of his time.

He retired in 1889, at the age of 78, and having been greatly interested in **geology** throughout his career he devoted much of his time to it in the ten years before he died.

Essential science

The Bunsen burner

Throughout his teaching career, Bunsen always emphasized the experimental side of science. He enjoyed designing new pieces of apparatus and developing and improving existing ones. Throughout the early 1850s he worked on the creation of a gas burner and, in 1855, developed one that built upon a burner originally designed by **Michael Faraday**. Among other experimental uses, Bunsen used his burner to identify metals and their salts according to their characteristic colored flames.

The Bunsen battery

Throughout the 1840s and 1850s Bunsen worked on and developed the **galvanic battery**. In 1841, he made a battery, known as the Bunsen battery, using heat-treated carbon as the negative pole, which was able to generate light of great intensity. He also made a battery with chromic acid instead of nitric acid as the **electrolyte**, and another with zinc and carbon plates in chromic acid.

Organic and inorganic chemistry

In his early years Bunsen completed some research in **organic** chemistry. Between 1837 and 1842 he published several papers on compounds of cacodyl, an arsenic-containing compound. In fact,

he lost the sight in one eye in 1843 after an explosion of cacodyl cyanide and, on another occasion, nearly died of arsenic poisoning. Through other studies on arsenic, he discovered that hydrated ferric oxide is an antidote to arsenic poisoning.

After abandoning research in organic chemistry, much of the rest of his life was devoted to **inorganic** chemistry. In this field, he developed a variety of analytical techniques for the identification, separation, and measurement of inorganic substances, and lectured on these techniques for much of his career.

Studies on gases

Between 1838 and 1846, Bunsen and Lyon Playfair (1818–98) investigated the industrial production of cast iron in Germany and demonstrated the inefficiency of the process. They discovered that over 50 per cent of the heat of the fuel was lost in escaping gases in charcoal-burning furnaces and 80 per cent was lost in coal-burning furnaces. In their paper, "On the Gases Evolved from Iron Furnaces with Reference to the Smelting of Iron", they proposed a series of techniques they had developed for recycling gases through the furnaces. Many of their discoveries in this area were collected in Bunsen's book *Gasometrische Methoden*, which also set out various methods for collecting, preserving, and measuring gases.

■ The Bunsen burner is a widely used piece of laboratory apparatus, which has transformed the practice of chemistry.

■ Bunsen developed and improved several other pieces of laboratory equipment, including the Bunsen battery, an ice calorimeter, a vapor calorimeter, a filter pump, and a thermopile.

■ During a scientific expedition to Iceland in 1846 Bunsen tested rocks from the volcano Hekla, concluding that volcanic rocks are mixtures of acidic **silica** rocks (trachytic) and basic rocks that are less rich in silica (pyroxenic). His assessment is no longer accepted but Bunsen's research in this field contributed greatly to the development of modern **petrology**.

■ In 1898, after his death, he was the first ever recipient of the Albert Medal, awarded by the English Society of Arts for his scientific contributions to industry.

Different flames produced by a Bunsen burner with different air valve settings, from closed (left) to fully opened (right).

The preservation and collection of gases is the first, and one of the most important operations in gasometry.

Gasometry: Comprising the Leading Physical and Chemical Properties of Gases (1857)

A chemist who is not a physicist is nothing at all.

Bunsen quoted in J. L. Partington (ed.), *A History of Chemistry* (1964)

Spectrum analysis

In the 1860s Bunsen and Gustav Kirchhoff (1824–87) employed the Bunsen burner in the field of **spectrum analysis**, using it to test the emission spectra from the flames of heated materials. Studies in this field led to their discovery of cesium, a new alkali metal.

Key dates

1811	Born in Göttingen, Germany.
1828	Graduates from the high school in Holzminden.
1830	Receives doctorate from the University of Göttingen with a thesis in physics.
1830–3	Travels in Europe.
1834	Discovers an antidote to arsenic poisoning.
1838	Appointed professor extraordinarius of chemistry at the University of Marburg.
1842	Appointed professor ordinarius of chemistry at the University of Marburg. Elected a foreign member of the Chemical Society of London.
1843	Loses sight in one eye after cacodyl cyanide explosion.
1845	Publishes "On the Gases Evolved from Iron Furnaces with Reference to the Smelting of Iron" (co-authored with Lyon Playfair).
1846	Scientific expedition to Iceland.
1852	Appointed professor at the University of Heidelberg.
1857	Publishes *Gasometrische Methoden* (English title: *Gasometry: Comprising the Leading Physical and Chemical Properties of Gases*).
1858	Elected foreign fellow of the **Royal Society of London**.
1860	Receives the Copley Medal from the Royal Society of London.
1877	Bunsen and Kirchhoff receive the first Davy Medal.
1889	Retires from teaching.
1899	Dies in Heidelberg in Germany.

James Dwight Dana

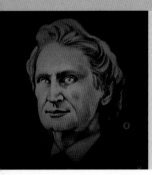

One of Dana's greatest contributions was to provide a classification system for minerals. As a geologist and zoologist he played a vital role in the United States Exploring Expedition (1838–42), collecting an extraordinary range of information on mountain building, volcanic islands, corals, and crustacea during the South Pacific voyage. Dana's *Manual of Mineralogy*, revised and republished, is still an important reference book today.

Dana was born into a devout Christian family in Utica, New York, where his father owned a hardware store. From an early age, Dana was fascinated with natural history, and collected plants, insects, and rocks. He entered Yale College (now Yale University) in New Haven, Connecticut, where one of his teachers was the prominent scientist and founder of the *American Journal of Science*, Benjamin Silliman (1779–1864). After graduating in 1833, his first important expedition overseas came with the post of mathematics teacher on a Navy ship. With Dana on board, the ship sailed to the Mediterranean and he was able to observe Mount Vesuvius erupting. His first scientific publication was a letter about the Vesuvius eruption, published in the *American Journal of Science*.

In 1836 Dana returned to Yale as assistant to Silliman. By studying his own and Silliman's mineral collections, Dana developed a mineral classification system and described it in a 580-page book called *A System of Mineralogy*, published when he was only 24 years old. Two years later he boarded the *USS Peacock*, one of a small fleet of US Navy ships that was sent by Congress to explore the Southern Seas. The main purpose of this United States Exploring Expedition was to chart the Pacific islands, and survey Antarctica. One of several scientists involved, Dana had the opportunity to study both the **geology** and **zoology** of unknown regions. The route of the expedition included the Andes, the coral islands (atolls) and reefs of the Pacific, and Kilauea, an active volcano in the Hawaian islands. His collection from the expedition included 300 fossils, 400 corals, and 1,000 crustacea. Afterwards, for more than ten years, Dana labored to write up his findings, which resulted in three detailed reports, published between 1846 and 1854: *Zoophytes, The Geology of the Pacific Area*, and *Crustacea*.

Around this time, and now much admired by many of his contemporaries and viewed as America's primary geologist, Dana was appointed Silliman's successor at Yale College. With his focus now firmly on geology, he continued to be a prolific writer, producing numerous skilfully illustrated reports and publications. They contained such theories as the age progression of young high islands to older atolls, and the "contracting earth" theory that explained mountain building. In 1848 his *Manual of Mineralogy* was published, a revised edition of which is in print today. He returned to Hawaii in his 70s for more studies in preparation for his work, *Characteristics of Volcanoes*, and continued to write and study until he died.

Essential science

System of Mineralogy

In the first editions of Dana's *System of Mineralogy*, the method he used for classifying minerals was similar to **Carolus Linnaeus'** classification system for plants and organisms (genus and species). The fourth edition of the book was revolutionary in using a chemical classification system in which minerals were classed according to their composition (for example, silicates, sulfates, and oxides) – a method still universally accepted today.

Contracting earth theory

Dana's "contracting earth" theory (also known as "global cooling") was that the once molten earth was still cooling. As its molten interior cooled and shrank, the rigid outer crust crumpled. Dana proposed that the crumpling effect was seen in features such as mountain ranges.

Atolls and reefs

A lively debate between **Charles Darwin**, naturalist Louis Agassiz (1807–73), and Agassiz's son Alexander was running in the nineteenth century as to the formation of coral reefs. On returning from his *Beagle* voyage in 1837, Darwin argued that most reef formations, such as atolls, were due to coral growth in shallow waters after the subsidence of oceanic islands. Dana, independently, worked out this same "subsidence theory", but added details from his own observations, such as how coral atolls eventually die and subside beneath the ocean surface. Agassiz's son argued against this subsidence theory – saying that some reefs grew where submarine mountains had risen as a result of accruing plankton debris. It was only in 1951, with evidence from drilling down into the ocean floor, that Darwin's and Dana's subsidence theory was proven.

Legacy, truth, consequence

■ Dana's system for the classification of minerals was a great leap forward for mineralogy, and his chemical classification system is still used universally, although it has been much extended with the ongoing discovery of new minerals.

■ His theory of a "contracting earth" was overtaken by the concept of **plate tectonics**, which had its beginnings in 1915, when scientist and meteorologist **Alfred Wegener** proposed the idea of continental drift (see pages 140–1).

■ Detailed observations of volcanic chains and coral formations led Dana to original and accurate predictions, such as the existence of drowned atolls, today called guyots. His understanding of the linear pattern of volcanic chains in the Pacific Ocean, and their age progression from active volcano to atoll, also holds true.

■ As one of the first trained scientists to study erupting volcanoes, Dana provided some of the earliest useful reports on volcanic activity. For example, he noted the difference between the viscous lava from Vesuvius and the more fluid nature of basaltic lava from Kilauea.

■ In order to display some of the nation's collection from the US Exploring Expedition, a first national museum of America was opened. With each publication reporting the findings of the Expedition, America's status in the world of science improved. Dana's works played a significant role and were described by German naturalist and explorer Alexander von Humboldt (1769–1859) as "*the most splendid contribution of science of the present day*".

The Forum, Pompeii, with Vesuvius in the Distance, a nineteenth century painting. On his first important expedition overseas, Dana observed Mount Vesuvius erupting.

I am really lost in astonishment at what you have done in mental labor. And, then beside the labor, so much originality in all your works.

Charles Darwin in letter to Dana about Dana's publication, *Crustacea*

United States Exploring Expedition

With six ships, nine scientists and artists, and a crew of over 300 men, the US Exploring Expedition was the largest voyage of its kind thus far, and was the first US-led exploration to gather geological information in a systematic way. Vast amounts of information and samples were garnered, adding greatly to America's scant scientific and **ethnographic** collections. Darwin was among those who admired Dana for the huge numbers of samples he gathered and outlined in *Crustacea*, and which included more than 500 newly discovered species of lobsters, crabs, barnacles, and shrimps.

Key dates

1813	Born in Utica, New York, US.
1830	Enters Yale College, New Haven, Connecticut.
1837	*A System of Mineralogy* is published.
1838	Joins the United States Exploring Expedition.
1843	*Manual of Mineralogy* is published.
1844	Marries Henrietta Silliman, daughter of his former professor, Benjamin Silliman.
1846	Becomes co-editor of the *American Journal of Science and Arts*.
1848	Publication of *Manual of Mineralogy*.
1849	Appointed professor of natural history at Yale College.
1852	First part of *Crustacea* is published.
1862	*Manual of Geology* is published.
1864	Appointed professor of geology and mineralogy at Yale College.
1872	*Corals and Coral Islands* is published.
1887	Revisits the Hawaiian Islands to continue his study of volcanoes.
1890	*Characteristics of Volcanoes* is published.
1895	Dies in New Haven.

Rudolf Virchow

German pathologist, anthropologist, archaeologist, and a prominent political figure and activist, Rudolf Ludwig Karl Virchow was the founder of cellular pathology and modern pathology. He was the first academic to confirm the theory that cells originate from cells, and that disease either originates from cells or is characterized by malfunction at the cellular level.

Born in Schivelbein, Pomerania, Prussia (now Swidwin, Poland), to a farming family, Virchow received a scholarship to attend the high school in Koslin at the age of 14. He read medicine at the Friedrich-Wilhelm Institute in Berlin, graduating in 1843, and went on to work at the Charité Hospital, where he became known for his series of lectures and demonstrations in **pathology**. In 1847 he became a permanent member of staff at the university, and editor of the *Journal for Pathological Anatomy and Clinical Medicine*, later known as *Virchow's Archiv*.

In 1848 Virchow conducted a study into the typhoid epidemic in Silesia, and argued that the country's poor health standards and disease-ridden populace were a direct result of their lack of freedom and democracy. This declaration was to be the starting point of his theories linking practical medicine and political legislation, which he supported in his weekly paper, the *Medical Reform*. This criticism of local government was to prove dangerous for Virchow, but he did not relent, usually writing most of the content for the journal himself.

Virchow's political views led him to leave the Charité for another position as professor of pathological anatomy at Würtzburg in

Bavaria. It was during this time that he met and married Rose Mayer, with whom he had six children.

It was also at Würtzburg that he formulated the basic biological law for which he is best remembered: that each cell originates from another cell (or cells multiply by division). This led to his theory that disease is not located in organs, tissues, vessels, or nerves, but originates from one cell, or one group of cells. He coined the term "cellular pathology", and in 1858 his most lauded work, *Cellular Pathology as Based upon Physiological and Pathological Histology*, was published: a collection of lectures he had given on the topic.

He was a popular academic and educator, and greatly admired for his sharp mind. He became known for encouraging medical students to use microscopes, embodying his enthusiasm for addressing disease at the cellular level. Due to his increasing popularity he was offered a place as chair of pathology at the University of Berlin in 1856, which he accepted under his own terms, requesting that the university provide him with a pathology institute – the place where he would spend the rest of his working life.

From 1859 his political career became more established. He was elected to the Berlin City Council, and then went on to be a

Essential science

Cellular theory

Virchow was one of the first physicians to promote the theory that "every cell originates from another existing cell like it", thus rejecting the current theories of spontaneous generation of disease. His theory was published in 1858 in *Cellular Pathology as Based upon Physiological and Pathological Histology*. Although the theory is not thought to originate with Virchow, many believe that it is his promotion of it in his publications and lectures that led to the acceptance of cells as the origin of other cells.

Oncology

Virchow's work on cellular origin led to pioneering research in related diseases, particularly in the field of **oncology**. He was not only the first to correctly describe a case of leukemia (cancer of the blood), but his research yielded important findings in other types of malignancy too. Virchow is one of two physicians who

simultaneously described the earliest signs of gastrointestinal malignancy (cancer of the stomach). One of these early symptoms is now known as "Virchow's node", and is characterized by an enlarged supraclavicular lymph node, usually on the left side. This is sometimes called "Troisier's sign", named after Charles Emile Troisier (1844–1919), who also described the association.

> **Medicine is a social science, and politics is nothing but medicine on a large scale. The physicians are the natural attorneys of the poor, and the social problems should largely be solved by them.**
>
> Attributed to Rudolf Virchow

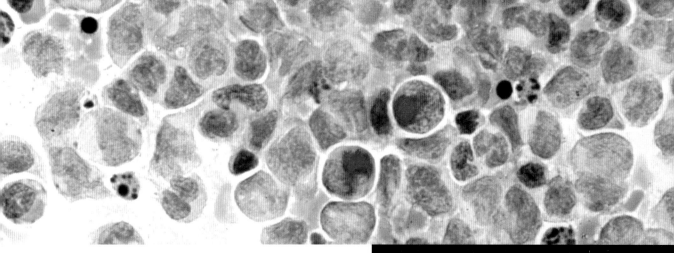

founding member of the German Progressive Party, as well as sitting as a liberal member of the Reichstag from 1880 to 1893. After the Franco-Prussian War (1870–1) he became involved in healthcare and emergency medicine, but nevertheless maintained his political interests for the rest of his life. He used both his popularity and his political influence as an activist, making considerable efforts to improve public health in Berlin by campaigning for better standards of care and advocating water and sewage purification.

Later on in life, Virchow also became involved in archaeology. One of his more popular science projects involved working with Heinrich Schliemann, the man who discovered ancient Troy. After accompanying Schliemann on digs in Troy and Egypt, Virchow published a book on Trojan graves and skulls. His interest in archaeology was closely linked to his passion for anthropology; he co-founded several anthropological societies during his life, such as the German Anthropological Society (1869). Virchow himself performed an anthropological study into German ethnicity, where he surveyed thousands of school children. The results were published in 1876 and called into question contemporary racist theories of the "Aryan race".

Legacy, truth, consequence

■ Virchow's ideas, that disease originates in cells or is represented by cells in an abnormal state, provided the foundations for modern pathology. His cellular theory replaced the existing theories of spontaneous generation, the notion of organisms arising from non-living matter. He is credited with founding "social medicine", the field of medicine based on the theory that disease is often socially and economically derived.

■ His descriptions of various types of malignancy led to better identification of early symptoms.

■ The terms "thrombosis", meaning the development of masses of blood vessels, and "embulus", meaning a part of a thrombus that can become separated and cause harm to surrounding vessels, were coined by Virchow, who was the first to disprove the common opinion that phlebitis (inflammation of a vein) was the cause of most diseases.

■ He introduced a standardized technique to perform autopsies, which is still used today.

Think microscopically

Virchow's persistent admonition to his medical students in his encouragement of their use of microscopes

Key dates

1821	Born in Schivelbein, Pomerania (now Swidwin, Poland).
1843	Graduates from the Friedrich-Wilhelm Institute, at the University of Berlin.
1845	Publishes the first report of a case of leukemia.
1847	Becomes editor of the *Journal for Pathological Anatomy and Clinical Medicine*.
1848	Launches the weekly paper *Medical Reform*.
1849	Following a temporary suspension from the Charité Hospital linked to his political views, he is appointed to a newly created position as chair of pathological anatomy at the University of Würzburg, Bavaria.
1856	Accepts a position specially created for him and returns to the University of Berlin.
1858	Publishes his key work, *Cellular Pathology as Based upon Physiological and Pathological Histology*.
1859	Becomes elected to the Berlin City Council.
1861	Becomes elected to the Prussian Diet.
1863–7	Publishes his book series on the pathology of malignant tumors.
1879	Accompanies Heinrich Schliemann on his excavation in Troy.
1880–93	Becomes an active member of the Reichstag.
1881	Goes on his own expedition to the Caucasus (and in 1894).
1873	Elected to the Prussian Academy of Sciences.
1894	After declining a noble title, is declared "Geheimrat" – privy councillor.
1902	Continues to research, write, edit, and lecture up to the year of his death.

Gregor Johann Mendel

An Austrian botanist and monk, Mendel is best known for his statistical experiments on generations of pea plants and his resulting theory of inheritance. The significance of his understanding of inheritance remained unrecognized for more than 30 years. Rediscovered in the early twentieth century, "Mendel's laws" provided the foundation for a revolutionary science – that of genetics. Today, Mendel is commonly called "the father of genetics".

Johann Mendel was born in the village of Heinzendorf, Austria (now Hyncice in the Czech Republic). His parents were farmers with limited resources, so to pay for his grammar schooling he had to teach fellow students. A loan from his sister then allowed him two years of study at the Institute of Philosophy in Olmütz (then Olomouc). In 1843, when money ran out, he entered the Augustinian abbey in Brno (then Brünne) as a novice, and took the name of Gregor. The Augustinians' emphasis on education and research provided Mendel with a golden opportunity to continue his studies.

Brno was the center of an agricultural region, and the abbey was involved in plant trials. The abbot, C. F. Napp, had already set up an experimental garden in the abbey grounds and saw that Mendel had a fascination for scientific research similar to his own. Mendel was ordained in 1847, and in 1851 Mendel's supportive abbot sent him to the University of Vienna to study mathematics, physics, and natural history. After returning to the abbey in 1853, Mendel began a 14-year period of teaching at a local school, and in 1854

was encouraged to begin the plant experiments that would lead to his major discovery.

Mendel's experiments on plant hybridization in the abbey garden were focused on common garden peas (*Pisum sativum*). By observing how characters or traits of the plants were transmitted from parent plants to their offspring he hoped to "deduce the law according to which they appear in successive generations". Over the course of ten years he completed the tedious work involving the cultivation and selective crossbreeding of nearly 30,000 plants, and the counting and categorizing of hundreds of thousands of peas. In 1865 he reported his results and theory of inheritance to the Association of Natural Research in Brno.

Although greatly disappointed by a lack of response from the scientific world, he attempted to test his theory on other plants, and on bees, but did not achieve satisfactory results. The last years of Mendel's life were largely taken up by administrative duties, which had become increasingly time-consuming since his election as abbot in 1868.

Essential science

Heredity

Heredity is the passing on of biological characteristics, or traits, from generation to generation of plants and animals. Before Mendel, the basics of heredity and how tiny units in cells, today called **genes**, carry these traits from one generation to the next was not understood. **Aristotle** taught that traits were carried through the blood. Likewise, scientists such as French biologist, Chevalier de Lamarck (1744–1829), mistakenly believed that traits acquired during a lifetime could be transferred to the next generation through the blood. For example, he thought that giraffes acquired long necks through generations of stretching to tree-top leaves, and that the trait was passed on in the blood. **Charles Darwin** held a similar theory: in 1886 he explained that particles in the blood, affected by activities in life, migrated to the reproductive cells and so were passed on. Before Mendel, few experiments were carried out to discover the truth about heredity, and Darwin, like other contemporary scientists, was unaware of Mendel's revolutionary theory.

Mendel's experiment

In Mendel's experiments he cross-pollinated purebred strains of the common pea plant, which had distinct traits. He chose to compare seven sets of contrasting plant traits in each generation of the plant, such as stem height (long or short), color of flower (purple or white), and seed/pea color (green or yellow). He found that the offspring plants always showed one or other of each trait, not a blend of the two; for example, the flowers were always purple or white, and not a different color blend. By cultivating further generations he found that one of each pair of traits was dominant. For example, in the first offspring the seeds were always yellow, and in the second generation the seeds were predominantly yellow in a 3:1 ratio. The ratio appeared in following generations, too.

Mendel's laws

Mendel's conclusions were published as an article entitled *Experiments with Plant Hybrids* in 1866, and came to be known as Mendel's laws.

Legacy, truth, consequence

■ It was not until 1900 that Mendel's laws were rediscovered, by three scientists independently: German botanist Carl Correns (1864–1933), Austrian agronomist Erich von Tschermak-Seysenegg (1871–1962), and Dutch botanist Hugo de Vries (1848–1935). Important **genetic** discoveries followed quickly, including the discovery that Mendel's law of independent assortment applies only to those genes that are located on different **chromosomes**.

■ As the "father of genetics" Mendel laid the foundations for modern sciences such as **genetic engineering**.

■ Mendel's theory of inheritance had a profound influence on scientists' understanding of many subjects, including **evolution**, **biochemistry**, medicine, and agriculture.

■ Mendel's meticulous experiments and use of mathematics, for example in his deduction of the 3:1 segregation ratio, marked the beginning of the systematic use of statistics in biology.

Key dates

1822	Born in Heinzendorf, Austria (now Czech Republic).
1843	Joins Augustinian abbey in Brno under Abbot C. F. Napp.
1846	Completes a course in agricultural studies at Brno Institute of Theology.
1847	Ordained a priest.
1851	Sent to University of Vienna.
1854	Begins teaching natural sciences at a local high school; also begins his experiments on garden peas in the abbey gardens.
1863	Publication of his meteorological observations.
1865	Gives lecture on his "Experiments with Plant Hybrids".
1866	Publication of his *Experiments with Plant Hybrids*.
1868	Elected abbot after the death of C. F. Napp.
1871	Begins experiments with bees.
1884	Dies in Brno (Czech Republic).

The value and utility of any experiment are determined by the fitness of the material to the purpose for which it is used, and thus in the case before us it cannot be immaterial what plants are subjected to experiment and in what manner such experiment is conducted.

Attributed to Gregor Mendel

• *The Law of Segregation* firstly says that inheritance of each trait is determined by "factors" (today called genes) that are passed on unchanged to offspring. Each plant must have a pair of factors for each trait, as demonstrated by the fact that first generation yellow-pea plants can produce both yellow- and green-pea plants. Secondly, the trait that is shown in the first generation plant is the result of a dominant factor, while the trait that is not shown is the result of the regressive factor. The paired factors must segregate (separate) during the division of sex cells and each sperm or egg gets only one trait of each pair. So the first generation plants can produce two types of sex cells, one with a factor for yellow peas and one with a factor for green peas. Mendel said that the random combination of sex cells at fertilization produced the ratios in the second generation.

• *The Law of Independent Assortment* says that each paired factor is separate, and not linked to other paired factors, so is inherited independently from one another. So, for example, a pea plant with purple flowers is not more or less likely to have yellow rather than green peas.

The results of Mendel's cross-pollination of yellow- and green-pea plants.

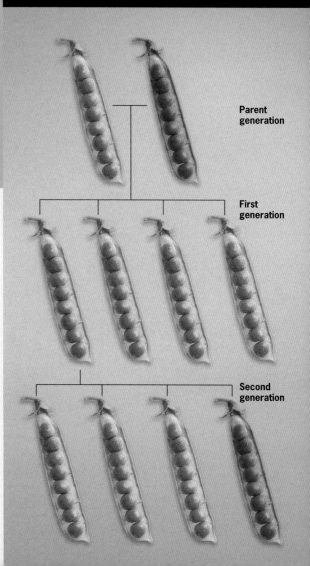

Parent generation

First generation

Second generation

Louis Pasteur

Louis Pasteur was a French chemist, credited as the "father of modern bacteriology", who promoted inoculation as a means of preventing disease and, through careful experiments with microbes, discovered hugely valuable vaccines for numerous diseases, including anthrax and rabies. He is also known for discovering pasteurization, the process that enables the preservation of certain liquids such as milk and wine.

Born in the French town of Dôle in the Jura Mountains in 1822, Louis Pasteur spent most of his childhood in Arbois, in the same region. His father was a sergeant major in Napoleon I's army and both his parents encouraged his desire to learn. From an early age, Pasteur was interested in the natural world. He would go out hunting birds after school and he enjoyed studying plants and animals in their natural habitats. He was a distinguished artist, a fact well known among the people of his hometown.

Throughout his formative years, Pasteur worked hard and excelled academically, eventually culminating in his receipt of a doctorate in physics. As an academic, he was a good lecturer and attracted many students.

His initial research focused on a combination of chemistry, **crystallography**, and **optics**. In 1847 he discovered that crystals are chemically the same but have different shapes because of how their molecules fit together. More generally, he showed that how the molecules fit together affects the chemical nature of a substance. After his famous studies on **fermentation**, he was made director of scientific studies at the École Normale in 1857, and in the following years, while working at a variety of different universities, he conducted research on pasteurization and a range of vaccines, especially for anthrax and rabies. With all these scientific breakthroughs, he became famous around the world and many sought his advice.

In later life, he founded the Pasteur Institute and served as its first director for several years. After suffering from the blood disease uremia for nearly a year, he died in 1895. He was given a state funeral and his body was buried in the crypt of the Pasteur Institute.

Essential science

Fermentation and pasteurization

Pasteur was often asked by local wine merchants and breweries to help with fermentation of their wine and beer. In particular, he was asked to investigate why the liquids went "off". Using a microscope, Pasteur discovered that cells in non-fermenting liquids were round whereas those in fermenting liquid grew longer, and that a certain type of cell was present in milk that had fermented and gone sour. Separating these cells out of the milk, he put them in another liquid and immediately the new liquid began fermenting. Following his investigations Pasteur was able to inform the local wine merchants that if wine is exposed to the atmosphere a fungus called *Mycoderma aceti* soon develops, turning the wine into acetic acid (vinegar), whereas if the wine bottle is hermetically sealed the wine will stay pure.

Pasteur also addressed a matter much discussed during the 1850s – the idea of spontaneous generation (living organisms springing into life from nothing). He devised an experiment, filling flasks with liquid that he had boiled to kill the **microbes**, and then sealing the flasks so that no air or dust could get inside. No fermentation occurred. But when the flasks were subsequently exposed to air and dust again, fermentation occurred in every flask. This proved that (1) microbes are present in the air and dust, and (2) fermentation will not occur in a liquid when it has been boiled and sealed, thus disproving the idea of spontaneous generation. Fermentation occurs because a liquid has come into contact with microbes present in the air and dust, and not because living organisms can spring into life from nothing.

To combat the problem of fermentation, Pasteur developed a method, called pasteurization, involving heat followed by rapid cooling, which destroyed the microbes that spoil food. In 1864 he used this process to prevent beer and wine from spoiling, heating samples to 122–140°F (50–60°C), followed by rapid cooling.

Anthrax and rabies vaccinations

Anthrax was a major problem in France during Pasteur's time. Pasteur discovered that microbes present in the blood of animals that had died from anthrax were the cause of the sickness, and that even if the dead were buried away from live animals, the earth where they had been became infected with the **bacteria** and the disease could be passed on to other animals through the earth or earthworms. This knowledge greatly helped farmers to protect their animals. Pasteur also worked on a vaccine containing anthrax bacteria. The vaccine was heated to 108°F (42°C) which weakened the bacteria. It was then injected into sheep, which gave them only a mild case

Legacy, truth, consequence

■ Pasteur's pioneering work in vaccination and sterilization has saved the lives of millions of animals and human beings, from the mid-nineteenth century onwards. It also paved the way for more sophisticated research. In addition to his anthrax and rabies vaccines, Pasteur used similar methods to discover vaccines for chicken cholera, swine erysipelas (skin infection), and silkworm diseases.

■ In the nineteenth century, many people died after operations. Pasteur recognized that the lack of cleanliness was a contributing factor because microbes could invade the body, causing disease and infection. In 1878 he wrote to the Academy of medicine in France:

> "If I had the honor to be a surgeon, aware as I am of the danger which microbes bring, not only would I make sure that the instruments were perfectly clean, but after washing my hands with great care and drying them quickly in heated air, I would only use lint, bandages, and sponges which had been exposed to air heated to a temperature of 130–150°C [266–302°F]. I would only use water which had been raised to a temperature of 110–120°C [230–248°F]."

■ With generous funding from various sources, he established the Pasteur Institute in Paris, dedicated to the science of biology. Today the Institute is an important center for the production of vaccines and serums and also a center for biological research, with fieldwork particularly on tropical diseases suffered in Africa.

Where observation is concerned, chance favors only the prepared mind.

Speech at the inauguration of the Faculty of Science, University of Lille (1854)

of anthrax, from which they recovered, while immunizing them against future attacks.

On May 5, 1882 Pasteur conducted a famous experiment that proved the worth of his vaccine once and for all, and garnered quite a crowd. He inoculated 25 sheep and left 25 without inoculation. Twenty-six days later he injected all 50 sheep with full-strength anthrax bacteria. Two days after that, all the sheep that hadn't been inoculated were dead and all the inoculated ones were alive.

To find a vaccination for rabies, Pasteur injected the rabies virus into the brain of a rabbit. He then removed a piece of the rabbit's spinal marrow and placed it in a sterilized heated flask, where it eventually dried out and lost its virulence. Decomposing the fragments in pure water he then injected the solution into dogs that had been bitten by rabid animals. The following day, he injected the same dogs with a much stronger dose of rabies virus. He continued the process until he injected a dog with spinal marrow of a rabbit that had died of rabies only a few hours earlier and found that this dog did not die. Thus the dog had developed a resistance to the disease and Pasteur had discovered a vaccination for rabies.

Pasteur was able to put his vaccine into action on a person in 1885, injecting an afflicted boy, Joseph Meister, with marrow that was only a day old. Ten days later the boy was well again.

Key dates

1822	Born in Dôle, France.
1843–6	Student at École Normale Supérieure, Paris.
1844	Studies the chemical composition of crystals.
1848	Appointed professor of physics, Lycée de Dijon.
1849–54	Appointed professor of chemistry, Strasbourg.
1854	Appointed professor of chemistry and dean of the Faculty of Sciences, Lille University.
1857	Conducts studies on fermentation. Appointed administrator and director of scientific studies, École Normale.
1863–7	Appointed professor of geology, physics, and chemistry, École des Beaux-Arts, Paris.
1864	Invents the process of pasteurization.
1867	Appointed professor of chemistry, Sorbonne, and director of the laboratory of physiological chemistry, École Normale. Lectures to the wine merchants of Orleans.
1874	Awarded the Copley Medal, **Royal Society of London** (for work on fermentation and silkworm diseases).
1877	Research on anthrax.
1878	Research on hospital cleanliness and infection.
1879	Discovers the principles of vaccination.
1881	Demonstrates a successful vaccination of rabies. Elected to the French Academy.
1882	Conducts anthrax experiment, which proves the effectiveness of his vaccine.
1885	Uses rabies vaccine on a person for the first time.
1888	Founds the Pasteur Institute in Paris and serves as its first director.
1895	Dies from uremia in Paris, France.

Louis Pasteur, portrait in *Vanity Fair*, 1887.

Ferdinand Cohn

Ferdinand Cohn, who began his professional studies as a botanist, is considered to be a founder of the science of bacteriology. His accomplishments include the description of different kinds of bacteria, the development of an early classification system for bacteria, and the discovery of the formation of heat-resistant spores. Cohn encouraged and supported the work of other scientists, including that of Robert Koch.

Ferdinand Cohn was born in the German Jewish ghetto of Breslau, Silesia (now Wroclaw, Poland). His father, a successful merchant, supported his education and career, both mentally and financially. Cohn was reportedly a child prodigy who was able to read before the age of two years. At the age of 14 he began studying at the University of Breslau, where he developed an interest in botany. Although he successfully completed four years of study, he was not awarded a degree because of his Jewish ancestry. He moved to Berlin in 1846 to study for a doctorate in botany, which he completed two years later, but was unable to obtain a teaching position in Berlin because of his liberal political views and probably again because of his Jewish roots.

When Cohn returned to Breslau in 1849 he worked at the university's Physiological Institute. At that time his father gave him an expensive, high-quality microscope, which became one of his most important research tools over the following years. He was appointed associate professor of botany in 1959 and initially devoted his time to the study of plant cells. In 1866

the university allowed him to develop an institute for plant **physiology**.

For a number of years he extensively studied fungi and algae, but he is probably most famous for his contributions to the science of **bacteriology**. His most important publications in this field include an early system of classification for bacteria, published in 1872, and the description of the life cycle of *Bacillus subtilis* (a rod-shaped bacterium, commonly found in soil), together with the discovery that these bacteria form heat-resistant **endospores** capable of germinating to form new bacilli, published in 1876.

Recognized as the foremost bacteriologist of his day, Cohn's lectures attracted many students and young scientists. He realized the importance of **Robert Koch**'s work, which he encouraged and supported. During his lifetime Cohn received numerous distinctions and was a foreign member of the **Royal Society of London** and the Linnean Society, the first society in the world for the study and dissemination of **taxonomy** and natural history.

Essential science

Algae and fungi

Cohn devoted a great part of his early career to the study of algae and fungi. Although his conclusion that algae and fungi belonged to the same class of organisms turned out to be wrong, his early efforts on the developmental and sexual cycles of algae and lower fungi are still considered important contributions to the biology and classification of these organisms.

Institute of plant physiology

In 1866 Cohn was provided with the facilities (a few empty rooms in a converted convent) to develop the world's first institute of plant physiology. He used these rooms for teaching and research, but the facilities soon became inadequate and there was no space for a museum of botany. Eventually, a proposal for a new building in the botanical gardens of Breslau was accepted and the new institute opened in 1888. The building contained a herbarium, a museum, a lecture room and the institute of plant physiology with

laboratories and a library. Cohn was the director of the institute until his death in 1898.

Foundation of the journal *Contributions on Plant Biology*

In 1872 Cohn founded his own journal, which contained the first essays on modern bacteriology. Not only was Cohn able to publish his own findings in this relatively new science, the journal also allowed other pioneers in the field to publish their research.

Classification of bacteria

Many of Cohn's contemporaries believed that all bacteria were variations of the same organism and that the differences observed only reflected the different stages of development. In addition, the naming of bacteria was extremely chaotic at that time, because the majority of scientists who microscopically examined bacteria named the organisms without regard for anybody else who had already named the same organism.

Legacy, truth, consequence

■ With his classification of bacteria, Cohn laid the foundation for subsequent systems of classification. The notion of defining different types of bacteria according to their appearance as well as according to their physiological properties still forms the basis of today's classification. Cohn's discovery that different kinds of bacteria have different properties was hugely important in establishing the concept that certain bacteria can cause particular infections. His work, combined with the findings of **Louis Pasteur** and the physicist John Tyndall (1820–93), eventually discredited the theory of spontaneous generation, which stated that living organisms could develop from nonliving matter.

■ Cohn was active in applied **microbiology**. He advised farmers on the diagnosis and treatment of fungal infections of crops. In addition, he recognized that water was capable of harboring and transferring infectious diseases to humans. A pioneer in water analysis, he investigated drinking water as a potential source of infection during cholera outbreaks, even though the pathogen causing cholera was only later described by Robert Koch.

■ Cohn recognized the importance of Robert Koch's work on *bacillus anthracis* and published Koch's findings in his journal, *Contributions on Plant Biology*. With Cohn's support Koch went on to discover the bacterial causes of cholera and tuberculosis.

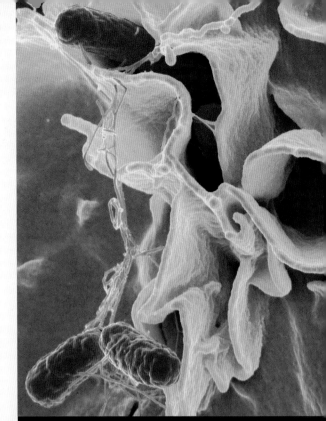

Salmonella bacteria (in red) invading cultured human cells. Cohn developed a system of classification for bacteria.

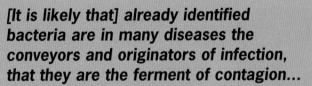

[It is likely that] already identified bacteria are in many diseases the conveyors and originators of infection, that they are the ferment of contagion...

Bacteria, the Smallest of Living Organisms (1873)

Cohn recognized the existence of different species of bacteria and, in 1872, he published the three-volume treatise *Researches on Bacteria,* which contained a system of classification for bacteria, dividing them into four groups: sphaerobacteria (round), micro-bacteria (short rods), desmobacteria (longer rods or threads), and spirobacteria (screw-like or spiral). He also recognized six genera, with at least one genus belonging to each group, and further divided some genera into subcategories (e.g. pigmented, fermenting, or contagious).

Description of *Bacillus subtilis* and discovery of endospores

In 1876 Cohn described the entire life cycle of *Bacillus subtilis*. He became the first person to demonstrate that these bacteria form endospores upon exposure to heat. Many bacteria can be killed by boiling, but spores are more resistant to heat and other physical agents than the vegetative forms of bacteria. When environmental conditions become more favorable again (e.g. by a return to room temperature), these spores are capable of germinating to form new bacilli.

Key dates

1828	Born in Breslau, Silesia, now in Poland.
1842	Admitted to the University of Breslau.
1846	Studies for a doctorate in botany in Berlin. He is introduced to the study of microscopic animals.
1848	Receives his doctorate at the age of 19 years.
1849	Awarded a position at the University of Breslau, where he remains for the rest of his working life. Given his own microscope by his father.
1859	Appointed associate professor of botany at the University of Breslau.
1866	Develops the world's first institute for plant physiology. Cohn is the director of the institute until his death.
1867	Marries Pauline Reichenbach.
1872	Founds the journal *Contributions on Plant Biology*, which contains the first essays in modern bacteriology. Publishes his first classification of bacteria in *Researches on Bacteria*. Becomes full professor of botany.
1876	Publishes essays outlining the life cycle of *Bacillus subtilis* and describes the formation of heat-resistant spores. Meets Robert Koch and supports Koch's work on *Bacillus anthracis*.
1885	Receives the Leeuwenhoek medal.
1895	Receives the Gold medal of the Linnean Society.
1898	Dies in Breslau, Silesia, Poland.

James Clerk Maxwell

James Clerk Maxwell is one of the greatest scientists the world has ever known, whose genius was manifested in his ability to understand physical phenomena in terms of mathematical equations. He is most famous for his comprehensive theory of electricity and magnetism, showing that electric currents can exist in any material, even empty space.

Born in Edinburgh, Scotland, in 1831, James Clerk Maxwell moved to Glenlair, in the Scottish area of Dumfries and Galloway, at the age of two, an estate that would intermittently remain his family home for the rest of his life. He was an only child and his mother's death when he was only nine had a profound effect on him. Likewise, an outing with his father, when he was 11, "*to see some electro magnetic machines*" was a defining event: it marked the beginning of his lifelong interest in science.

Educated at home, the Edinburgh Academy, and then at the University of Edinburgh, he went on to Cambridge University in 1850 and came second in their Mathematical Tripos (mathematics course) of 1854. At Cambridge he engaged in not only physical studies of natural phenomena but also religious, philosophical, and theological questions. During these years, he had something of a reputation for eccentricity. Unusually for a man of such practical genius, he was interested in mysticism and had a playful personality. He was also a lover of music, although as a boy he had been unable to gain enjoyment from listening because of persistent ear inflammation.

From 1856 to 1860 he was professor of **natural philosophy** at Aberdeen, during which time he married Katherine Mary Dewar, daughter of the principal of his college. In 1860, after the suppression of his professorship due to the merging of Aberdeen's two colleges, Maxwell accepted a position at Kings College, London. He then left London in 1865, retiring to Glenlair for six years.

Maxwell returned to Cambridge in 1871 and was appointed to the newly founded chair of experimental physics. After a major donation from the Duke of Devonshire, of the Cavendish family, he was entrusted to build, furnish, and organize the new Cavendish Laboratory, a feat for which he is well remembered. In 1879 he died from abdominal cancer, at the age of 48.

Essential science

Theory of electromagnetic radiation

Maxwell's research on **electromagnetism**, as expounded in his 1864 publication "A Dynamical Theory of the Electromagnetic Field", marks a continuation of **Michael Faraday**'s breakthroughs in the subject. Where Faraday sought to understand the relationship between electricity and magnetism, Maxwell formulated Faraday's laws of **electromagnetism** into a mathematical form, now known as the "Maxwell equations". The equations showed the following:

1. Unlike charges attract each other; like charges repel (also called Coulomb's law).
2. There are no single, isolated **magnetic poles** (if there is a north, there will also be an equivalent south pole).
3. Electric currents can cause magnetic fields.
4. Changing **magnetic fields** can cause electrical currents.

For many physicists these equations constituted a vision of great beauty. **Ludwig Boltzmann** was led to quote the writer Johann Goethe when he said "*was it God who wrote these lines …?*" Maxwell demonstrated that electrical and magnetic effects were distinct manifestations of a single electromagnetic force, thus unifying electricity and magnetism under the electromagnetic field, and showed that light is a form of electromagnetic radiation. He describes it as:

"*An electromagnetic disturbance in the form of waves propagated through the electromagnetic field according to electromagnetic laws.*"

("A Dynamical Theory of the Electromagnetic Field", 1864)

Shortly after this, other forms of electromagnetic radiation were discovered and a whole electronic **spectrum** encompassing such phenomena as visible light, radio waves, and **X-rays** was conceived. These electromagnetic waves were all conceived of as disturbances in the electromagnetic field and were characterized by their wavelength.

Color vision and optics

Maxwell was a pioneer in the study of color vision and **optics**. He created the science of quantitative colorimetry, proving that mixtures of up to three spectral stimuli may match all colors: combinations of the primary pigments, red, yellow, and blue, yield any desired hue when mixed together in various combinations. Among many contributions to physiological optics, he

■ Before Maxwell, electricity and magnetism were understood in terms of particles exerting forces upon one another. After Maxwell, electricity and magnetism were understood in terms of space-filling fields as defined by the Maxwell equations. Maxwell's equations are employed today in countless ways, including electrical power generation, television, radio, radar, and particle physics.

■ Areas of the planet Venus were named after Maxwell because his electromagnetic theories led to the invention of the radio telescope that first enabled pictures of those areas to be taken.

■ On the basis of Maxwell's theory of light, **Albert Einstein** was able to determine the connection between energy, mass, and the speed of light, which is the fundamental structure underpinning his **theory of relativity**. Einstein was a huge admirer of Maxwell, claiming that one scientific epoch ended and another began with Maxwell.

■ He demonstrated the principle by which we see colors and produced the world's first color photograph.

■ He introduced statistical methods into physics, which have become standards of the discipline.

■ His *On Governors* is frequently cited as an early example of **control theory**.

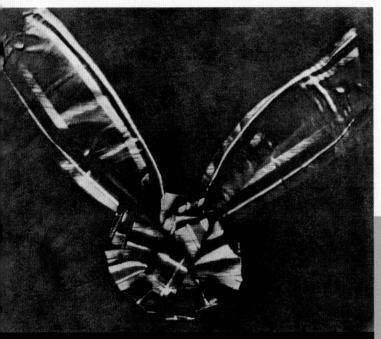

This picture of a tartan ribbon, presented by James Clerk Maxwell in 1861, is considered to be the world's first color photograph. It was made using red, green, and blue filters.

> **To know absolutely where we are, and in what direction we are going, are essential elements of our knowledge as conscious beings.**
> Matter and Motion (1876)

demonstrated that color blindness is due to the ineffectiveness in one or more receptors and, in 1861, he projected the first color photograph, using red, green and blue filters, in front of an audience at the Royal Institution that included Faraday.

Maxwell's demon

Maxwell's demon was the first effective scientific **thought experiment**. Introduced in *A Theory of Heat* (1871), Maxwell's demon is a hypothetical molecule-sized creature who can make molecules flow from one container to another, thereby showing that the second law of thermodynamics is only probabilistic and not necessarily true. The second law of **thermodynamics** states that two bodies of equal temperature in an isolated system will not arrive at a state where the temperature of one is significantly higher than the other. But if Maxwell's demon is able to release one of the faster-than-average molecules from one chamber into another, this will significantly increase the temperature of the recipient container, thus contravening the second law of dynamics. Many have interpreted Maxwell's demon as a rejection of "Laplace's demon" (1814), which was an argument for **determinism**.

Key dates

1831	Born in Edinburgh, Scotland.
1850	Enters the University of Cambridge, England.
1854	Comes second in the Mathematical Tripos at Cambridge.
1855	Gains Fellowship at Trinity College, Cambridge. Publishes *On Faraday's Lines of Force*.
1856	Becomes professor of natural philosophy at Marischal College, Aberdeen.
1858	Marries Katherine Mary Dewar.
1860	Appointed professor of natural philosophy at Kings College, London.
1861	Produces the first color photograph.
1861–2	Publishes "On Physical Lines of Force" in four parts in the *Philosophical Magazine*.
1864	Reads "A Dynamical Theory of the Electromagnetic Field" to the **Royal Society**.
1865–71	Retires to Glenlair, Scotland.
1868	Publishes *On Governers*.
1871	Returns to Cambridge to organize the new Cavendish Laboratory. Publishes *A Theory of Heat*.
1873	Publishes *A Treatise on Electricity and Magnetism*.
1874	Completes work on the Cavendish Laboratory.
1876	Publishes *Matter and Motion*.
1879	Dies in Cambridge from abdominal cancer.

Alfred Bernhard Nobel

Alfred Bernard Nobel was a Swedish chemist and industrialist who pioneered work in explosives, most notably inventing dynamite, and was at the forefront of the explosives industry, owning many factories worldwide. He is best known for founding the Nobel Prizes, the annual awards for services to mankind in a variety of disciplines.

Born in Stockholm, Sweden, in 1833, Alfred Bernhard Nobel descended from a line of ancestors that already boasted some very talented individuals. His father, Immanuel Nobel the younger, was a builder, industrialist, and inventor, who invented plywood, and his great-great-great-grandfather, Olaus Rudbeck (1630–1702), was one of the most important Swedish scientists of the seventeenth century.

After Nobel attended Saint Jakob's Higher Apologist School, Stockholm, he and his family moved to Saint Petersburg, where Nobel and his brothers were privately tutored. In 1850 Nobel embarked on a two-year trip to Germany, France, Italy, and North America to further his studies, particularly in science and languages.

During the Crimean War (1853–6), he worked at his father's firm, a torpedo works, in Saint Petersburg before the family returned to Sweden in 1859 when the business went bankrupt. Both Nobel and his father were very interested in the study of nitroglycerin, an oily, colorless, explosive substance, and worked on it independently over the following years. In 1863 Nobel's first breakthrough came when

he invented the Nobel patent detonator, a blasting cap using a strong shock rather than heat combustion to ignite the nitroglycerin. By 1865 Nobel was able to open the first of what would ultimately be many factories producing nitroglycerin. The growth of a worldwide empire marked Nobel's ascension as a renowned inventor and industrialist.

After a long period studying the properties of nitroglycerin, often gaining further knowledge as a result of accidents in his factories, Nobel patented dynamite in 1867 in Sweden, England, and the Unites States. He continued to work on dynamite in the subsequent years and, by 1875, had produced an improvement on his initial invention. One advantage of dynamite over the volatile liquid nitroglycerin was that in the form of a malleable paste dynamite could be employed, among other things, to insert into the drilling holes used for mining.

With both his patents on explosives and his ability as an industrialist, Nobel amassed a large fortune, and by the time he

Essential science

The invention and production of dynamite

Nobel was the first scientist to transform nitroglycerin, discovered originally in 1847 by Ascanio Sobrero (1812–88), into a useful explosive. His dynamite was a solid and ductile explosive consisting of nitroglycerin absorbed by *kieselguhr*, a very porous diatomite (a chalky sedimentary rock). This dynamite, known as guhr dynamite, had several weaknesses and Nobel continued to develop it.

In 1875 he created blasting gelatin, a colloidal solution of **nitrocellulose** (also known as guncotton) in nitroglycerin, which proved to be a better explosive. It possessed greater force than pure nitroglycerin, was less sensitive to shock, and was strongly resistant to moisture and water. Known variously as Nobel's extra dynamite, express dynamite, blasting gelatin, saxonite, and gelignite, it was soon put into production in Nobel's many factories.

Although most famous for dynamite, Nobel also produced other explosives. In 1863 he invented the Nobel patent detonator, which introduced the "initial ignition principle" into the

technique of blasting. In 1879 he invented the nearly smokeless blasting power known as ballistite or Nobel's blasting powder, a mixture of nitroglycerin, nitrocellulose, and 10 per cent camphor (a clear waxy solid).

The Nobel Prizes

In his will, Nobel laid out instructions for awarding a set of annual prizes to individuals who had served mankind. The Nobel Prizes were to be awarded in the fields of physical science, chemistry, physiology or medicine, literature, and peace. The Nobel Foundation would be the owner of the fund capital, from which the recipients received their prize money.

In addition, it was stipulated that various institutions – some of which had not been consulted prior to the opening of the will – would award the prizes. The Royal Swedish Academy of Science would award the prizes for physics and chemistry; the Royal Caroline Medical Institute (Karolinska Institute) would award the prizes for medicine or physiology; the Swedish Academy would award the prize for literature; and a committee of five from the

Detail of Nobel's application for a patent regarding the "principles for initial ignition of nitroglycerine", 1867.

Legacy, truth, consequence

- The Nobel Prizes have grown in stature over the years and been awarded to some of the finest minds in the world. They are widely regarded as the most prestigious awards in each of the respective categories.
- Having patented dynamite in 1867, Nobel was able to corner the market in explosives and the production of dynamite. The fortune he amassed has been used for purposes intended to benefit mankind.
- As well as his research and inventions within the field of explosives, Nobel also contributed work in electro-chemistry, optics, biology, physiology, and various methods of telecommunication.

The whole of my remaining realizable estate shall be dealt with in the following way: the capital ... shall constitute a fund, the interest on which shall be annually distributed in the form of prizes to those who, during the preceding year, shall have conferred the greatest benefit on mankind.

From Alfred Nobel's will (1895)

died he was a multi-millionaire. In his will he left his wealth to a foundation that would award annual prizes to those who had benefited mankind.

Although he never married, he had an 18-year relationship with a Viennese girl named Sophie Hess, who was 23 years his junior, although it was over by the time he died. Despite his riches, throughout his life he suffered from loneliness and melancholia. Always traveling, he described himself as "*the wealthiest vagabond in Europe*" and claimed "*My home is where I work – and I work everywhere.*"

Norwegian parliament would award the prize for peace. (Note that the prize for economic sciences is a separate entity established later by the Swedish Riksbank.)

The Nobel Prizes are still awarded to this day. Prize presentation takes place annually on December 10, the anniversary of Nobel's death. Recipients receive prize money, the Nobel gold medals, and diplomas. The only obligation on a laureate is that the recipient must deliver a Nobel lecture, which is then published in the Nobel Foundation's annual publication "Les Prix Nobel".

All the prizes stemmed from Nobel's hopes for the future of mankind. The scientific prizes reflected his own professional interests and the prize for literature was a response to his lifelong interest in the subject and his own efforts to write poetry and prose. However, he was not solely responsible for the introduction of the peace prize. Nobel had a lasting friendship with Baroness Bertha von Suttner (1843–1914), a pioneer in the peace movement. Although Nobel himself was a pacifist who abhorred wars between nations, von Suttner is thought to have played a role in the foundation of the peace prize.

Key dates

1833	Born in Stockholm, Sweden.
1841–2	Attends Saint Jakob's Higher Apologist School, Stockholm.
1842	Nobel family moves to Russia.
1843–50	Tutored privately in Saint Petersburg.
1850	Embarks on trip to Germany, France, Italy, and North America.
1853	Goes to work for his father in Saint Petersburg.
1863	Invents the Nobel patent detonator.
1865	Opens his first explosive factory.
1867	Patents dynamite in Sweden, England, and the United States. Conducts demonstrations of his explosives in Redhill, England.
1875	Invents blasting gelatin and puts it into factory production. Publishes *On Modern Blasting Agents*.
1879	Invents the smokeless blasting powder ballistite.
1895	Signs his will, outlining the establishment of the Nobel Prizes.
1896	Dies in San Remo, Italy.

Robert Koch

The German physician and bacteriologist Robert Koch is considered to be one of the founders of microbiology. He devised techniques for culturing bacteria outside the body and he was able to isolate the causative organisms for a number of diseases, including anthrax, tuberculosis, and cholera. In 1905 Koch was awarded the Nobel Prize for Physiology or Medicine for his research on tuberculosis.

Robert Koch was born in Clausthal, Germany, in 1843. The son of a mining engineer, he was one of 13 children. He reportedly taught himself to read with the help of a newspaper at the age of five years, before attending his local school. He went on to study medicine at the University of Göttingen and obtained his degree in 1866. Having worked as a general practitioner and as a field surgeon during the Franco-Prussian war, Koch became a district medical officer in Wollstein (now Wolstyn, Poland), where he also started his research on bacteria.

One of the first diseases he studied was anthrax and, in 1876, he became the first person to conclusively demonstrate that a living microorganism, *bacillus anthracis*, was the causative agent of an infectious disease. In 1880 he was appointed to the Imperial Health Office in Berlin, where he continued to develop his methodology and techniques for bacterial examination.

Koch is probably most famous for his research on tuberculosis, for which he was later awarded the Nobel Prize. In 1882 he announced that he had isolated and grown the tubercle bacillus (a bacterium from a lesion of tuberculosis), and he demonstrated that this was the causative agent of tuberculosis.

Koch then went on to investigate cholera outbreaks in Egypt and India, and was again able to determine the causative agent, *vibrio cholerae*. In 1885 he was appointed professor of hygiene at the University of Berlin. Six years later he became the director of the new Institute for Infectious Diseases, which had been founded for him, and which still bears his name today.

Many countries sought Koch's advice on infectious diseases and he spent much of the later years of his career traveling. He visited India and Italy, as well as countries in Africa, where he investigated a number of diseases, including cattle pest, malaria, and plague.

On his death in 1910 he was cremated and his ashes were buried in a mausoleum that had been erected for him at his institute.

Essential science

Anthrax

Koch carried out his initial research on anthrax at his home in Wollstein, at a time when anthrax was a highly prevalent disease among cattle. Although the anthrax bacillus had already been described by the French physician Casimir Davaine (1812–82), no advances had been made in the prevention and treatment of the disease and researchers were unable to explain the fact that cattle not only contracted the disease from other infected cattle, but also from grazing in pastures where infected animals had been kept years earlier. Koch was able to isolate and culture the bacilli and to observe their entire life cycle. He noted that the bacilli formed resistant **endospores** when conditions around them were unfavorable (for example, due to a lack of oxygen). These spores were able to remain dormant for long periods and then, under the right conditions, give rise to bacilli again. This explained the recurrence of the disease in pastures that had been unused for grazing for a number of years.

Tuberculosis

When Koch started his work on tuberculosis, researchers already suspected that the disease was caused by an infectious agent. The causative organism, however, had not yet been isolated and identified. In 1882 Koch announced his discovery of the tubercle bacillus in a lecture in Berlin. He was able to isolate and culture the organism – a remarkable achievement in itself, because of the microorganism's fastidious food requirements and slow growth. Koch demonstrated that the organism was the causative agent of tuberculosis. Once the bacillus had been identified, finding a vaccine or treatment became a priority. Koch was able to extract a protein component, which he called tuberculin, from cultures of tubercle bacilli. After experimentation he concluded that he might have developed a cure for the disease, but tuberculin proved to be ineffective in the treatment of tuberculosis. It did, however, turn out to be a valuable diagnostic tool.

Legacy, truth, consequence

■ Koch developed and perfected various methods and techniques of bacteriological research. **Louis Pasteur** initially introduced the idea that a microorganism might be cultured outside the body, but it was Koch who perfected the technique of pure culture that is necessary for doing so. With this technique, a sample containing many different species of microorganisms is manipulated to spread and dilute the cells on the surface of a culture medium. The aim is to eventually obtain a laboratory culture containing only a single species of microorganism. Koch also developed new staining techniques, making bacteria more visible and easier to identify. His slide technique still forms the basis of routine laboratory examination of bacteria, and his methods and techniques have laid an important foundation for bacteriological research.

■ Although tuberculin did not have the healing properties that Koch had anticipated, a tuberculin derivative is still used today in the diagnostic skin test performed to identify a tuberculosis infection.

■ Koch recognized the importance of hygiene and perfected methods for disinfection and sterilization. These concepts are not only important in laboratory research, but are also essential when trying to limit the spread of infectious diseases.

> *In the future the battle against this plague of mankind [tuberculosis] will not just be concerned with an uncertain something but with a tangible parasite, about whose characteristics a great deal is known and can be explored.*

Robert Koch (1882), quoted in *The Indian Journal of Tuberculosis* (2001)

Cholera

In 1883 and 1884 Koch investigated cholera outbreaks in Egypt and India. He managed to identify the causative organism, *vibrio cholerae*, and its transmission via drinking water. On the basis of these findings, he was able to formulate rules for the control of cholera epidemics. He also greatly influenced plans for the conservation of water supplies.

Koch's postulates

Together with his colleagues, Koch determined the following principles used to establish the causal relationship between a particular microorganism and a disease: the microorganism has to be present in all cases of the disease; the microorganism can be isolated and grown in pure culture; the cultured microorganism produces the disease when transferred to a healthy animal or human host; the microorganism can then be isolated from the newly infected host. These four criteria are usually referred to as Koch's postulates.

Key dates

1843	Born in Clausthal, Germany.
1862–6	Studies medicine at the University of Göttingen.
1867	Marries Emmy Fraatz. Starts work as a general practitioner.
1868	His daughter Gertrud is born.
1870–1	Serves as a field surgeon in the Franco–Prussian war.
1872–80	Works as district medical officer in Wollstein, where he begins his bacteriological research.
1876	Demonstrates for the first time that a microorganism (anthrax bacillus) is the causative agent of an infectious disease.
1880	Appointed member of the Imperial Health Office in Berlin.
1882	Announces his discovery of the tubercle bacillus.
1883	Undertakes his first expeditions to Egypt and India to investigate outbreaks of cholera. Discovers the causative agent of cholera.
1885	Appointed director of the newly founded Institute for Hygiene in Berlin.
1890	Discovers tuberculin.
1891	Appointed director of the Institute for Infectious diseases in Berlin.
1893	Divorces Emmy Fraatz and marries Hedwig Freiberg.
1896	Begins to study several tropical diseases, including malaria, sleeping sickness, and cattle pest.
1905	Awarded the Nobel Prize for his research and achievements in the field of tuberculosis.
1910	Dies in Baden-Baden, Germany.

A photomicrograph of *bacillus anthracis*, the bacteria Koch showed to be the cause of anthrax.

Ludwig Boltzmann

An Austrian physicist, Ludwig Boltzmann is most noted as a pioneer of quantum mechanics, specifically in the fields of statistical thermodynamics and statistical mechanics. His ideas of atomic theory were so controversial that, despite an eminent professional academic career, it wasn't until after his death – by his own hand – that their true worth were accepted.

Ludwig Boltzmann was born into a secure middle-class family in Vienna, then the capital of the Austrian Empire. His parents, in particular his mother, offered every encouragement to the young and eager Ludwig in pursing his chosen study, at that time a general interest in the world of nature. When he was 15, however, he had to cope with the death of his father. He became prone to bouts of depression, though it is unlikely that his father's death was the original catalyst for his condition.

He received his doctorate from the University of Vienna, where his chief mentor was the much revered physicist, Josef Stefan (1835–93), whose studies in **radiation** inspired Boltzmann to continue the work himself when he became a professor. Boltzmann was itinerant in his professorships, holding positions in mathematics and physics at the universities of Vienna, Graz, Munich, and Leipzig in his lifetime. He also found time to work in Heidelberg and Berlin. He once jestingly put this restlessness down to being born during a Mardi Gras ball.

Boltzmann was one of the few physicists at the time who supported the **atomic theory** pioneered by **John Dalton**, and the view that **atoms** and **molecules** were part of reality and not merely theoretical constructs. In his thirties he published a succession of papers outlining his hypothesis that the second law of **thermodynamics** (stating that physical systems involving energy exchange tend to drive irreversibly towards a state of disorder – a process known as **entropy**) could be better illustrated by applying the theory of **probability** and the laws of **mechanics** to the movements of the atoms. He clarified that the second law was basically statistical, and after further research on the distribution of energy this led to his part in the conception of the Maxwell–Boltzmann distribution law (see below).

His other main contribution to statistical mechanics was the Boltzmann equation. This formula was based on his conclusion that the entropy of a system (its level of disorder) is in proportion to the probability of the configuration of its integral particles.

In 1876 Boltzmann married a fellow teacher of mathematics and physics called Henriette von Aigentler. They met when he successfully advised Henriette in her battle to be permitted to scrutinize lectures in Graz at a time when women were not admitted to Austrian universities. They had two sons and three daughters.

Boltzmann was honored at home and abroad during his lifetime, but opposition to some of his theories was fierce from some quarters, most notably from a fellow professor in Vienna, Ernst Mach (1838–1916), with whom he also had personal differences. Allied to these difficulties, his lifelong depression deepened and, 30 years after his marriage, Boltzmann hung himself during a family holiday in Italy.

Essential science

The Maxwell–Boltzmann distribution law

On completing his studies in Vienna, and with a successful doctorate on the **kinetic** (motion) theory of gases to his credit, Boltzmann was encouraged by Josef Stefan to familiarize himself with the **electromagnetism** work of the English physicist, **James Clerk Maxwell**. Boltzmann improved his English and followed his mentor's advice. Working separately from each other they derived the Maxwell–Boltzmann distribution law: first presented by Maxwell in 1859, it was generalized by Boltzmann in 1871.

Starting from basic principles, the temperature of any physical system is the result of the motions of the molecules and atoms that make up the system. These particles have a range of different velocities, and the velocity of any single particle constantly changes due to its collisions with other particles. Since all molecules of a particular chemical, element, or compound have the same mass, their kinetic (motion) energy is entirely dependent on the velocity of the particles. The Maxwell–Boltzmann distribution law shows how, at a certain temperature, the speeds (and hence the energies) of the moving particles of a mixture vary and how the average motion energy of any given molecule is the same in each different direction. This important principle of science, describing how energy is distributed in a system, is particularly useful in statistical mechanics, the branch of physics that applies statistical principles to the mechanical behavior of large numbers of small particles (such as molecules or atoms) in order to explain the overall properties of the matter composed of such particles.

Legacy, truth, consequence

■ It was the older, established scientists who were the main opponents to Boltzmann's statistical mechanics theories – many of the younger mathematicians championed him. Atomic theory was still highly controversial at the time and one reason for their attitude was that they feared a public outcry and condemnation of physical science, as the crux of Boltzmann's research, and atomic theory in general, seemed to deny that God existed. However, the studies in atomic physics that were carried out around the time of Boltzmann's death were eventually responsible for vindicating his misinterpreted theories. It was finally accepted that the kinetic energy in gases could only be explained by statistical mechanics.

■ In his lectures on natural philosophy, Boltzmann anticipated aspects of the theory of **special relativity**, specifically the equal treatment of space coordinates and time.

■ In 2006, one hundred years after Boltzmann's death, a conference was held in his honor to recognize and discuss his contributions to **thermodynamics**, statistical mechanics, and kinetic energy. It highlighted the many recent developments in mathematics and physics that were influenced by Boltzmann's work.

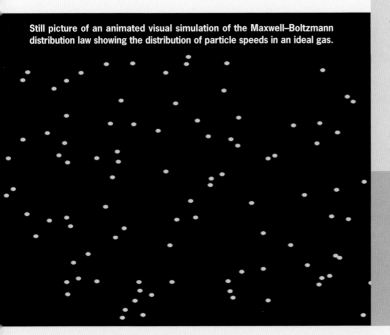

Still picture of an animated visual simulation of the Maxwell–Boltzmann distribution law showing the distribution of particle speeds in an ideal gas.

I see no reason why energy shouldn't also be regarded as divided atomically.

Comment to Wilhelm Ostwald and Max Planck
at the Halle Conference (1891)

The Boltzmann equation

The famous equation named after Boltzmann was developed to describe the dynamics of an **ideal gas** and is applied generally to determine the entropy of a system. It is the logarithmic connection between the entropy and the probability of the configuration of its integral particles. Written as "$S = k \log W$", where k is Boltzmann's constant, it connects entropy (S) with molecular structure (W). Although he was first to make this link, the equation itself was written down by **Max Planck**. Boltzmann, according to Planck, *"never gave thought to the possibility of carrying out an exact measurement of the constant"*. Nevertheless the equation is inscribed on Boltzmann's tombstone.

Evolution

Later in life, Boltzmann came to admire **Charles Darwin**'s ideas. He wanted to extend the theory of biological evolution to cultural evolution, which he saw also as a physical process but taking place in the brain. For Boltzmann, the development of living organisms represented a perfect example of thermodynamics. As he said in 1900:

"The overall struggle for existence of living beings is therefore not a struggle for raw materials – the raw materials of all organisms are available in excess in the air, water, and ground – nor for energy, which in the form of heat is plentiful in every body, but rather a struggle for entropy, which becomes available in the flow of energy from the hot sun to the cold earth."

Wilhelm Conrad Röntgen

Wilhelm Conrad Röntgen was a German physicist who is most famous for his discovery of X-rays while experimenting in his laboratory. For this achievement, he became the first person to be awarded the Nobel Prize for Physics.

Born in 1845 in Lennep, Prussia (present-day Germany), Röntgen, or Roentgen in English, was the son of a cloth manufacturer and merchant. After receiving his early education in Holland he moved on to Zurich Polytechnic, Switzerland, in 1866. Although not particularly studious at school, preferring outdoor activities, at Zurich he diligently applied himself to mechanical engineering, for which he received a diploma and a doctorate in the late 1860s.

Röntgen worked as an assistant to August Kundt (1839–94), then professor of physics at Zurich, and began to focus on physics. Eventually, after a brief stint teaching at Hohenheim and Strasbourg, Röntgen became professor of physics at the University of Giessen, a position he kept for nearly ten years, and later professor of physics at Würzburg, eventually becoming director of the Physical Institute there.

In November 1895, partly by chance, Röntgen made his momentous discovery, observing a mysterious form of **radiation**, which he called **X-rays**. Röntgen was so astonished and puzzled by his discovery that he was unable to tell anyone all day; even his wife wondered what was wrong. However, after fixing his findings on metallic plates, he was certain of his achievement.

He made the news public early the following year and later demonstrated it to the kaiser. There was an immediate reaction, both inside and outside academia. Among his fellow scientists, he correspondended with **Ludwig Boltzmann**, Friedrich Kohlrausch (1840–1910), Lord Kelvin (1824–1907), and Henri Poincaré (1854–1912), most of whom ran to their laboratories to repeat Röntgen's experiment. X-ray equipment was developed and installed in hospitals throughout Europe and the US. As a man of great personal integrity, Röntgen refused to benefit financially from his discovery, believing that it should be made freely available to all. His achievement was greeted by a standing ovation at the Physical–Medical Society. News spread around the world, reporters came to visit him, and he was invited to lecture in many prestigious universities. But he declined all offers, preferring to continue to study X-rays. To his concern, however, this left him unable to address the spread of rumors about the possible harmfulness of X-rays. A congressman in New Jersey, US, tried to pass a law to forbid the future manufacture of X-ray glasses to see through people, and publications in the media included this poem:

> The Röntgen Rays, the Röntgen Rays
> What is this craze
> The town's ablaze
> With the new phase
> Of X-ray ways.

Essential science

The discovery of X-rays

In 1895 Röntgen was investigating the properties of **cathode rays** emitted by high-vacuum **discharge tubes**. While experimenting on a Crookes tube, a type of discharge tube, Röntgen discovered that a screen coated with a thin layer of the chemical barium platinocyanide became **fluorescent** when the tube was operating, emitting light. Objects placed between the tube and the screen cast shadows that could be recorded on a photographic plate. The denser the object, the darker the image, so when a hand was positioned there, the bones cast darker shadows than the flesh.

Moving the screen to an adjacent room, Röntgen found that it still produced a luminescent glow when the tube was activated. This high power led Röntgen to conclude that the radiation was entirely different from cathode rays. Since he was unable to establish the exact nature of the newly discovered rays that came from the glass walls of the tube, he called them "X-rays". Before announcing his major discovery he conducted further experiments, establishing that X-rays pass unchanged through cardboard and thin plates of metal, travel in a straight line, and are not deflected by electric or magnetic fields.

Röntgen publicly announced his discovery early the following year, illustrating his lecture with an X-ray photograph of a man's hand. Although it is likely that X-rays had been produced before Röntgen, because experiments with cathode rays had been performed for many years, it was Röntgen who first noticed and highlighted their existence and investigated their properties.

Röntgen discovering, partly by chance, a mysterious form of radiation.

- X-rays were used in Vienna hospitals within only a few weeks of Röntgen's discovery and were soon made widely available for medical purposes. They have transformed medical science. More uses for X-rays were also found, such as **metallography**, **crystallography**, detection of hidden objects, and advances in **atomic physics**.

- The discovery of the X-ray served as an impetus to other scientists of the period. In particular, in the same year it led **Henri Becquerel** to the discovery of **radioactivity**, which in turn led to much research on the **atom** and atomic structure.

- A Röntgen or Roentgen (r) is now the international measurement of X-rays or gamma-rays.

I'm full of daze
Shock and amaze
For nowadays
I hear they'll gaze
Through cloak and gown – and even stays
These naughty, naughty, Röntgen Rays.

Despite the scepticism, shortly after taking up a position at the University of Munich, Röntgen was awarded the first ever Nobel Prize for Physics in 1901.

During World War I, Röntgen was persuaded to sign the proclamation of the "ninety-three intellectuals", signatories to the famous "appeal to the civilized world" to believe that Germany had not caused the war. However, he later came to regret this decision.

Röntgen retired in 1920, a year after his wife died following a long illness, and spent much of his time in the library and enjoying walks in the Bavarian mountains near his country house in Weilheim, near Munich. He died at the age of 78, sadly impoverished following the great inflation in Germany and suffering from ill health due to prolonged overexposure to X-rays.

Electricity

In 1888 Röntgen made an important contribution to the scientific understanding of electricity. He showed that a convection current is the same as a conduction current; that the current obtained by moving charges is the same as the current in a wire. This was a major discovery because it constituted further support for **Michael Faraday**'s claim that there is only one kind of electricity.

Other work

As well as his famous discovery of the X-ray, Röntgen also worked on elasticity, heat conduction in crystals, specific heat capacities of gases, and the rotation of plane-polarized heat. He wrote 58 papers throughout his career, some with collaborators.

> *For brevity's sake I shall use the expression "rays"; and to distinguish them from others of this name I shall call them "X-rays".*
>
> "On a New Kind of Rays" (1895)

Key dates

1845	Born in Lennep (now part of Remscheid), Prussia (modern Germany).
1866	Attends Zurich Polytechnic.
1868	Receives diploma in mechanical engineering.
1869	Receives doctorate in mechanical engineering.
1871	Moves to Würzburg as Kundt's assistant.
1872	Marries Anna Bertha Ludwig. Moves to Strasbourg as Kundt's assistant.
1875	Appointed professor of physics and mathematics, Agricultural Academy of Hohenheim.
1876	Teaches physics at Strasbourg University.
1879–88	Professor of physics at Giessen.
1888	Appointed professor of physics at Würzburg. Confirms that magnetic effects are produced by the motion of electrostatic charges.
1895	Discovers X-rays. "On a New Kind of Rays" is published by the Physical-Medical Society of Würzburg.
1896	Publicly announces his discovery of X-rays. A translation of "On a New Kind of Rays" is published in *Nature*.
1900	Appointed professor of physics and director of the Physical Institute, Munich.
1901	Awarded the Nobel Prize for Physics.
1923	Dies in Munich, Germany.

Thomas Edison

A prolific inventor of unsurpassed productivity, Thomas Edison is responsible for many of the commodities that the modern world takes for granted, from the electric light bulb to the film industry. He worked in self-financed laboratories with a small team of co-workers, often discovering momentous principles by chance as a consequence of his haphazard, but unflinchingly positive approach to experimentation. He never invented anything unless it was necessary.

Edison was born in Milan, in the American state of Ohio. Like many great scientists before him, he had a strong sense of curiosity from a young age. He devoured books on chemistry and gave an indication of his practical and hard-working approach to life when, aged only ten, he grew and sold vegetables in order to finance chemical experiments in his cellar. By the age of 15 he was printing his own newspaper on a printing press located in the luggage wagon of a train.

In a developing country as vast as America the telegraph, as well as the train, played a hugely important role in the growth of the country's economy. To learn about telegraphy, Edison worked for a number of telegraph offices, finally settling in a job in New York. Within a year he had set up his first workshop.

Edison was a fervent experimenter. He fed this passion by establishing two laboratories – firstly for eight highly successful years at Menlo Park in New Jersey and then at the grander, but less productive, Edison laboratory in New Jersey.

One invention would lead to another. His phonograph (a device for recording messages), for example, needed a power source to operate, yet most homes didn't have electricity, so he began developing the alkaline storage battery. When he subsequently became an automobile enthusiast, he saw this kind of battery as the most likely to power cars. By the time it was developed, gasoline had superseded it, so the battery was used for other things such as train signaling and for lighting miners' lamps and submarines.

Edison's lack of good managerial or organizational skills actually instigated a fearless and open-minded approach to work that insulated him from the disenchantment of failure. He didn't believe that he was particularly clever – he was an advocate of hard graft. He begrudged wasteful tasks such as eating and resting that would impinge on his work time and was known to sleep fully dressed on his workshop tables.

He was married twice and had six children. The first two were nicknamed "Dot" and "Dash" after the telegraph code terms.

Essential science

Telegraphy

Telegraphy, the communication system involving the transmission of electric signals through wires that translated into a message, was developed in the nineteenth century, most successfully by Samuel Morse (1791–1872). Morse's system sent pulses of current through a wire, which deflected an **electromagnet**, which in turn embossed a strip of paper with dots and dashes – known as Morse code. Edison's contribution was to develop a printer that turned the electric signals into printed letters. He also patented a duplex telegraph (allowing communication in both directions), which was able to send up to two messages together over the same wire without getting them mixed up. Later he put two of these machines together to create the appropriately named quadruplex telegraph.

Carbon-button transmitter

Between 1877 and 1878 Edison invented a carbon-button transmitter or microphone that would prove to be a crucial component in the next development of telegraphy – the "speaking telegraph" or telephone – along with Alexander Graham Bell's

receiver. Edison's device used a diaphragm to convert sound to electrical signals and consisted of two metal plates connected by an electric current and separated by granules of carbon. One plate acting as a diaphragm would vibrate in response to a speaker's voice, which would change the pressure on the granules, and this in turn changed the electrical resistance between the plates (a higher pressure would push the granules closer together and lower the resistance). The changing resistance resulted in a changing electrical current between the plates, which could then be fed into a telephone system as an electrical signal.

The phonograph

Edison saw that the new invention of the telephone had a major drawback: the transmission of messages was too rapid for people to write down what was being said. His answer was the invention of the phonograph: a device that could record and play back a vocal message. He came to this idea after playing the tape of a telegraph transmitter at high speed and hearing a sound similar to a human voice. The idea that the paper tape moving through the machine could produce a

Legacy, truth, consequence

- Principles discovered by the "Edison effect" led to the development of the **electron tube** and form the bedrock on which the electronics industry is laid.
- Edison's carbon-button transmitters used in telephones until recently have only become less common since the growth of cordless phones.
- He set up the first industrial research laboratory, in Menlo Park, New Jersey, US, employing a team of researchers under his direction with the purpose of developing new and improved technology.
- 1,093 patents are recorded in his name. As a compulsive experimentalist, not everything that Edison touched was liable to turn into gold. One of his ideas was to make things out of cement – from cupboards and pianos to houses. It didn't catch on quite as he had hoped, but he did receive a contract to build the New York Yankees' stadium with cement.

Key dates

1847 Born in Milan, Ohio, US.
1869 Moves to New York. Receives first patent.
1870 Opens first workshop in Newark, New Jersey.
1876 Moves to his Menlo Park laboratory.
1877 Invents the phonograph.
1882 Opens a commercial electric station in New York.
1883 The electric light bulb is patented.
1888 Meets Eadweard Muybridge (1830–1904), an expert in photographic motion analysis. Develops an interest in the moving picture, which culminates in the founding of the motion picture industry.
1889 Edison General Electric Company is formed. Develops talking dolls.
1893 Completes construction of his first film studio, called the Kinetographic Theater.
1896 Forms National Phonograph Company.
1913 Launches the Kinetophone which attempts to synchronize phonographic cylinder records with moving pictures.
1928 Awarded the Congressional Gold Medal.
1931 Dies in West Orange, New Jersey, US. During his funeral the American public dim their lights for one minute.

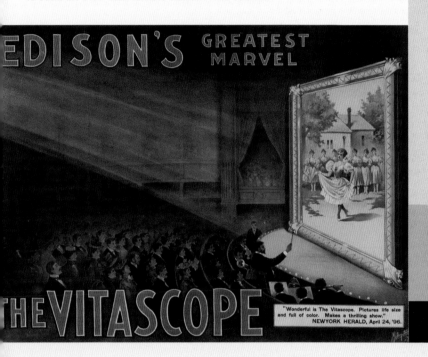

What man's mind can create, man's character can control.
Newspaper interview (1921)

An advertisement for one of Edison's inventions, c.1896.

---

noise resembling spoken words made him think that by combining the technologies of the telephone and the telegraph he might succeed in both recording and playing back a spoken message. First he attached a stylus taken from a telegraph machine to the diaphragm in a telephone receiver (the mouthpiece), with the loose end of the stylus placed so that it would indent a sheet of paper as the diaphragm vibrated. These indentations could then be played back by another stylus and diaphragm unit, reproducing the original recorded vibrations, or sound. He soon replaced the paper with tinfoil wrapped around a rotating cylinder, and achieved a clear result: after speaking into the machine he could rotate the cylinder to play back his recorded message. Thus it was that, with some astonishment, he heard his own voice reciting, "*Mary had a little lamb*".

Electric light bulb

A problem for previous experimenters had been the overheating and disintegration of the bulb itself. Edison intended not only to sort that out, but ultimately to produce an incandescent light that was safe, practical, and cheap – something that could be used in the home. A year and a half of experimentation, characterized by Edison's trial and error approach, produced a bulb in which a reduced current of electricity, operating in a more efficient vacuum, flowed from a fine, carbonized wire filament to a plate fixed inside, and which, crucially, burned for over 13 hours.

By the time the bulb was patented, in 1883, it was noted that in a vacuum bulb the wire and the bulb itself blackened at the negative pole, but at the positive pole a blue glow was observed. This became known as the "Edison effect".

Electric light distribution

The invention of the light bulb necessitated further electrical development, the most significant being Edison's seven-point program for electricity distribution, the components of which are the parallel circuit, a durable light bulb, an improved dynamo, the underground conductor network, devices for maintaining constant voltage, safety fuses and insulating materials, and light sockets with on-off switches. Each of these components had to be separately invented and developed into producible units.

Ivan Petrovich Pavlov

A Russian physiologist, surgeon, and psychologist, Ivan Petrovich Pavlov is best known for his study of conditioned reflexes in dogs. His experimental work on animal and human behavior, and its links with the nervous system, provided clues to the working of the brain, and greatly influenced behavioral and learning theories.

Pavlov was born into a large and impoverished family in the small village of Ryazan in Russia. His father was the village priest, and at first Pavlov pursued the same career. He went to the church village school, and then trained as a priest at a theological seminary. But exciting new ideas from **Charles Darwin** and Russia's eminent physiologist, Ivan Sechenov (1829–1905), inspired him to change direction – from religion to science.

In 1870 he went to the University of Saint Petersburg to study natural science. There, his lasting interest in **physiology** began, as well as his drive to succeed in spite of a lack of money. He soon showed promise as a researcher, sharing an award for a treatise on the physiology of pancreatic nerves. After receiving the degree of Candidate of Natural Sciences, he went on to the Academy of Medical Surgery. From 1879 to 1890 he worked in the laboratory of a famous Russian clinician, Sergei Botkin, researching cardiac physiology and acquiring excellent surgical skills. His skills and achievements were recognized in 1890 by his appointment as professor of physiology at the newly founded Imperial Medical Academy in Leningrad (Saint Petersburg), a post he held for 34 years. His studies soon turned to the physiology of digestion, and he discovered the vital role of the nervous system. This led to his key paper on the function of the principal digestive glands, which was to become the basis of modern physiology of digestion.

Pavlov's main area of research was on the secretion of fluids from digestive glands. By experimenting on dogs he could observe the links between salivation and digestion. He found that stimuli other than food could make the dogs salivate. This led him in 1903 to the theory that salivary secretion was of a "conditional reflex nature".

Pavlov's work was widely recognized and highly acclaimed. He won the Nobel Prize for Physiology in 1904, was elected Academician of the Russian Academy of Sciences in 1907, and received an honorary doctorate from Cambridge University in 1912. Throughout Pavlov's remaining years, he continued his research, applying his theories to human psychoses. Although he was critical of the Bolshevik government in Russia (1917–22), the government continued to support him and provide resources for his research.

Essential science

Pavlov's method

In his reflex experiments, Pavlov used a method established by his former colleague, D. D. Glinskii, which involved putting fistulas (small holes) in the ducts of the salivary glands so any secretions could be collected and measured. The animal involved did not need to be anesthetized, which meant that its nervous system was not inhibited during the experiment and any link between, for example, stimulation of a sense organ and the reflex, could be monitored. Previously, dissection had been the most common method of investigating the digestive system.

Conditioned reflexes

Pavlov noted that salivation was a reflex action in a dog when it was offered meat (just as a hungry person's mouth might water at the sight of a tasty dish). He rang a metronome each time he gave food to the dog. Then he removed the food and simply rang the metronome, and he found that the dog still salivated – it had learned to respond to a "conditioned stimulus". Pavlov called this response a "conditional" reflex. In the dog, a normal unconditioned reflex (UR) had been replaced by what came to be known as a "conditioned reflex" (CR). He repeated the experiment by associating other stimuli with food, including lights, sounds, and touch, all of which produced the same newly conditioned reflex.

Further experiments showed that the conditioned responses could be "unlearned", for example if the metronome was repeatedly rung and no food was supplied to the dog, the dog would cease to salivate at its sound.

Classical conditioning

The learning of new stimuli in Pavlov's experiments is known as classical conditioning. To achieve it, before conditioning, the new "conditioned stimulus" (such as Pavlov's metronome) must attract the animal's attention but should not produce the studied reflex action (salivation). The reflex action should only occur when the original "unconditioned stimulus" is provided (such as giving food to the dog). Once this has been established,

Legacy, truth, consequence

- In measuring the amount of saliva secreted in unanesthetized animals, Pavlov was able to carry out experiments on the reflex of salivary secretion. This was a major step forward from Sechenov's former subjective and theoretical interpretation of "psychic" salivary secretion. Pavlov's systematic experimental work inspired experimental research into the functions of the nervous system in physiological institutions all around the world.

- Pavlov's understanding of the brain and its responses to stimuli from the outside environment provided a foundation for **behaviorism** and **behavioral psychology**. However, **psychiatrists** today recognize the limitations of Pavlov's work, including his reliance on conditioned reflex as an explanation of behavior. Also, advances in **neuroscience** have since provided a greater understanding of the workings of the nervous system.

- Classical conditioning has become a major technique in behavior therapy, such as aversion therapy. It is also widely used in analyzing nerve structures and mechanisms in learning and memory.

Mankind will possess incalculable advantages and extraordinary control over human behavior when the scientific investigator will be able to subject his fellow men to the same external analysis he would employ for any natural object, and when the human mind will contemplate itself not from within but from without.

Scientific Study of So-Called Psychical Processes in the Higher Animals (1906)

Key dates

1849	Born in Ryazan, Russia.
1870	Leaves his religious career to enrol in a natural science course at the University of Saint Petersburg.
1875	Receives the degree of Candidate of Natural Sciences. Continues his studies at the Military-Medical Academy.
1879	Completes a third course at the Academy of Medical Surgery and is awarded the gold medal. Becomes director of the Physiological Laboratory at S. P. Botkin's clinic.
1881	Marries Seraphima (Sara) Vasilievna Karchevskaya, a teacher and friend of Russian novelist Fyodor Dostoyevsky.
1883	Presents his thesis on "The centrifugal nerves of the heart" for his Doctor of Medicine degree.
1890	Takes the role of organizing and directing the Department of Physiology at the Institute of Experimental Medicine.
1891	Begins major research on physiology of digestion.
1895	Made chair of physiology at the Institute of Experimental Medicine.
1897	Publishes his "Lectures on the function of the principal digestive glands".
1898	Begins his work on unconditioned and conditioned reflexes.
1904	Wins the Nobel Prize for Physiology or Medicine *"in recognition for his work on the physiology of digestion"*.
1915	Awarded the Order of the Legion of Honor.
1921	Government decree, signed by Lenin, acknowledges the significance of his work.
1936	Dies in Leningrad (now Saint Petersburg), Russia.

Stuffed model of one of the dogs used by Pavlov in his researches: a container to catch the saliva was surgically implanted in the dog's muzzle.

the new conditioned stimulus is then associated with the unconditioned stimulus (the metronome is sounded at the same time as the food is given). Through association, the animal learns the new stimuli and salivates at the sound of the metronome.

Conditioned reflexes in human beings

Pavlov thought that a conditioned reflex was caused by a physiological event – the formation of a new reflexive pathway in the brain's cortex. His theory led to more investigations on the brain, and a greater understanding of how it reacts to stimuli. He tried to apply his theories to the subject of human psychoses, for example by investigating whether unpleasant stimuli could condition a person to develop a phobia. In the 1930s Pavlov proposed that conditioned reflexes play an important part in the way human beings adapt to their environment, and went on to say that human language was based on long chains of conditioned reflexes involving words.

Karl Ferdinand Braun Guglielmo Marconi

Italian physicist Guglielmo Marconi invented the first successful radio telegraphy system, and provided the foundation for long-distance radio communications. German physicist Ferdinand Braun improved Marconi's telegraphy system, and is also known as the developer of the cathode-ray tube, forerunner of the television tube.

As a boy, Marconi was fascinated by his mechanical toys, and often made gadgets of his own. However, home-educated until 13 years of age, his first formal lessons on physics and electricity were not until 1887, when he began to attend a technical school in Leghorn.

In 1894, inspired by the work of physicist **Heinrich Hertz**, Marconi first started to experiment with radio waves at his father's villa. He sent radio signals without wires across an attic room, then experimented outdoors and discovered that equipment on the ground, with high antennae, extended the range of the signals. In 1895 he successfully sent and picked up radio waves a mile away on the other side of a hill – he had invented a wireless telegraph system that was unaffected by physical obstacles.

In 1896, unable to get funding from the Italian government, Marconi went to England where he impressed the Post Office's chief engineer with his new system of telegraphy without wires. He was given the world's first wireless patent and the opportunity to develop his apparatus. After further experiments he achieved an event that made him famous – in 1901 he successfully sent wireless signals across the Atlantic Ocean. In 1909 Marconi shared the Nobel Prize in Physics for achievements in wireless telegraphy with Ferdinand Braun, another brilliant physicist.

Braun had been a precocious child and had quickly achieved academic success. After taking several teaching posts, he became director of the Physical Institute and professor of physics at the University of Strasbourg in 1895. In 1897 he developed the cathode-ray tube, the potential of which was not immediately recognized. However, his work in 1898 on improving radio wave transmitters was noticed.

At the time, Marconi had only so far achieved transmission distances of about nine miles (14 km) without using a disproportionate amount of energy. Braun solved the problem by producing a circuit that greatly improved the broadcasting range.

Essential science

Telegraphy

From the mid-1800s, telegraphy involved sending messages, usually in Morse code, down wires. The message passed along the wires in the form of pulses of electric current. As the popularity of telegraphy spread, networks of cables had to be laid at great expense. It was Marconi, and others, who pioneered a revolutionary "wireless" system that did not require cables.

Wireless telegraphy

Work on wireless radio waves began 30 years before Marconi. Marconi would have studied the work of Scottish professor **James Clerk Maxwell**, who predicted the existence of radio waves and how they could be reflected, absorbed, and focused. German scientist Heinrich Hertz who first produced and transmitted radio waves in 1887, also inspired Marconi.

Marconi's wireless system involved transmitting Morse code over long distances using radio waves that travel through the air. Radio waves, like light and infrared, are a type of **electromagnetic** radiation. Electromagnetic waves vary in wavelength, the longest of which are radio waves. In his first wireless demonstration Marconi used Hertz's spark coil as a transmitter and Edouard Branly's (1844–1940) coherer as a radio receiver (the coherer, invented, around 1890, recovered information contained in a **modulated waveform**). By gradually improving his equipment, Marconi was able to achieve longer transmission distances. Marconi's system only sent Morse code signals, and did not transmit speech. In the early 1900s, however, other inventors were developing radio telephony – through which speech could be transmitted and heard.

Cathode-ray tube

Cathode-ray tubes are vacuum tubes in which streams of **electrons** are produced. Electrons had been discovered in 1897 by Joseph John Thomson (1856-1940) and scientists noted that the beams of electrons could not be controlled. Braun, however, found a way of producing a narrow beam of electrons using alternating voltage, which could be controlled like a torch beam and focused to trace patterns on a **fluorescent** screen. The oscilloscope or "Braun tube" became an important scientific instrument, and led to the television tube.

Operators copying messages transmitted by wireless telegraphy from ships at sea, c.1912.

It was patented in 1899. He also found a way of sending waves in definite directions, which led to the development of improved radio receivers. Braun noted that patents filed by Marconi in 1900 and 1901 were very similar to Braun's earlier patent filed in 1899. Braun and Marconi met to discuss the matter and Marconi admitted that he had "borrowed" Braun's ideas.

In 1914, when World War I began, Braun moved to New York, which brought an end to his work. When the US declared war on Germany in 1917, Braun knew he was unable to return home. Marconi, however, went on to achieve many other innovations and patents. He improved the transmission and reception of long-distance radio systems, and from 1916 began to experiment with shorter wavelengths. In 1932, using very short wavelengths, his team built the first microwave telephone system.

Legacy, truth, consequence

■ Marconi's work provided the basis of modern-day wireless radio and television communications. His first transatlantic transmission in 1901 made him famous worldwide and led to the rapid development of radio communications. In 1920 transmitting stations started the world's first public broadcasts, and in 1936 the world's first television broadcasting service began.

■ Marconi was certainly a pioneer of wireless telegraphy but he admitted using Braun's and others' ideas. His patent No. 7777 was overturned in 1945 as inventors, such as Serbian-American physicist Nikola Tesla (1856–1943), appeared to have been first in developing radio-tuning apparatus.

■ Braun is sometimes considered a "forgotten pioneer of radio" and is often remembered only for his work on cathode-ray tubes. However, Braun's wireless inventions, including his crystal diode rectifier and tuning patents, were used by Marconi and played a significant part in the evolution of radio.

■ Braun's development of the cathode-ray tube oscilloscope (also called the "Braun tube") was a forerunner of the television tube and radarscope. The cathode-ray tube (CRT) is only now being replaced by new technology in flat-screen televisions and computer monitors.

Key dates

1850	Braun born in Fulda, Hesse-Kassel, now in Germany.
1874	Marconi born in Bologna, Italy.
1894	Marconi begins long-wave research.
1896	Marconi obtains first patent for a system of telegraphy using "Hertzian waves" (radio waves) and gives first formal demonstration of new wireless telegraphy system, in London, sending a Morse code message from one Post Office building to another.
1897	Braun develops the cathode-ray oscillograph. Marconi establishes the Wireless Telegraph and Signal Company (renamed Marconi's Wireless Telegraph Company in 1900 and The Marconi Company in 1963). Marconi International Marine Communication Company is set up to install and operate wireless telegraphy services on ships.
1898	Braun demonstrates his new wireless transmission circuit. Application for a patent on Braun's new circuit is filed. Marconi opens world's first "wireless" factory, in England.
1900	Marconi files an application for what becomes known as his No. 7777 tuning patent for transmitters.
1901	Marconi transmits across the Atlantic Ocean from a transmitting station at Poldhu in Cornwall, England, to a receiving station at St John's, Newfoundland.
1907	Marconi opens world's first transatlantic commercial service between Glace Bay, Nova Scotia, and Clifden, Ireland.
1909	Braun and Marconi both awarded the Nobel Prize for Physics for development of wireless telegraphy.
1918	Braun dies in Brooklyn, New York, US. Marconi sends first radio message from England to Australia.
1932	First microwave radio telephone invented by Marconi's team.
1937	Marconi dies in Rome, Italy, following a series of heart attacks.

Marconi's original system had its weak points. The electrical oscillations sent out from the transmitting station were relatively weak … It is due above all to the inspired work of Professor Ferdinand Braun that this unsatisfactory state of affairs was overcome.

H. Hildebrand, President of the Royal Swedish Academy of Sciences, Nobel Prize Presentation Speech (1909)

William Morris Davis

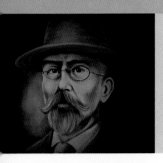

An eminent American geographer, geologist, and meteorologist, William Morris Davis proposed theories of how landscapes were formed. The now well-established discipline of geomorphology (the study of landforms) originated from his work.

Born in Philadelphia in the US to a Quaker family, Davis studied geology and mining engineering at Harvard University, graduating with a masters in engineering in 1870. He became an instructor in **geology** at Harvard in 1879, teaching physical geography, and gained a professorship there in 1890. At the start of his career, little was understood about how the landscape evolved and how its characteristic appearances developed according to the structure of the land and environmental conditions. It was Davis' descriptions of, in particular, the "cycle of erosion" that sparked the development of geomorphology as a science, and by the end of his career he had contributed much to the establishment of geography as a profession in its own right through his research and advocacy.

Davis was among the founders of the Association of American Geographers and was heavily involved in the National Geographic Society. After both his first and then second wife died, Davis remarried, to Lucy L. Tennant of Milton, Massachusetts, who survived him after his death in Pasadena, California.

> *... many pre-existent streams in each river basin concentrated their water in a single channel of overflow, and that this one channel survives – a fine example of natural selection.*
>
> William Morris Davis (1883)

Essential science

Davis' predecessors believed that the shape of a landform was determined purely by its structure, or was created by the biblical flood. By contrast, Davis often described the slow development of landforms as having parallels in evolutionary theories. His theory of landscape development – the geographical cycle (or "cycle of erosion" or "geomorphic cycle") – was his major contribution to the field of geomorphology. It was published in his article "The Rivers and Valleys of Pennsylvania" in *National Geographic* in 1889, and reiterated in subsequent works.

The cycle began with uplift of land to form mountains, followed by erosion and weathering to form at first V-shaped and then wider valleys, and rounded hills. Davis proposed such geographical cycles for different environments, according to factors such as humidity and latitude. He described three **variables** contributing to the appearance of landforms: (1) *structure* (resistance of rock to weathering and erosion, and the shape of the rock strata or layers), (2) *process* (actions such as weathering, erosion, and deposition by streams), and (3) *stage* (youth, maturity, and old age, which indicated how long the processes had been continuing). Although now considered simplistic, Davis' ideas remained highly influential until around 1950.

Legacy, truth, consequence

- Influenced by the evolutionary theories of **Charles Darwin**, the erosion cycles described by Davis started a new era in the understanding of landforms.
- Theories of stream development are still based on the systematic approach set up by Davis.
- Many of Davis' students at Harvard went on to become famous geographers themselves; among these were Albert Perry Brigham (1855–1932), Isaiah Bowman (1878–1950), Richard Elwood Dodge (1868–1952), Mark Jefferson (1863–1949), and Ellsworth Huntington (1876–1947).

Key dates

1850	Born in Philadelphia, Pennsylvania, US.
1889	Publishes his article "The Rivers and Valleys of Pennsylvania" in *National Geographic*.
1890	Gains a full professorship in physical geography at Harvard.
1904	Founds the Association of American Geographers, and becomes its first president.
1911	Leads a nine-week geographical pilgrimage from Wales to Italy.
1912	Appointed emeritus professor at Harvard.
1912	Organizes an eight-week transcontinental expedition across the US, sponsored by the American Geographical Society.
1934	Dies in Pasadena, California, US.

Emil Fischer

In the course of his long studies into chemical compounds, the German organic chemist Hermann Emil Fischer made many important discoveries of the structures of sugars and proteins and of the nature of purines (certain compounds sharing an organic base). His descriptions of carbohydrates and amino acids helped found the discipline of biochemistry.

Known by his middle name Emil, Hermann Emil Fischer was the son of a lumber merchant who insisted the boy should join the firm after leaving school. Fischer showed no business ability at all, so in 1871 he was allowed to go to university.

In his second year he met his mentor, Adolf von Baeyer (1835–1917), with whom he studied and worked for years. Fischer's doctoral research included the synthesis of the compound phenylhydrazine, which probably gave him cancer. He also suffered from mercury poisoning.

Fischer refused all job offers from chemical companies, and during World War I he was an advisor to the government. In 1919, depressed by the deaths of two of his sons in the war and diagnosed with intestinal cancer, Fischer killed himself. His surviving son, Hermann Otto Fischer, went on to become a professor of biochemistry himself.

> ### ... the structure will only be finally elucidated ... by building up the molecule ... by what is termed synthesis.
>
> Nobel Prize Lecture (1902)

Essential science

Purines

Fischer spent years on one subject before deciding he could do no more. His work on purines lasted 17 years, beginning in 1882 when he proved that several seemingly unrelated natural compounds were linked. Some were animal products, such as uric acid and guanine, and others, including caffeine and theobromine, were from plants, but they all shared a common chemical basis.

"*They contain a common atomic group ... it consists of five carbon atoms and four nitrogen atoms so arranged that two cyclic groups with two common atoms are formed ...*"

He named this common link purine, proved that all purines could be derived from each other, synthesized many of them, including caffeine, and in 1898 managed to synthesize purine itself.

Sugars

Before Fischer's research, little was known about sugars. He used new analytical techniques to discover their structure, purify them,

Legacy, truth, consequence

- Fischer's work helped advance **physiology**. He pointed out that some purines are part of a cell's nucleus so "*Knowledge of their chemical constitution and of their transformation into one another will make it easier for physiological research.*"
- By creating his synthetic compounds Fischer was not only proving their chemical structure, but he also hoped to be able to supply cheap medical substances or even cheaper food products. Barbiturate drugs are one result of his work.

Key dates

1852	Born in Euskirchen, Germany.
1882	Appointed director of a chemical institute at the University of Erlangen.
1882	Begins research on purines.
1884	Begins study of sugars.
1885	Moves to the University of Würzburg.
1892	Appointed professor of chemistry at the University of Berlin.
1899	Begins research into proteins.
1902	Awarded the second Nobel Prize in chemistry for his studies of sugars and purines.
1919	Dies in Berlin.

and identify **isomers**, eventually creating synthetic glucose, fructose, and mannose out of glycerol.

His old friend phenylhydrazine proved particularly useful, since he discovered it could crystallize sugar, making it easier to isolate.

In order to represent the shapes of the isomers, Fischer developed a new projection system now named after him. And in identifying sugar isomers, he also made a discovery about **enzymes**. He found that yeast enzymes only eat some sugar isomers, and since they vary only in shape, he concluded that enzyme activity can be determined by molecular structure, not content.

Proteins

Fischer discovered the peptide chain linking amino acids together. He also identified many new amino acids, and succeeded in synthesizing several.

Henri Becquerel

Antoine Henri Becquerel was a French physicist who is most famous for his discovery of radioactivity. Together with Marie and Pierre Curie, he won the Nobel Prize for Physics in 1903 for his pioneering work in this field, which is credited as bringing about the advent of nuclear physics.

Known as Henri, Antoine Henri Becquerel was born in Paris in 1852 into a renowned scientific family. He was the son of physicist Alexander Edmond Becquerel (1820–91) and the grandson of electrochemist Antoine-César Becquerel (1788–1878), both members of the Academy of Sciences and each in turn professor of physics at the Natural History Museum, Paris (Muséum National d'Histoire Naturelle).

Influenced by his father and grandfather, Becquerel focused on science and engineering, first at the École Polytechnique and then at the École des Ponts et Chaussées, where he entered the Administration of Bridges and Highways with the rank of *ingénieur*.

On completing his engineering training, he married Lucie-Zoé-Marie Jamin, daughter of J.-C. Jamin, professor of physics at Paris University. He began to conduct private research, in particular on polarized light in magnetic fields and crystals, before taking up a position as teacher and researcher at the École Polytechnique.

In 1878, a few weeks after the birth of their son Jean, his wife died. The same year he moved to the Natural History Museum, Paris, where he eventually succeeded his father as professor.

Although much of Becquerel's original research was in the field of **optics**, he also continued the work of his father and grandfather on **fluorescence** and **phosphorescence**, especially after learning of **Wilhelm Conrad Röntgen**'s discovery of **X-rays** in 1895. His thoughts on the matter were greatly stimulated by a famous meeting of the French Academy of Sciences, at which mathematician and scientist Henri Poincaré (1854–1912) gave a demonstration of Röntgen's X-rays and showed the audience – of which Becquerel was a member – the first X-ray images, depicting the bones in a human hand. Only months later, Becquerel was to stumble upon the discovery which would make his name.

Becquerel was a modest man and always played down his own achievements, seeing his work very much in the context of his family line. Nevertheless, he received worldwide acclaim for his discovery, which came to be known as "**radioactivity**" – named by Pierre and **Marie Curie**, who did not participate in Becquerel's discovery but further investigated the properties of the phenomenon. In 1903 Becquerel shared the Nobel Prize with the Curies for "*their joint researches on the radiation phenomena*", and later he became a foreign member of the **Royal Society of London**, and of Academies of Science in Berlin, Rome, and Washington.

In 1908 having recently been appointed president of the Academy of Sciences, Becquerel, died suddenly of a heart attack at the age of 55. Notably, Henri was not the last Becquerel to hold the position of professor at the Natural History Museum. His son Jean succeeded him, making it four generations of Becquerels that held the esteemed academic post.

Essential science

The discovery of radioactivity

Towards the end of February 1896, inspired by Röntgen's discovery of mysterious rays which had been labeled "X-rays", Becquerel began to investigate the possibility of X-ray emission from fluorescent crystals and, in doing so, accidentally discovered radioactivity in uranium salts.

Henri Poincaré had informed Becquerel that X-rays were emitted from a fluorescent spot on the glass **cathode ray** tube used by Röntgen. Immediately this suggested to Becquerel that X-rays may be produced naturally from fluorescent crystals. He positioned the fluorescent crystals on top of a photographic plate wrapped in black paper and made the substance fluoresce through exposure to a light source. He knew that if the substance did emit X-rays under these conditions then the plate would be affected.

Initially, certain crystalline substances yielded unexciting results, but it was when he introduced crystals of potassium uranium sulphate to the photographic plate that he made his momentous discovery. Positioning these crystals on the plate, he placed a silver coin beneath one of them. The whole setup was then exposed to sunlight for several hours.

On developing the photographic plate he found a silhouette of the crystals and coin. At first he thought the crystals had emitted X-rays but, on reflection, considered that this could not have been the case. A good critical thinker, he made up more arrangements of crystals and photographic plates in order to test his findings.

Unfortunately, poor overcast weather halted the proceedings, removing Becquerel's natural light source. He became impatient and, on March 1, decided to develop the plates. Expecting to find

Legacy, truth, consequence

- Becquerel's discovery of radioactivity and the subsequent developments in this field by the Curies caused a revolution in physics. It paved the way for **nuclear physics**, which showed that **atoms**, and nuclei within atoms, are composed of even smaller particles.

- The development of **radium therapy** as a treatment for cancers came about, in part, because of an experiment by Becquerel. Observing that early workers handling radium suffered from skin burns, in 1901 Becquerel carried a tube containing radium in his waistcoat pocket for six hours while at a conference. Nine days later he noticed a red area on his skin adjacent to where the radium had been positioned. His doctor advised him that it was an X-ray-like burn. Based on this finding, a number of pioneering surgeons came to see that the radiation from radium could be used to treat cancers and other diseases.

- Becquerel's discovery led to Pierre and Marie Curie's search for other radioactive materials: they discovered polonium and radium in 1898.

- The unit of measurement of radioactivity, the becquerel (Bq), is named after him. It is defined as the activity of a quantity of radioactive material in which one nucleus decays per second.

- There are craters named after Becquerel on both the moon and Mars.

Key dates

One wraps a Lumière photographic plate with a bromide emulsion in two sheets of very thick black paper ... One places on the sheet of paper, on the outside, a slab of the phosphorescent substance, and one exposes the whole to the sun ... When one then develops the photographic plate, one recognizes that the silhouette of the phosphorescent substance appears in black on the negative. If one places between the phosphorescent substance and the paper a piece of money or a metal screen, one sees the image of these objects appear on the negative ... One must conclude ... that the phosphorescent substance ... emits rays which pass through the opaque paper and reduce silver salts.

"Sur les Radiations Émises par Phosphorescence" (1896)

only weakly defined images because the crystals had been exposed to so little sunlight, he found to his surprise that the silhouettes were all remarkably clear.

Still he felt compelled to verify his results and he conducted more experiments, this time discovering, by placing the crystals in a drawer, that the silhouettes were just as clear with no sunlight at all. With these findings, he abandoned the idea that fluorescence and X-rays were connected. Evidently the uranium compound was emitting some penetrating **radiation** from within itself.

Studying this radiation, he found that it behaved like X-rays, in that it could penetrate matter and **ionize** air. Conducting further tests, he proved this was due to the presence of uranium in the crystals and subsequently discovered that pure uranium is highly radioactive.

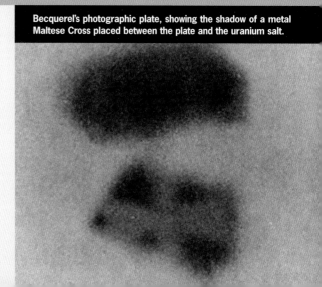

Becquerel's photographic plate, showing the shadow of a metal Maltese Cross placed between the plate and the uranium salt.

Albert Abraham Michelson

Albert Abraham Michelson was an American physicist who had a lifelong passion for precision measurement in experimental physics. He was the first person to accurately determine the velocity of light and, for this feat, became the first American to win the Nobel Prize for Physics. With physicist Edward Morley, he performed the famous Michelson–Morley experiment for light waves.

Michelson was born into a Jewish family of modest means in 1852, in Strzelno in German-occupied Poland. The family emigrated to the US while Michelson was still a young child, settling in the mining town of Murphys in Calaveras County, California, where his father established himself as a merchant.

The young Michelson was sent to high school in San Francisco, after which he spent several years in the US Naval Academy. As a naval officer he studied celestial navigation, went to sea for two years, and was later appointed an instructor in physics and chemistry at the naval academy.

His fascination with science was growing, and after marrying Margaret Hemingway in 1877 he traveled to Berlin to study under the physicist Hermann von Helmholtz (1821–94), and later to Paris. It was during this time that he began his experiments to measure the velocity of light.

On his return to the US, he was appointed professor of physics at the Case School of Applied Physics in Cleveland, Ohio, where he developed apparatus for measuring light velocity. In 1885 he began a collaboration with Edward Morley (1838–1923), working on ether-drift experiments that culminated in the famous Michelson–Morley experiment of 1887. To their disappointment, this failed to produce evidence of the existence of ether (or aether) – a hypothetical medium that was believed at the time to occupy all space and through which light and **electromagnetic** waves travel.

Michelson was appointed head of physics at the new University of Chicago in 1893, and in 1899, by then a world-renowned physicist, he gave the Lowell Lectures at Harvard University, later published as *Light Waves and Their Uses*. In 1907 he became the first American citizen to win the Nobel Prize in one of the science categories, honored *"for his precision optical instruments and the spectroscopic and metrological investigations"*.

Throughout the 1920s, Michelson spent as much time as he could in California, conducting further research and enjoying his lifelong passions of watercolor painting, billiards, chess, and tennis. In 1929 he retired from his post at Chicago and moved to Pasadena, California, where he repeated the Michelson–Morley experiment at Mount Wilson. Two years later, at the age of 79, he died from a cerebral hemorrhage during an elaborate test of the velocity of light in a partial vacuum over a mile-long course.

Essential science

The speed of light

In 1850 Léon Foucault (1819–68) had succeeded in measuring the speed of light to within a surprisingly accurate value using his rotating mirror method. This involved shining a sharply focused beam of light onto a rotating mirror, from where it would travel to a fixed mirror, then bounce back to the source at a slightly different angle due to the rotational movement of the mirror. From this he measured the angle between the original light source and the reflected beam, and taking into account known constants (the distances between the various surfaces and the speed of the mirror's rotation), he calculated the speed of light at 185,168 miles/second (298,000 km/second) – remarkably close to the modern measurement of 186,282 miles/second (299,792 km/second).

Michelson was able to refine the rotating mirror technique to achieve an even closer measurement, by constructing an interferometer, consisting of an octagonal drum of mirrors spinning at 550 revolutions per second. The device was used to detect differences in the velocity of light in two directions at right angles to one another, by employing a split beam of light: each beam of light was directed towards a mirror, and the reflected beams were made to interfere with each another. The pattern of the interference was then recorded, from which he was able to calculate the velocity of light.

Over the years he worked on the precision of his measurements and measuring devices: in the early 1920s he measured light over a 22-mile (35-km) path between two mountain peaks in California, and later, in 1926, he obtained the value 186,285 miles/second (299,796 km/second), his most precise measurement.

The Michelson–Morley experiment

After **James Clerk Maxwell**'s theory of electromagnetism, there were many attempts to prove the existence of the mysterious

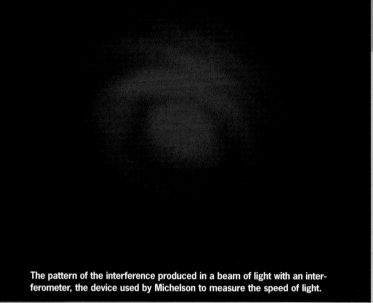

The pattern of the interference produced in a beam of light with an interferometer, the device used by Michelson to measure the speed of light.

Legacy, truth, consequence

■ In 1907 Michelson became the first American to win the Nobel Prize for Physics. Other honors included the presidency of the American Physical Society and Copley Medal of the **Royal Society**.

■ Albert Einstein once remarked, "*I always think of Michelson as the artist in Science. His greatest joy seemed to come from the beauty of the experiment itself, and the elegance of the method employed.*"

■ Despite its failure to produce the result Michelson desired, the Michelson–Morley experiment is generally considered to be the primary cause and justification for Einstein's first work on the theory of relativity. Michelson, however, remained sceptical of these developments in special relativity.

■ In 1920, using his interferometer, Michelson was able to announce the size of the giant star Betelgeuse, making him the first scientist to measure a star.

If a poet could at the same time be a physicist, he might convey to others the pleasure, the satisfaction, almost the reverence, which the subject inspires.

Light Waves and Their Uses (1903)

substance known as ether. Michelson's research on the subject had begun as early as 1881 in Berlin, but it was not until 1887 that he, along with Morley, carried out the definitive experiment.

At the time it was believed that the earth traveled through the ether, a medium through which light waves propagated. Michelson and Morley used the interferometer to attempt to detect such a phenomenon.

The two scientists hypothesized that if the earth were moving through the ether, then light would travel slower in the direction of the earth's motion than at right angles to it. Therefore, the degree of change exhibited by the interference pattern produced in the interferometer should indicate how fast the earth was moving through the ether.

A number of interferometers were constructed, with ever-increasing sensitivity for measurement, but to no avail. Nothing that could be described as a conclusive effect due to the ether was ever found.

Many regard the Michelson–Morley experiment as a turning point in theoretical physics. Its negative result demonstrated that the velocity of light was constant whatever the motion of the observer. There could be two possible explanations for this: either the ether did exist but moved with the earth, which was disproved by Olive Lodge (1851–1940) in 1893, or the ether did not exist and moving objects contracted slightly in the direction of their motion.

The latter view was put forward by George Fitzgerald (1851–1901) and became part of **Albert Einstein**'s **special theory of relativity**. Michelson's further failed attempts at the experiment only served to verify Einstein's view.

Santiago Ramón y Cajal

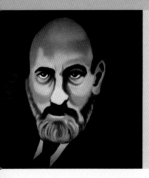

Santiago Ramón y Cajal, a Spanish histologist, physician, and Nobel laureate, is widely regarded as Spain's greatest and most distinguished scientist. He conducted pioneering studies into the function and organization of the nervous system and is commonly credited as one of the founding fathers of neuroscience.

The son of a county surgeon, Santiago Ramón y Cajal was born in the mountain village of Petilla de Aragón in Navarra, Spain, and raised in one of the poorest districts of upper Aragón. He was an intelligent yet rebellious and headstrong child who did not respond well to the discipline of school. At the age of eleven, he was imprisoned for using a homemade cannon to destroy the town gate.

Throughout childhood his first love was art not science but his unsympathetic father forbade his son from painting and drawing at home, fearing that it might divert him from a career in medicine. However, Cajal was later able to exhibit his artistic talents once his scientific career began, with his famously meticulous pen-and-ink illustrations of the organization of the nervous system.

After a year working as a shoemaker's apprentice, his father decided Cajal should enter college, and later, medical school at Zaragoza. While in Cuba on compulsory military service, Cajal suffered from malaria and tuberculosis, but recovered enough to follow a course in anatomy on his return. His interest in science was growing. In 1877 he gained a doctorate, and then took up his first academic position in a microscopy laboratory, though recurring ill health hindered his development during this time.

As a professor of anatomy at Valencia in his early thirties, he wrote and illustrated the first original **histology** textbook in Spain. Published after he took up a post at Barcelona – where he experienced some of his most productive years in scientific research – this book remained the standard text on the topic throughout his lifetime.

At the age of 40, he was appointed to the chair of histology and pathological anatomy at the University of Madrid, where he enjoyed the interdisciplinary intellectual scene. A keen debater in the capital's cafés, he was often known to contemplatively gather the table's food crumbs into a pile, and then sweep them onto the floor with a flourish in order to emphasize his point.

Throughout this time at Madrid, he conducted studies on the retina, **sympathetic ganglia**, the **cerebral cortex**, and the nervous system, and began work on his well-known book, *Advice for a Young Investigator*. He remained at Madrid for the rest of his career and accumulated many awards, including the Moscow Prize with its 6,000 francs in 1900, which enabled him to buy his first microtome (a device for cutting and preparing specimens for microscopic examination). In 1906 he shared the Nobel Prize for Medicine or Physiology with Italian histologist and physician Camillo Golgi (1843–1926), who was, as Cajal put it in his autobiography, *"the originator of the method with which I accomplished my most striking discoveries."* Despite this connection, the two men met for the first time when they were introduced at the ceremony in Stockholm. With these triumphs came international recognition and public work. He was constantly consulted on medical and educational matters, and was even invited to be government minister for education, although he never took the post. He was ill at ease with much of the publicity and could be fairly reclusive, preferring calmer pursuits like chess and photography. He died in Madrid in 1934.

Essential science

The central nervous system and the "neuron doctrine"

Cajal did his most important work on the central nervous system. Before this time, the path of the nervous system was unknown. Building on the histological staining techniques pioneered by Camillo Golgi, he used an inorganic chemical reagent, potassium dichromate, and the soluble chemical compound silver nitrate to stain sections of brain tissue.

Using this procedure he showed that the nerve fibers (axons) of **nerve cells** (neurons) end in the gray matter of the central nervous system and never join the endings of other axons or the cell bodies of other nerve cells. He concluded that the nervous system consists entirely of independent units and, contrary to the popular belief at that time, in particular Golgi's view, is not to be conceived of as a network. This redefined human understanding of brain circuits, presenting histological evidence that the central nervous system is not a continuous reticulum (or web) of interconnected cells. Rather, it consists of individual neurons that conduct information in just one direction. He discovered the electrical synapse and later explained the major neural systems in terms of chains of independent neurons. His "neuron doctrine" states that neurons are the basic structural and functional units of the nervous system.

Legacy, truth, consequence

■ Along with British **neurophysiologist** Sir Charles Sherrington (1857–1952), Ramón y Cajal is credited as the founding father of **neuroscience**. Modern **neurology** has its foundations in his research because he pioneered the contemporary understanding of the role of the nerve cell in the nervous system, and of the nervous impulse.

■ Much of his research in histology laid the foundation for subsequent studies of tumors of the brain and spinal cord. In 1913 he invented a gold sublimate to stain nerve structures, which is now used in the study of tumors of the central nervous system.

■ An inspirational teacher, many of his students went on to accomplish great achievements. Perhaps the most well known is neurosurgeon Wilder Penfield (1891–1976).

■ Cajal published more than 100 scientific works in various languages.

■ The asteroid 117413 Ramonycajal is named after him.

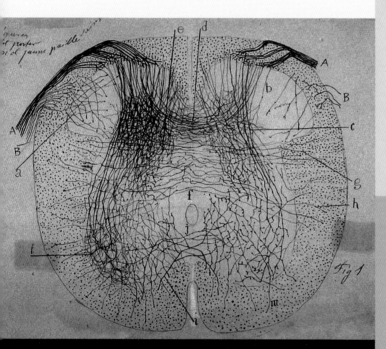

One of Cajal's drawings of his observations.

The cerebral cortex

In 1897, using methylene blue and Golgi's silver nitrate stain, Cajal investigated the human **cerebral cortex**, describing several types of nerve cell and discovering distinct structural patterns in different parts of the cortex.

Neurofibrils

In 1903 he found that silver nitrate-stained structures within the cell body were neurofibrils (long, thin filaments extending through the body of a neuron and into the nerve fibers and **dendrites**), and that the cell body itself was concerned with the conduction of information.

Key dates

1852	Born in Petilla de Aragón, Navarra, Spain.
1873	Gains his license to practice.
1874–5	Expedition to Cuba.
1877	Graduates as a doctor of medicine from Madrid, Spain.
1879	Marries Silveria Fananás García. Appointed director of the Zaragoza Museum.
1884	Appointed professor of anatomy at Valencia.
1887	Appointed professor of normal and pathological histology at Barcelona.
1889	Publishes his textbook *Histologia y Técnica Micrographica* (*Manual of normal histology and micrographic technique*).
1892	Appointed professor of anatomy and histology at Madrid.
1894	Publishes *New Ideas on the Structure of the Nervous System*. Delivers the Croonian Lecture to the **Royal Society**.
1899–1904	Publishes the three-volume *Histology of the Nervous System in Man and Vertebrates*.
1900	Awarded the Moscow Prize. Appointed director of the newly founded National Institute of Hygiene at Madrid.
1905	Awarded the Helmhotz Gold Medal of the Imperial Academy of Sciences.
1906	Shares the Nobel Prize in Physiology or Medicine with Camillo Golgi.
1909	Elected a foreign member by the Royal Society.
1922	Founder of the Cajal Institute. Retires from Madrid.
1934	Dies in Madrid.

A marvelous field of exploration opened up before me [with the microscope], full of thrilling experiences. A fascinated spectator, I examined the red blood corpuscles, the cells of the skin, the muscle fibers, the nerve fibers, pausing here and there to draw or photograph the more intriguing scenes in the life of the infinitely small.

Ramón y Cajal quoted in Dorothy Cannon, *Explorer of the Human Brain* (1949)

Nerve fiber generation and degeneration

In later years he conducted research into the generation and degeneration of nerve fibers, demonstrating that nerve fiber regeneration occurs because it grows from the stump of the fiber still connected to the cell body.

Sigmund Freud

The Austrian doctor and philosopher Sigmund Freud created the modern discipline of psychoanalysis as both a theory of the mind and a form of therapy. Introducing concepts such as the power of the unconscious mind and the sexual origin of psychological problems, he revolutionized our idea of the mind and its neuroses.

Aged 41 when Sigmund was born, and 20 years older than his wife, Freud's father was a remote, authoritarian figure, while Sigmund's mother was caring and nurturing. The oldest child of seven, Freud was favored because of his intellectual brilliance. His early family circumstances played a major part in the theories of the mind that he later formulated.

Unable to make a decent living as wool merchants, the family moved to Vienna. Freud studied medicine and in 1885 won a scholarship to study in France under the renowned **neurologist** Jean Martin Charcot (1825–93), who used hypnotism to treat hysterical disorders.

Back in Vienna, Freud specialized in neurology, using the then standard treatments of electrotherapy or hypnotism. He soon realized that these were ineffective, and instead experimented with what was called the "talking cure", encouraging

> ## The ego is not master in its own house.
>
> *A Difficulty in the Path of Psycho-Analysis (1917)*

patients to talk about and release their problems. Freud used cocaine to expand his mind, and at some points in his life he was clearly addicted to the drug.

In his forties Freud spent a period of time intensively exploring his own psychology. He reached several universal conclusions from this, particularly that the sexual impulse is the source of many neuroses. The wider scientific community reviled his explorations of sexuality, especially his belief that even infants are driven by it, and for a time Freud had to work in isolation. But by 1906 he had gathered a group of followers, including Carl Jung (1875–1961) and Alfred Adler (1870–1937), and in 1908 the first psychoanalytic conference took place in Salzburg, shortly followed by the establishment of the International Association of Psychoanalysts in 1910.

When Hitler's Nazis took power in Germany in 1933, Freud's books were among the first to be thrown on the public bonfires. Five years later the Nazis took over Austria and began to harass everyone with Jewish ancestry, such as Freud, even though he himself was an atheist. Freud decided he would prefer to "die in

Essential science

Psychoanalysis
Freud's psychoanalysis covered three different areas: a therapeutic technique; a theory of the mind and its related human behavior; and a philosophy.

Psychoanalytical method
Freud developed the technique of analytical discussion between patient and psychotherapist, through which patients will eventually bring their problems into the open, be able to confront them rationally, and thus make any necessary changes in their behavior. He asked patients to lie on a couch, which he thought would make them relaxed and open-minded, and approached their neuroses through free association of ideas and the interpretation of dreams, which he thought offered an insight into the unconscious – "the royal road to the unconscious". He also introduced the idea of transference in a psychotherapeutic relationship, whereby the patient projects feelings and ideas onto the analyst.

The unconscious
Freud's fundamental theory was that the mind has several layers and levels of functioning. Most of the time these layers are unaware of each other, and may in fact act in opposition to each other. They are:
• The id – the primitive, selfish, infantile, demanding lower level. Sometimes it can sabotage the other layers by its impulses, or reveal itself through "Freudian" slips.
• The super-ego – the superior moral psychological code.
• The ego – the part of the psyche supposedly in control of everyday matters, balancing the other two parts. This ego may try to repress the id and any painful memories of the past.

Freud concluded that the adult personality was largely created by childhood experiences, even though many formative experiences might have been forgotten by the conscious mind.

Sexuality
Freud speculated that much of our unconscious motivation is driven by our sexual drives, and that neuroses often originate from

> **The unconscious is the larger circle which includes within itself the smaller circle of the conscious, everything conscious has its preliminary step in the unconscious ...**
>
> *Psychoanalysis for Beginners (1920)*

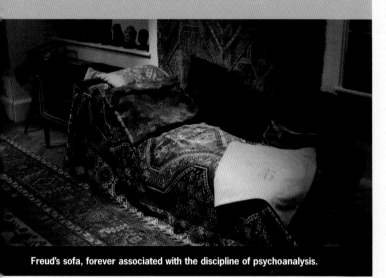

Freud's sofa, forever associated with the discipline of psychoanalysis.

Legacy, truth, consequence

- Freud initiated the whole field of psychoanalysis. Other giants of psychotherapy such as Carl Jung owed their inspiration to Freud. Jung launched a new approach involving concepts such as mythological archetypes, the collective unconscious, extroversion, and introversion, but was set on his path by Freud's original work.

- A controversial figure from start to finish, Freud has always had critics. Many psychologists think that his theories are unproven speculation, and some think that his ideas of infantile erotic instincts are obscene and shocking. On the other hand, there have always been enthusiastic supporters of his ideas, particularly of the fact that the unconscious can drive us in unsuspected ways.

- Regardless of whether we agree or disagree with him, he changed the way people think about themselves and about their behavior.

- Many terms and concepts he introduced such as id, ego, Oedipus Complex, libido, repression, the death drive, defense mechanism, penis envy, and Freudian slip are now not only accepted in mainstream psychology, but are also commonplace ideas and parts of everyday conversation.

freedom", even if it meant he had to leave Vienna, so he and his family left for London in 1938.

Freud endured many unsuccessful operations for throat cancer, but eventually he could take no more suffering. In September 1939 he persuaded his friend the doctor Max Schur to help him die – *"You promised me then not to forsake me when my time comes. Now it is nothing but torture ..."* Schur gave him three doses of morphine, and Freud died peacefully in his home in north London.

an abused, frustrated, or complicated sexuality. Probably his most famous example is the Oedipus Complex, when he argued that the attraction that young boys feel towards their mother is partly sexual, and subconsciously they hate their father out of jealousy because he can have sexual relations with her. He named this theory after the ancient Greek prince Oedipus who was given away as a baby, and as a grown man unknowingly killed his real father and married his real mother. He wrote: *"I found in myself a constant love for my mother, and jealousy of my father. I now consider this to be a universal event in childhood."*

Freud developed a similar theory for girls, and is also famous for suggesting that objects that pop up in the subconscious or in dreams have sexual meaning. For example, linear shapes such as swords and pens are supposed to be phallic symbols, but he also pointed out, *"Sometimes a cigar is just a cigar!"*

Sexual instincts are part of what Freud called life instincts, the drive to sustain life and reproduce. He contrasted this with death instincts, the unconscious self-destructive or aggressive impulse.

Key dates

1856	Born in Freiberg, Moravia, Austro-Hungary (now Pribor in the Czech Republic).
1885	Studies under famous neurologist Jean Martin Charcot in Paris.
1886	Opens a medical practice in Vienna specializing in "brain disorders".
1895	Publishes his first work on psychoanalysis, *Studies in Hysteria*, with Joseph Breur.
1895	Embarks on a four-year period of self-analysis.
1896	Introduces the term "psychoanalysis".
1900	Publishes *The Interpretation of Dreams*, outlining the first approach to psychoanalysis.
1901	Publishes *The Psychopathology of Everyday Life*, suggesting that slips of the tongue ("Freudian slips") are the unconscious at work.
1902	Appointed a professor at the University of Vienna, teaches, and founds a psychoanalytical society.
1905	Publishes *Three Essays on the Theory of Sexuality*.
1923	Publishes *The Ego and the Id*.
1923	Diagnosed with cancer of the jaw, possibly due to cigar-smoking.
1930	Receives prestigious German award, the Goethe Prize, for contributions to literature and culture.
1938	Flees the new Nazi regime for London, UK.
1939	Dies from assisted suicide in London.

Heinrich Hertz

Heinrich Rudolf Hertz was a German physicist who is most famous for discovering radio waves and for producing evidence for James Clerk Maxwell's electromagnetic theory. Sadly, his tragic early death meant he was unable to witness the radio communications industry that burgeoned from his momentous discovery. A "hertz", the unit of frequency, was named after him.

Born in Hamburg, Germany, in 1857, into a prominent local family, Heinrich Rudolf Hertz was a very practical child who enjoyed working in his makeshift workshop, conducting optical and mechanical experiments.

He was schooled at the Johanneum Gymnasium in Hamburg and had a brief spell studying engineering at Dresden Polytechnic, before undertaking his compulsory military service. With the intention of becoming a scientist, he studied mathematics and practical mathematics in Munich and then gained a doctorate from Berlin University.

Early in his career, Hertz worked as assistant to Herman von Helmholtz (1821–94) at Berlin. This was a very profitable relationship for the young Hertz as Helmholtz recognized his aptitude for research and gave him much encouragement. Their collaborative work during these years produced a prolific 14 published papers on a variety of topics in physics.

Hertz began his teaching career at Kiel before moving on to Karlsruhe where, in 1887, at the age of 30, he discovered radio waves, his most important contribution to physics. During this time he wrote ten important papers and outlined a definitive confirmation of **James Clerk Maxwell's** **electromagnetic** theory. In 1889 he succeeded Rudolf Clausius (1822–88) as professor of physics at Bonn, and throughout the next few years he developed an interest in **mechanics**.

The recurrent ill health from which Hertz suffered began as far back as 1887, shortly after he married Elizabeth Doll, with whom he had two daughters. During this time, a painful toothache was an early manifestation of an ensuing case of bone disease. His physical condition deteriorated steadily over the years and, having made a concerted effort to complete the research for his posthumously published book, *The Principles of Mechanics*, he eventually died of blood poisoning in 1894, at the age of 36. Tragically, he did not live to see his famous discovery used as the platform for a worldwide method of communication.

> ## The domain of electricity extends over the whole of nature.
> On the Relations between Light and Electricity (1889)

Essential science

Electromagnetism and radio waves

In 1879, under the instruction of Helmholtz, Berlin University offered a prize for a solution to a problem associated with James Clerk Maxwell's theory of electricity: whether or not electricity moved with inertia. Unfortunately, Hertz was unable to take part in the task at the time in part due to the demands of his doctoral research. It was not until the mid-1880s, once he was at Karlsruhe, that he was able to perform the experiments that would address the problem.

It was during these experiments that Hertz proved the existence of electromagnetic waves with the aid of a simple table-mounted apparatus, which included a circuit containing an induction coil, a wire loop, and a spark gap. At the other end of the table, he arranged another circuit with only a spark gap. He then observed that a discharge from the **induction coil** across the first gap was accompanied by a weaker spark across the gap in the receiving circuit, thus proving the existence of electrical waves.

Performing further experiments on these waves – later to be known as "radio waves" – Hertz was able to determine their velocity. He discovered that the velocity of radio waves is the same as that of light, and he devised more experiments to show that radio waves could be reflected, **refracted**, and **diffracted**.

Mechanics

During his later work in the field of mechanics, Hertz developed a system with only one law of motion. The law stated that the path of a mechanical system through space is as straight as possible and is traveled with uniform motion.

Legacy, truth, consequence

- As the discoverer of radio waves, Hertz was a pioneer in the field of radio communication. Although his equipment allowed detection of radio waves over a distance of only 60 feet (18 meters), Hertz's research in this area laid the groundwork for Italian inventor **Guglielmo Marconi** (1874–1937) to transmit across the Atlantic.

- The unit of frequency bears his name: the International Electrotechnical Commission officially adopted the term "hertz" in his honor in 1933. One hertz is equal to one complete revolution, or cycle, per second.

- A portrait of Hertz displayed in Hamburg City Hall was later removed when the Nazis came to power because of Hertz's Jewish ancestry. His widow and two daughters fled Nazi Germany in the 1930s and settled in England. Hertz has no direct descendents as his daughters never married or had children.

- Hertz's nephew, the experimental physicist Gustav Ludwig Hertz (1887–1975) shared the Nobel Prize for Physics with James Franck (1882–1964) in 1925 for their discovery of the laws governing the impact of an **electron** on an **atom**; and one of Gustav's sons, the physicist Carl Helmuth Hertz (1920–90), invented medical ultrasonography, an ultra–sound based diagnostic imaging technique.

Key dates

1857	Born in Hamburg, Germany.
1872–4	Privately tutored.
1874	Enters Johanneum Gymnasium, Hamburg.
1876	Studies engineering at Dresden Polytechnic.
1876–7	Completes compulsory military service.
1877–8	Studies at Munich University and Polytechnic.
1880	Gains doctorate in physics from Berlin University.
1880–3	Becomes an assistant to Helmholtz at Berlin.
1883	Appointed professor of physics at the Technische Hochschule, Kiel.
1885	Appointed professor of physics at Karlsruhe University.
1886	Marries Elizabeth Doll.
1887	Proves the existence of electromagnetic waves.
1889	Appointed professor of physics at Bonn University. Gives famous address in Heidelberg "On the Relations between Light and Electricity".
1890	Awarded the Rumford Medal from the **Royal Society of London**.
1893	Publishes *Electric Waves*.
1894	Dies from blood poisoning in Bonn, Germany His *Principles of Mechanics* is published (English translation 1899).

The most direct, and in a sense the most important, problem which our conscious knowledge of nature should enable us to solve is the anticipation of future events, so that we may arrange our present affairs in accordance with such anticipation.

The Principles of Mechanics (1899)

Two figures from the 1893 edition of *Electric Waves* showing configurations of electromagnetic waves as calculated by Heinrich Hertz.

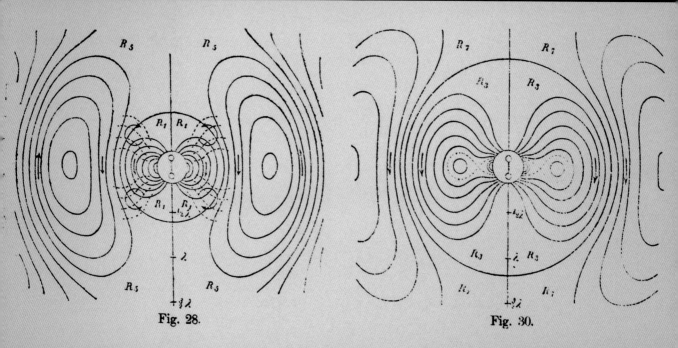

Fig. 28.   Fig. 30.

Jagadis Chandra Bose

Sir Jagadis Chandra Bose was instrumental in developing modern experimental science in India. Originally trained as a physicist, he was a prolific inventor, ahead of his times in many ways since some of his ideas and discoveries were only widely adopted years after his death. In particular, he pioneered studies into microwaves, radio optics, and the physiology of plants.

Jagadis (or Jagdis) Chandra Bose first attended the local village school before going to Calcutta (Kolkata) to finish his schooling. His father, a deputy magistrate, stretched his finances to send Jagadis to England for further education, where he attended the University of London to study medicine, then transferred to Christ's College, University of Cambridge, to study natural sciences.

He gained a master's degree before returning to India, where he was appointed acting professor of physics at the Presidency College, Calcutta, partly on the recommendation of his Cambridge lecturers. He was the first native Indian to hold the post, and suffered some racial discrimination, not least a pay package that was less than that of Europeans doing an equivalent job. When the university turned down his appeal for a pay rise in line with Europeans, Bose refused to accept his salary. He worked for three years with no pay, fully supported by his wife Abala Das, doing such an excellent job that the university eventually agreed to increase his salary and give him back-pay.

Bose was one of the scientists – often from the Bengal area – who gave an intellectual edge to Indian nationalism. He argued strongly that India needed to build up a skilled, modern scientific base, and he was opposed to caste distinctions and the religious arguments between Hindus and Muslims that bedeviled India.

As well as teaching, Bose carried out extensive experiments into radio waves and plant **physiology**, and his achievements are all the more impressive because he started out with practically no scientific equipment at all in his college. He had to buy equipment himself and work in a tiny room after his lecturing duties were over.

Bose firmly believed that knowledge should benefit the whole of humanity, so he refused to file for patents until 1901, when his colleagues persuaded him to register one. Previously, after giving a lecture to the **Royal Institution** in 1897 on the application of short radio waves, he had turned down an offer by the business partner of **Guglielmo Marconi** to share the profits from a patent.

As well as his practical science, Bose was one of Bengal's first science-fiction writers, producing a story about taming cyclones in 1896. For his science he was awarded several honors, and was a member of the League of Nations' Committee for Intellectual Cooperation.

Essential science

Radio communications

Following on from his first work, on the transmission of electrical energy, Bose began to study radio waves, conducting experiments in radio communication at the same time as the Italian physicist Guglielmo Marconi, who is generally held to be the inventor of the wireless telegraph in 1896.

In 1895 in Calcutta, Bose gave the first ever public demonstration of wireless **electromagnetic waves**, transmitting a signal across nearly a mile (1.6 km) to ring a bell and explode some gunpowder.

Microwaves

Bose discovered the existence of very short wavelengths – only a few millimeters (a fraction of an inch) in size – and in experimenting on these "millimeter waves" he built an improved coherer, or early radio detector. He also built several microwave components, whose value was not recognized at the time, although they are now commonplace. It was nearly 50 years before

other scientists made use of his discoveries of the quasi-optical properties of short radio waves.

Semiconductors

Like his work on millimeter wavelengths, Bose's studies of semiconductors lay unused for decades. He was the first scientist to identify the use of semiconducting crystals in detecting radio waves, but it took another 60 years before work began on modern semiconductors and scientists began to build on Bose's original discoveries.

Plant physiology

Bose was one of the few physicists who successfully crossed over into the biological sciences. He created his own highly sensitive instruments with which to measure plants' rates of growth as well as their tiny movements or responses to external stimuli, such as light, touch, sound, temperature, electric currents, and also deliberately unpleasant stimuli such as cuts or the application of

■ Bose never pushed to make money from his experiments and did not even seek public recognition. As a result he was often overshadowed by other scientists. However, in recent years his work has been reassessed. He is now acknowledged for his pioneering discoveries in the fields of millimeter waves, wireless detection equipment, and plant physiology.

■ At the time many scientists did not accept Bose's findings that plants behaved in the same way animals do when stimulated. Today, this behavior as a response of the nervous system is fully accepted, although most scientists do not agree that this proves any consciousness on the part of plants. Instead, the response to stimuli is considered to be a "knee-jerk reaction".

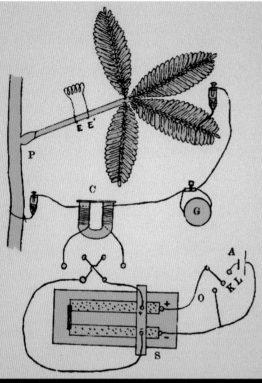

Sketch of the "complete apparatus for investigation of the variation of conducting power in plants" used by Bose for his experiments.

The invisible light can easily pass through brick walls, buildings, etc. Therefore, messages can be transmitted by means of it without the mediation of wires.

Jayant Narlikar, *The Scientific Edge* (2003)

Key dates

1858	Born in Mymensingh, East Bengal, India (now Bangladesh).
1882	Studies science at University of Cambridge, UK, then returns to India.
1885–1915	Professor of physics at Presidency College, Calcutta.
1895	Gives first ever public demonstration of wireless electromagnetic waves.
1897	Refuses to take out a patent on his electromagnetic discoveries, arguing that scientific advances should be free for the benefit of all.
1901	Persuaded to file for his first ever patent, for a "detector for electrical disturbances".
1903	Publishes paper concluding that plants' responses to stimuli are similar to those of animals.
1917	Knighted for services to science.
1917	Founds the Bose Research Institute for Science in Calcutta.
1920	Elected fellow of the **Royal Society of London**.
1926	Publishes one of his most significant books, *The Nervous Mechanism of Plants*.
1927	President of the Indian Science Congress.
1937	Dies in Bengal.

harmful chemicals. Bose was able to prove that plants' reactions to stimuli were electrical in nature, not chemical, as had previously been thought. His "crescograph" magnified the movements of plants up to ten million times, and he even measured the electrical force produced at a plant's death by its "death-spasm".

Further experiments showed that noise can have an effect on plant growth: they grow faster and stronger when exposed to calming and pleasant music, but their growth is retarded by harsh, discordant sounds.

Bose's overall findings were considered at the time to be extraordinary, showing as they do that plants react to nervous stimuli in exactly the same way that animals do, but he himself accepted the results as quite natural, interpreting them in a mystical, religious way as an expression of the Hindu belief that the whole universe is essentially united as one.

In 1927 he proposed a new theory to explain how sap rises in plants, arguing that it was the electromechanical pulsation in plant cells that caused sap to ascend.

Cybernetics

Bose's studies of human memory resulted in a model of memory as an information storage and retrieval device, the foundation of modern **cybernetics** and another area that Bose effectively anticipated long before it became a commonplace discipline.

Max Planck

A theoretical physicist, Planck created the quantum theory and, probably without realizing it, laid the foundations for a complete revolution in the scientific view of matter and of the very nature of reality. His ideas were as important to modern science as Einstein's ideas of relativity in space-time.

Planck was in the lucky position of having several academic disciplines to choose to study at university. In particular, he was a gifted musician, but he settled on mathematics and physics, and during his career he attended or taught at just three institutions: the universities of Munich, Kiel, and, finally, Berlin, where he was to remain for more than 30 years.

His earliest work was on **thermodynamics**, which led him on to the study of radiation, and from there to the problems of explaining blackbodies – objects that absorb and reemit all energy radiated upon them.

In describing the properties of blackbodies, Planck had to overturn one of the standard models of **classical physics** – that energy moves in a continuous flow. Instead, he hypothesized that energy comes in tiny discrete packets, which he named quanta (singular = quantum), from the Latin for "amount".

In one small, logical theoretical step, Planck therefore created **quantum physics**, the science that was later applied to light and to subatomic matter, and revolutionized the scientific world view.

Planck was much respected not only for his groundbreaking work but also for his moral integrity, and he had a preeminent position in German science. An able administrator, he held senior positions in the German Physical Society, the Prussian Academy of Sciences, the Kaiser Wilhelm Society for the Advancement of the Sciences, and the Society of German Naturalists and Physicians.

He was born at a time when duty to the empire was emphasized, and science was thought to have enormous national cultural value. With this background, during World War I he signed an open letter from German intellectuals titled "An Appeal to the Civilized World" in which they argued that German militarism was essential to safeguard German culture. He later became much less militant and tried to maintain international scientific relations regardless of politics.

As a loyal German, he was torn by conflicting emotions during the rise of Nazism, and he went directly to Hitler to protest against the anti-Semitism which was destroying the careers of Jewish scientists. As the Nazis tightened their grip, Planck tried to keep the scientific community focused on true science, and, though he remained in Germany in a public position, he worked behind the scenes to help his Jewish colleagues.

In 1944 Planck's house in Berlin, containing priceless scientific records, was destroyed in an Allied bombing raid, and later that same year his second son was executed for his part in the failed plot to assassinate Hitler, the fourth of Planck's five children to die before him. It is thought that he lost the will to live after this, and after helping to rehabilitate German science into the international community, he died in 1947.

> *We have no right to assume that any physical laws exist, or if they have existed up to now, that they will continue to exist in a similar manner in the future.*
>
> The Universe in the Light of Modern Physics (1931)

Essential science

Thermodynamics

Planck made an important contribution to physical chemistry, exploring the transitions from solid, liquid, and gaseous states and outlining the theory of chemical equilibrium based on thermodynamic potential.

Planck's radiation law and quantum theory

At the end of the nineteenth century physicists could not explain why the spectrum of radiation emitted by a blackbody did not match expectations according to standard **electromagnetic** theory. After some years studying this problem, Planck identified the first demonstrable failure of classic physics models when he proposed that the unexpected spectrum could only be explained if energy did not come in a continuous flow, as suggested by accepted electromagnetic theories, but instead came in separate, tiny packets or quanta. The singular, quantum, is a discrete amount, the smallest possible packet which cannot be divided any further.

His public announcement of this theory in 1900 marked the birth of quantum theory or quantum physics, a profoundly new way of looking at the underlying principles of reality.

Legacy, truth, consequence

- Planck's theories caused an about-turn in physics by showing that some observed phenomena could not be explained by classical physics. Instead, a whole new physical description was required, which came to be called **quantum physics**.
- He thought at first that his theory of quantization applied only to the emission of energy. **Albert Einstein**, however, showed that the quantum theory and the Planck constant can also be applied to light radiation (with a light quantum coming to be called a photon), and **Niels Bohr** then applied the theory to atomic structure. Quantum physics therefore overturned not only electromagnetism as proposed by **James Clerk Maxwell**, but also Newtonian mechanics (see pages 58–9), the other main foundation of classical physics.
- Quantum theory has since been applied so widely that a great

deal of modern physics – such as lasers, electronics, photoelectric cells, quantum computing – is now based on Planck's original concept.
- Like Albert Einstein and **Erwin Shrödinger**, Planck's instincts were to reject the indeterministic, relative worldview he had himself helped to introduce. He intuitively thought that the universe existed independently of humanity, whereas the Copenhagen Interpretation of quantum physics put forward by Niels Bohr, **Werner Heisenberg**, and Max Born (1882–1970) suggested that reality depends upon the observer.
- Planck was convinced that an explanation for quantum theory would soon be found within the scope of classical physics. Instead, further work on quantum theory simply extended it further beyond the range of the "old" physics.

I can tell you as a result of my research about atoms this much: there is not matter as such. All matter originates and exists only by virtue of a force ...

"The Nature of Matter", speech (1944)

Max Planck sitting at his desk.

The Planck constant

In order to mathematically describe his theory, Planck developed the equation that the energy of a vibrating molecule, E (measured in joules), is equal to its frequency, v (measured in hertz), multiplied by a new constant value, h (measured in joule-seconds):

$$E = hv$$

The new constant he identified came to be called the Planck constant, although it is sometimes called the elementary quantum of action. He calculated it to be the tiny figure of 6.626×10^{-34} joule-seconds.

Key dates

1858	Born in Kiel, Germany, the son of a law professor.
1880–92	Publishes a series of important papers on thermodynamics, later collected together as the book *Thermodynamics*.
1889	Takes a position at the University of Berlin.
1900	Announces to the German Physical Society his hypothesis of quantum radiation law, "Planck's radiation law", containing the quantum theory and the Planck constant.
1918	Wins Nobel Prize for Physics.
1926	Retires but continues to head German scientific societies.
1947	Dies in Göttingen, Germany.

Florence Bascom

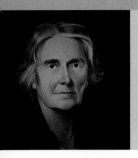

Florence Bascom was an American geologist who was a pioneer for women in both science and academia. She was an authority on the rocks formed by crystallization in the Piedmont region of Pennsylvania, and an inspiration for a generation of female geologists who came to study under her and who followed in her path.

Florence Bascom grew up in Williamstown, Massachusetts, where her father was professor of oratory and rhetoric. Both her parents supported women's rights and education. In 1874 the family moved after her father became president of the University of Wisconsin.

Bascom herself also joined the university, where she studied geology. After receiving her master's degree, she left Wisconsin and took a teaching position at Rockford College, Illinois. Part of her reason for this decision was that she experienced some problems with the **geologist** Roland Irving, one of her instructors at Wisconsin, who objected to coeducation. But opportunities for women were very gradually opening up in America. Johns Hopkins University in Baltimore, Maryland, began taking women at its graduate school and Bascom went there to study **petrology**. She gained her doctorate in 1893 with a thesis on "A Contribution to the Geology of South Mountain, Pennsylvania". However, this was not achieved without difficulty. The university still had a strongly conservative attitude and, in order to study there a woman had to prove that she could not receive equivalent instruction elsewhere. A friend of Bascom's father intervened on her behalf and she was admitted, although she was not enrolled as a regular student, did not pay tuition except laboratory fees, and had to sit behind a screen so male students would not know she was there. Even on her graduation, the degree had to be granted "by special dispensation". Johns Hopkins University did not admit women *officially* until 1907.

In the year she completed her doctorate, Bascom took her first academic position, as instructor and associate professor in geology and petrography at Ohio State University. Two years later she moved to Bryn Mawr College, Pennsylvania, where she worked as a geological assistant for the US Geological Survey – the first woman to do so. Throughout the summer months at Bryn Mawr, she mapped rock formations in Pennsylvania, Maryland, and New Jersey, and during the winters she analyzed microscope slides. The results were published in folios and bulletins of the US Geological Society.

Never married, she retired from teaching in 1928 but maintained her work for the US Geological Survey until 1936. She died of a cerebral hemorrhage in Northampton, Massachusetts, in 1945, a month before her eighty-third birthday.

Essential science

Crystalline rocks in Pennsylvania

In her work for the United States Geological Survey, Bascom was assigned the mid-Atlantic Piedmont region in Maryland, Pennsylvania, and parts of Delaware and New Jersey as her area of geological investigation. Her specialty field was petrology, the study of how present-day rocks are formed, and for many years she spent the summers mapping the **crystalline rocks** of this area and slicing thin sections of rock which she would then study during the winters. As a result, she became an authority on the complex, highly metamorphosed, crystalline rocks of the Piedmont area between the Susquehanna and Delaware rivers.

Her results were published in a series of comprehensive reports, often coauthored with colleagues, in US Geological folios and bulletins. She became widely known and respected among fellow geologists, and students came from around the world to study under her. Her work mapping the crystalline rock formations in the region became the basis for many later studies.

The Wissahickon controversy

Bascom was always very keen to encourage her graduate students to carry on the work she had begun. However, a problem emerged between Bascom and her students in 1930. Two of her graduate protégés, Anna Jonas (1881–1974) and Eleanora Bliss (1893–1974), who collaborated on a series of papers, broke with their mentor and offered a conflicting interpretation of the age of the Wissahickon schist – a rocky outcrop in Pennsylvania. Jonas and Bliss proposed an age of **Precambrian** for the schist and insisted on the presence of the Martic overthrust (supposedly an extensive sheet of older rock lying over younger rock, though many geologists at the time argued over whether the Martic overthrust interpretation was correct). Their proposal was in conflict with Bascom's view that the schist was **Paleozoic**. Eventually, with some modifications, Bascom's view prevailed, but the situation led to a good deal of bitterness, especially from Bascom who resented her students' unwillingness to discuss their work before publication.

The fascination of any search after the truth lies not in the attainment ... but in the pursuit, where all the powers of the mind and character are brought into play and are absorbed by the task. One feels oneself in contact with something that is infinite and one finds joy that is beyond expression in sounding the abyss of science and the secrets of the infinite mind.

From Bascom's writings discovered after her death

A picture from the US Geological Survey Folio 225, published in 1929 and coauthored by Florence Bascom, shows some "typical rock ledges of schistose Weverton sandstone" in Pennsylvania.

It is an interesting manifestation of the attitude of certain public critics towards change, that when the collegiate training of women was first on trial there were clamorous complaints that the health of young women was being wrecked; now the same class of public critics are loudly complaining that college women are "Amazons".

Florence Bascom, "The University in 1874–87" (1925)

Legacy, truth, consequence

■ As the first professional woman geologist in the US, Bascom was a true pioneer for women, both in science and academia. She gained admission to Johns Hopkins University when it was very difficult for women to do so and she was the first woman to gain a doctorate there. She was the first woman to serve as a geologist for the US Geological Survey, the first woman to present a scientific paper to the Geological Society of Washington, the first woman to be elected to a fellowship in the Geological Society of America, and the first woman to become vice-president of that organization.

■ Possibly her greatest achievement is that, as a university lecturer, she was mentor to an entire generation of young women geologists. Three of her students, Eleanora Bliss Knopf, Anna Jonas, and Julia Gardner (1882–1960), went on to join the US Geological Survey.

■ Bascom provided an important description and interpretation of rocks in major parts of Pennsylvania and the surrounding region, the areas she studied for the US Geographical Survey. Her bibliography contains more than 40 titles.

Key dates

1862	Born in Williamstown, Massachusetts, US.
1884	Completes her master's thesis on "The Sheet Gabbros of Lake Superior".
1893	Gains a doctorate from Johns Hopkins University. Becomes instructor and associate professor at Ohio State University.
1895	Leaves Ohio State University.
1896	Becomes geological assistant for the US Geological Survey. Publishes *The Ancient Volcanic Rocks of South Mountain*, Pennsylvania.
1898	Appointed reader at Bryn Mawr College.
1901	Ends her work for the US Geological Survey.
1903	Appointed associate professor at Bryn Mawr College.
1906	Appointed professor at Bryn Mawr College.
1928	Becomes professor emerita.
1929	Publishes "Cycles of Erosion in the Piedmont Province of Pennsylvania" in the *Journal of Geology*.
1938	Publishes "Geology and Mineral Resources of the Honeybrook and Phoenixville Quadrangles, Pennsylvania" in *US Geological Survey Bulletin*, coauthored with George Willis Stone.
1945	Dies in Northampton, Massachusetts.

Marie Curie

Physicist, chemist, and the recipient of two Nobel prizes, Marie Curie was one of the most renowned scientists of her time. The first female professor of the University of Paris, she and her husband Pierre were pioneers in the field of radioactivity, discovering not only the concept of radioactive material, but also being the first to identify new radioactive elements.

Marie Sklodowska-Curie was born in Warsaw in 1867, and was one of four children. She attended local schools and was taught advanced science by her father, a high school teacher. Because of a turbulent Polish uprising against the Russian Empire, Marie was prevented from attending university in Warsaw and was forced to work as a governess. Involved in a student revolutionary organization, Marie left Warsaw for the relative safety of Cracow, then a part of Austria.

In early childhood, Marie and her sister Bronislawa had formed an agreement under which Marie would help fund her sister's education in exchange for the same favor; as a result, in 1891, Marie was able to join Bronislawa in Paris to pursue her own studies at the Sorbonne, the University of Paris. Curie obtained degrees in physics and mathematics. She also met her future husband and colleague, Pierre Curie, then the Head of the Physics Laboratory at the Sorbonne and renowned for his work in **crystallography** and **magnetism**.

Together with Pierre, Marie Curie pioneered the field of **radioactivity**. Fascinated by the recent observations of **Henri Becquerel** on the unique properties of uranium, the Curies investigated similar properties in other elements. They studied large amounts of pitchblende (uranium ore) and found that the unique activity observed by Becquerel was maintained even when the uranium had been extracted. Further research resulted in the discovery of polonium and radium and, subsequently, to a purified form of metallic radium. Combined, these findings comprised the discovery of the phenomenon known today as "radioactivity", earning the Curies and Henri Becquerel the Nobel Prize for Physics in 1903.

Marie and Pierre Curie had two children together, before Pierre's tragic death in 1906, struck by a vehicle in the street. Although crushed by her loss, Marie was appointed to Pierre's professorship, making her the first female to become a lecturer at the Sorbonne. In recognition of her isolation of radium in its pure metallic form, Marie Curie received the Nobel Prize for Chemistry in 1911, making her the first woman to ever receive two Nobel Prizes and the first person to win two Nobel Prizes in two different fields of science. Despite these accomplishments and multiple accolades from other countries, Marie Curie faced opposition and prejudice against female scientists in her adopted France, and failed to be elected to the French Academy of Sciences by one vote.

Sadly, as a pioneer of the field of radioactivity, neither Marie nor her contemporaries were aware of the risks of radioactive exposure. She died in 1934 of a form of leukemia, presumed to be a result of her increased exposure to radioactive material. Studies of the notebooks and laboratory tools she left behind show her notebooks are radioactive to this day.

Essential science

Radioactivity and the new elements: polonium and radium

Henri Becquerel was the first to observe the "unique activities" of uranium (see pages 120–1). Trying to identify a focus for her postgraduate study, Marie Curie decided to investigate this "unique activity" in other elements. She and Pierre Curie studied waste obtained from uranium plants, and found that this activity was still present in the uranium-free material. They concluded that there must be some other elements responsible for the observed activity, and then the subsequent isolation and identification of polonium, named after Marie's native Poland, and, later on, radium, named after their newly coined term for the unique activity – radioactivity – because of its intensely radioactive properties.

Radioactivity in medicine

In their early research days, Pierre had been the first to observe that the "unique activity" of radium, later identified by the pair as radioactivity, was a chemical property directly affecting organic tissue. It was this fundamental discovery that led Marie Curie to research the application of radioactivity in medicine, and in 1915 she began to train doctors in the use of radium to treat scar tissue, arthritis, and some types of cancers. Later on she initiated research into radioactive therapeutics and their medical application. During World War I, she worked with X-ray scientists to bring mobile radiography units (popularly known as *petites Curies* or "Little Curies") to the field, to help remove shrapnel from the wounds of soldiers.

Legacy, truth, consequence

- The curie (Ci), the unit used to measure radioactivity, and the element curium are so named in honor of Marie and Pierre Curie as pioneers in the field of radioactivity.
- Marie Curie pioneered the use of the radioactive properties of radium in medicine. Her research into the therapeutic potential of radioactive material was crucial for the development of X-rays in surgery.
- Radium was used in luminous paints for watch dials, and even as a food additive, until the discovery of the associated serious adverse health effects in the 1930s.
- Marie Curie is a female icon in the scientific world and has been a major influence on subsequent generations of nuclear physicists and chemists.

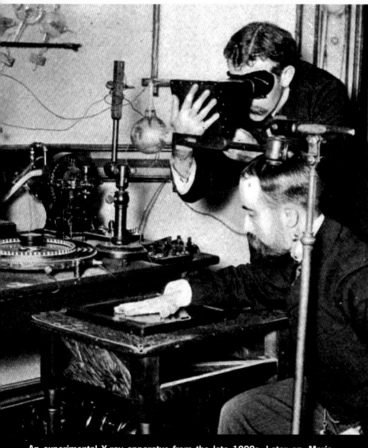

An experimental X-ray apparatus from the late 1800s. Later on, Marie Curie's research would be crucial for the development of X-rays in surgery.

Key dates

1867	Born in Warsaw, then part of the Russian Empire.
1891	Studies mathematics, physics, and chemistry at the University of Paris.
1893	Graduates first in her class from the University of Paris.
1892	Obtains a masters degree in mathematics from the University of Paris.
1895	Marries Pierre Curie.
1898	With Pierre Curie publishes an article describing their discovery of a new chemical element, polonium, and names it in honor of her native Poland. Later that year, the Curies publish their discovery of a second new element, named radium.
1902	Refines radium chloride.
1903	Along with Pierre Curie and Henri Becquerel, Marie Curie is awarded the Nobel Prize in Physics for their research on **radiation**, making her the first female Nobel laureate.
1903	Under the supervision of Becquerel, becomes the first female in France to receive a doctorate degree from the University of Paris.
1906	Pierre Curie is killed in a street accident.
1909	Appointed the first female professor at the University of Paris.
1911	Awarded the Nobel Prize in Chemistry for her discovery of radium and polonium and the isolation of radium.
1921	Receives a gift of a single gram of pure radium to use in her research from the United States President W. G. Harding.
1925	Founds the Warsaw Radium Institute and appoints her sister Bronislawa as Director.
1932	Founds the Radium Institute, now the Maria Sklodowska-Curie Institute of Oncology, in Warsaw, Poland.
1934	Dies in Sallanches, Savoy, of aplastic anemia most likely caused from her increased exposure to radiation.
1935	Curie's eldest daughter, Irene Joliot-Curie, wins the Nobel Prize in Chemistry for her discovery that aluminium can be radioactive and emit neutrons when treated with alpha rays.
1955	In honor of Marie's and Pierre's lifetime achievements, the remains of the couple are moved to the Pantheon in Paris.

We must not forget that when radium was discovered no one knew that it would prove useful in hospitals. The work was one of pure science. And this is a proof that scientific work must not be considered from the point of view of the direct usefulness of it. It must be done for itself, for the beauty of science, and then there is always the chance that a scientific discovery may become like the radium a benefit for humanity.

Lecture at Vassar College, Poughkeepsie, New York (1921)

Albert Einstein

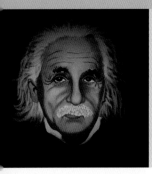

A theoretical physicist, Albert Einstein is considered one of the greatest scientists and intellects of all time. He answered fundamental scientific questions and revolutionized ideas on matter, energy, gravitation, light, space, and time. On completing his general theory of relativity, he advanced Newtonian physics, became world famous, and received a Nobel Prize that formally acknowledged his *"services to theoretical physics"*.

Einstein's fascination for invisible forces began at an early age when he first saw the effects of a magnetic compass. A book on **Euclidean geometry**, read when he was 12 years old, also had a lasting impression. He was inspired by an uncle's interest in mathematics, and another uncle's interest in science, but was less impressed by his schooling. Strict classroom discipline and Einstein's own boredom led him to leave his secondary school in Munich at the age of 15, with poor grades. He joined his parents in Switzerland, finished his schooling, and then entered the highly acclaimed Swiss Federal Institute of Technology, Zürich. But he still preferred to read in the library than go to classes. After graduating and a short period as a mathematics teacher, he took a job at the Swiss Patent Office in Bern.

The new job allowed time for Einstein to pursue his own ideas and in 1905 he completed a dissertation for his doctorate. The same year he published four revolutionary papers. One was a **quantum theory** of light and an explanation of the **photoelectric effect** (see below); another was an analysis of **Brownian motion**; the third was his **special theory of relativity**, notes for which he had begun when only 16 years old; the fourth paper stated that energy and mass are interchangeable and provided the famous equation $E = mc^2$.

Einstein was soon admired by other physicists, and the doors to academia opened. In 1909 he took the post of associate professor at the University of Zurich, and in 1914 moved to Berlin to become the director of a new research center, the Kaiser Wilhelm Institute for Physics. World fame came in 1916 with the publication of his **general theory of relativity**. But Einstein, who was of Jewish descent, faced verbal attacks as anti-Semitism grew prevalent in Germany, and he preferred to travel abroad.

During the 1920s he worked on the theory of **quantum mechanics**, but began to doubt its truth by the middle of the decade. In his later years, Einstein focused on establishing a **unified field theory**, which would describe the universal properties of matter and energy in a single formula or sentence. It was never completed. With Hitler's rise to power in 1933 Einstein renounced his German citizenship and later joined the Institute for Advanced Study in Princeton, New Jersey, US, to continue his research. A lifelong pacifist, his concerns became focused on the atomic bomb, and how to control the spread of nuclear technology. Days before his death he signed the Russell–Einstein Manifesto, which highlighted the dangers of nuclear weapons and called on international leaders to use peaceful means to resolve conflicts.

Essential science

The photoelectric effect

Einstein's first important paper published in 1905 extended physicist **Max Planck**'s theory that energy is emitted in tiny packets, or units, each one called a **quantum** (plural quanta). Einstein explained that light is made up of quanta (now called **photons**), and that this explains the photoelectric effect – how **subatomic particles** (called **electrons**) are emitted from some solids (such as metals) when struck by light (photons).

Brownian motion

Brownian motion is the random movement of microscopic particles suspended in a fluid, or gas, caused by the random bombardment of molecules of the fluid, or gas. The movement was first noticed by botanist Robert Brown (1773–1858), in 1827. Working independently, Einstein in 1905, and Polish scientist Marian Smoluchowski

(1872–1917) in 1906, were the first to explain it in terms that proved the existence of **atoms**.

Special theory of relativity

In his paper "On the Electrodynamics of Moving Bodies", Einstein presented his special theory of relativity. This states that all motion is relative and there is no stationary reference frame from which to take measurements (we stand on an earth which orbits around the sun, other planets move in relation to one another, and so on), effectively overthrowing **Isaac Newton**'s notions of absolute space and time.

His theory incorporates the principle that the speed of light is constant (which he called "c"). If two people, one sitting at the back of a train and one watching the train go by, were to measure the speed of a light shining along the train from its last carriage to its first, they would note exactly the same speed. The speed of light

Legacy, truth, consequence

- The year 1905 became known as Einstein's "miracle year" because of the publication of his four papers, each of which provided a major contribution towards an understanding of the universe. These theories and his general theory of relativity have become the foundations of modern physics.

- Ironically, proof of Einstein's equation $E = mc2$ came with the creation of atom bombs. It firmly linked him to the atomic age although he pleaded for the control of nuclear technology and worked to prevent the use of the bomb.

- After predictions in Einstein's general theory of relativity were verified in 1919, scientists around the world became profoundly impressed by his work. Though few fully understood his general relativity theory, it marked a step beyond Newton's view of the universe.

- Einstein's hope for a unified field theory was viewed by fellow scientists as impossible to achieve, largely because of quantum theory, which showed that there was an **uncertainty principle** in measurements of the motion of particles. Einstein continued to hope and became increasingly unsure about quantum theory.

- Elements of Einstein's theories were proven before and after his death. For example, **Edward Hubble**'s discovery in 1929 that the universe is expanding proved Einstein's equations that showed a dynamic universe; and overwhelming evidence that a star Cygnus X-1 is a black hole began to emerge in 1971.

EINSTEIN EXPOUNDS HIS NEW THEORY

It Discards Absolute Time and Space, Recognizing Them Only as Related to Moving Systems.

IMPROVES ON NEWTON

Whose Approximations Hold for Most Motions, but Not Those of the Highest Velocity.

The New York Times, 1919.

1879	Born in Ulm, Württemberg, Germany.
1896	Enters the Swiss Federal Polytechnic School, Zürich (renamed the Swiss Federal Institute of Technology in 1911).
1901	Granted Swiss citizenship and successfully applies for a post as technical assistant at the Swiss Patent Office.
1905	Publishes four ground-breaking papers, including his theory of special relativity.
1914	Appointed director of the Kaiser Wilhelm Institute for Physics, Berlin.
1916	Publication of his general theory of relativity.
1919	Observations taken during a solar eclipse prove Einstein's prediction of starlight bending near the sun.
1921	Begins first of many world tours. Awarded the Nobel Prize for Physics.
1932	Leaves Germany for the last time and joins the Institute of Advanced Studies in Princeton, US.
1936	Publishes, with other physicists, an article criticizing the quantum theory.
1939	Writes a letter to US President Roosevelt, warning him that Germany may be constructing atom bombs.
1940	Becomes a United States citizen.
1952	Offered and declines the post of president of Israel.
1955	Dies in Princeton, US.

The eternal mystery of the world is its comprehensibility ... The fact that it is comprehensible is a miracle.

Einstein, "Physics and Reality", Franklin Institute Journal (March 1936)

is the same for all observers, regardless of their motion relative to the source of the light, so long as the observers are still or in uniform motion and not accelerating.

This contradicts the fact that to the person watching the passing light, the light speed would seem to be faster than to the observer on the train. Einstein explained this in terms of time and space being relative to the observer: time and space are *perceived* differently by observers in different states of motion. For example, scientists have since shown that an atomic clock traveling at high speed in a jet plane ticks more slowly than if stationary on the ground.

$E = mc^2$

Einstein explained that his special theory of relativity led to the now famous equation $E = mc^2$, which states that a body's energy (E), equals the body's mass (m) multiplied by the speed of light (c)

squared. Because the speed of light is extreme, the conversion of even a tiny amount of mass releases a vast amount of energy.

General theory of relativity

A main part of this theory explained that gravitation is not a force as Newton described, but is a curved field caused by the presence of mass. Einstein said that this could be proven through a study of the way the sun's gravitation bends light rays from stars. The astronomer Arthur Eddington (1882–1944) did just this during a total **solar eclipse** in 1919. His calculations showed how starlight was bending, as Einstein had predicted.

Einstein's explanation of gravitation was based on how time and space are not separate, and that the effects of gravity are equivalent to the effects of acceleration. The theory led to Einstein's prediction of the existence of **black holes**.

Alfred Wegener

Alfred Wegener was a German meteorologist and geophysicist. He is the originator of the controversial idea of continental drift, which came to be viewed as the founding theory at the heart of a scientific revolution. He is also remembered for his perilous expeditions to Greenland's ice cap.

Alfred Lothar Wegener completed his secondary education in Berlin before moving on to the universities of Heidelberg and Innsbruck. From early on in his life, he enjoyed wide scientific interests. One of his passions was to use kites and balloons to study the upper atmosphere, which led to a world record in an international hot-air balloon contest when he and his brother stayed aloft for more than 52 hours. He had a keen interest in both **meteorology** and **geology**, and a year after presenting his thesis on astronomy at Berlin, he was delighted to join a Danish expedition to Greenland's unmapped northeast coast as the group's meteorologist.

On his return from Greenland, he was appointed lecturer in meteorology at the Physical Institute in Marburg, where he remained for four years, proving to be a popular lecturer. During this time he worked on his grand vision of continental drift (see below), which he publicly announced in a lecture at Frankfurt in January 1912. In the same year, he set off on his second expedition to Greenland, led by the Danish Captain J. P. Koch, to study **glaciology** and **climatology**.

During World War I, Wegener served as a junior military officer and endured long periods of sick leave after being wounded twice. Following the war, he worked at the meteorological experimental station of the German Marine Observatory at Gross Borstel, near Hamburg, and then went on to gain a professorship in meteorology and geophysics at the University of Graz.

In 1929 and 1930 he led two more expeditions to Greenland. On his fiftieth birthday, while on his fourth expedition, Wegener went to check on a supply drop but he never returned. His frozen body was later found and his death was attributed to heart failure.

Essential science

Continental drift

Wegener first considered the idea of continental drift in 1910 when he observed the correspondence in shape between the coastlines of the countries on either side of the Atlantic Ocean, in particular South America and Africa.

His theory was supported when, in 1911, he learned of evidence of **paleontological** similarities on both sides of the Atlantic – evidence that was being used to support the claim that a "land bridge" between Brazil and Africa had once existed.

After extensive study of the paleontological and geological evidence, such as similarities in fossils, land features, animals, and plants on both sides of the Atlantic, Wegener announced his theory in 1912, arguing that rather than having originated in their present positions, the separate continents were once a single landmass and had actually moved thousands of miles apart over geological time. During his wartime sick leave, Wegener wrote *The Origin of Continents and Oceans*, an extended account of the theory, which was published in 1915. In the book, he claimed that near the end of the **Permian period** there was a single super-continent, which he called "Pangaea". This super-continent split into several pieces, or sub-continents, which began to move, mostly westwards, with some bound for the equator. Then, at the beginning of the

Quarternary period, in the aftermath of drifting continents, smaller pieces detached, forming islands such as Greenland, the Antilles, Japan, and the Philippines.

He also described mountain formulation along similar lines, claiming that mountains are formed by the compression of an advancing front of a moving continent against the resistance of the ocean floor.

According to Wegener's theory, continental drift could be explained by the operation of a mechanism of centrifugal forces, which was generally moving the continents westwards, and moving the larger continents towards the equator and away from the poles.

In subsequent years, Wegener sought further evidence to support his theory, particularly through **geodetic** support and in his studies of **paleoclimatology**, but he had mixed success and could produce no significant results.

Reaction to continental drift

The theory of continental drift caused much controversy in the years following Wegener's initial announcement, especially when his mechanism of centrifugal forces designed to explain the phenomenon was shown to be untenable. The controversy came to a head in 1928 at an international conference attended by 14

The basic "obvious" supposition common to both land-bridge and permanence theory – that the relative position of the continents ... has never altered – must be wrong. The continents must have shifted.

The Origin of Continents and Oceans (1915)

Legacy, truth, consequence

- After the ascension of geological developments such as seafloor spreading and plate tectonics, Wegener's theory of continental drift was shown to be at the foundation of a scientific revolution. The theory of plate tectonics, in particular, is a direct descendent of the theory of continental drift and Wegener was recognized as the founding father of the movement to explain the global distribution of geological phenomena.

- His theory of continental drift offered a solution to an apparent problem with **Darwinism**. Before Wegener, the only way to reconcile **Charles Darwin**'s theory of evolution with the generally agreed fact that there are stark similarities between species that live on different continents was by positing the one-time existence of "land bridges". As an alternative to the problematic "land-bridge theory", continental drift (in the sense that there were not bridges between the continents but rather the continents were at one time part of a single landmass) strengthened the case for Darwinism.

Wegener was part of several expeditions to Greenland's ice cap.

eminent geologists who voted on the theory. Five supported it unreservedly, two with reservation, and seven opposed it.

Due to the worries regarding Wegener's mechanism and the controversy over the notion of continental drift among geologists of the time, Wegener's theory received little attention for several years. However, in the early 1950s the new science of **paleomagnetism** was developing, which rejuvenated interest in continental drift. Later on, with such developments in geology as **seafloor spreading** and **plate tectonics**, scientists began to take Wegener's theory very seriously again.

Key dates

1880	Born in Berlin, Germany.
1889	Completes secondary education in Berlin.
1899–1904	Attends the universities of Heidelberg, Innsbruck, and Berlin.
1904	Receives doctorate from the University of Berlin.
1905	Presents his thesis on astronomy at Berlin.
1905–06	Becomes an assistant at the Aeronautic Observatory, Lindenberg.
1906–08	First expedition to Greenland.
1909	Appointed lecturer in meteorology at the Physical Institute, Marburg.
1911	Publishes *The Thermodynamics of the Atmosphere*.
1912	Publicly announces his theory of continental drift.
1912–13	Second expedition to Greenland.
1913	Marries Else Köppen, daughter of meteorologist Wladimir Köppen.
1914–19	Serves as a junior military officer in World War I.
1915	Publishes *The Origin of Continents and Oceans*.
1917	Becomes a professor at Marburg.
1919–24	Works at the German Marine Observatory, Hamburg, and becomes professor at the University of Hamburg.
1924	Appointed professor of meteorology and geophysics at the University of Graz. Publishes *The Climates of the Geologic Past*, coauthored with Wladimir Köppen (1846–1940).
1929–30	Third expedition to Greenland.
1930	Fourth expedition to Greenland. Dies from heart failure while crossing Greenland's ice cap.

Scientists still do not appear to understand sufficiently that all earth sciences must contribute evidence towards unveiling the state of our planet in earlier times, and that the truth of the matter can only be reached by combining all this evidence.

The Origin of Continents and Oceans (1915)

Alexander Fleming

A Scottish biologist and bacteriologist, Alexander Fleming made one of the greatest discoveries in medicine when, in 1928, he identified the bacteria-killing properties of penicillin, the first ever antibiotic. It was hailed as a "miracle drug" when it first appeared, and probably no other group of drugs has saved as many lives as antibiotics.

His first name usually shortened to Alec, Fleming did not come from a scientific or even a professional family. He was the son of a Scottish farmer, and his first job was as a clerk in a shipping office in London. But in 1901 he inherited the then large sum of £250 from an uncle, and decided to use it to start a new career as a doctor.

After graduation Fleming worked as a **bacteriologist** in the inoculation department of his medical school in London, St Mary's, and also became a surgeon and a medical writer. His quiet sense of fun led him to grow some bacteria cultures not for work, but just for their pretty colors and patterns. He called them his "germ paintings".

During World War I he and his colleagues went to France as part of the Army Medical Corps, working in battlefield hospitals, but with peace restored in 1918 he returned to the lab. He received promotion in 1928 to professor of bacteriology, and just a few weeks later went off for a family summer holiday.

It was when he returned to his lab that he made his legendary discovery, purely by accident. He had forgotten to cover a dish of **bacteria**, and during his absence several molds had grown on it.

Some of this mold, he noticed, seemed to have killed the surrounding bacteria. This was the source of penicillin, the world's first antibiotic and one of the greatest medical discoveries of all time.

After further tests, Fleming announced his results to the Medical Research Council in 1929. A shy, retiring man, his presentation was poor, and his discovery was ignored by other scientists for years. Undeterred, he continued with researches, and when other scientists did look again at his work and refined the drug, he received full credit for the discovery.

Publicized as a "miracle drug", pure penicillin was eventually mass-produced in 1943, and it immediately saved the lives of thousands of soldiers in World War II. Fleming was catapulted into world-wide fame, becoming the first global medical celebrity.

Awards and honors from around the world were suddenly heaped on him. In all he received 26 medals, 18 prizes, 14 decorations, and 25 honorary university degrees, culminating in the Nobel Prize in 1945, shared with Howard Florey (1898–1968) and Ernst Chain (1906–79).

Essential science

Antiseptics

While treating injured soldiers during World War I, Fleming experimented with antiseptics, showing that they should be used sparingly so as not to destroy the body's natural defenses. He also completed pioneering work on blood transfusion.

Lysozyme

In 1921 Fleming discovered that a chemical substance present in egg whites, nasal mucus, and human tears could inhibit bacteria. He called it "lysozyme" because it had the properties of an **enzyme** and dissolved or "lyzed" **microbes**. But, at the time most scientists thought that microbes were destroyed by mechanical, not chemical, means, and his discovery was not followed up for some time.

Penicillium

Fleming's job was to search for new ways to protect against germs; it involved growing bacteria, then killing or weakening them with chemicals, and testing the resulting vaccines.

His great discovery came about by accident, when he left a petri dish of live bacteria uncovered in his lab while he went on holiday. When he returned, there were many patches of different molds growing on the dish, and he was observant enough to spot that one in particular seemed to have killed the bacteria around it. This led to his investigations.

Molds grow from microscopic, seed-like spores, which float in the air and are difficult to guard against. Researchers normally throw away mold-contaminated specimens, but fortunately Fleming kept the sample and soon identified the white, fluffy growth as a type of penicillium, common on soil, rotting fruit, and rotting bread. He extracted a liquid from the mold, which he called "mold-juice" or penicillin. Many tests later, he found that penicillin killed or stopped the growth of even the most harmful bacteria.

Comparison tests

For comparison, Fleming did the same tests on all the molds that he could find. His colleagues began to worry about him as he

Legacy, truth, consequence

■ Before the discovery of antibiotics, even the tiniest scratch could prove fatal. A child falling over and grazing a knee, a housewife cutting herself with a kitchen knife, a farmer scraping himself with a scythe, and, of course, every wounded soldier, was in danger of dying not necessarily from the injury, but from the bacteria which got into the wound and caused an infection. More soldiers died from infections than from the wounds themselves, and despite Joseph Lister (1827–1912) having persuaded doctors to use antiseptics to kill germs on hands and medical equipment, infection after an operation was still a serious threat. Fleming's discovery completely changed the lethal risk of everyday accidents and cuts.

■ There are now more than 8,000 different antibiotics available to fight infections, every one due to Fleming's original discovery of penicillin.

■ Antibiotics have had such an impact on health that some doctors divide the history of medicine into the "pre-antibiotic age" and the "antibiotic age".

■ Fleming never received any royalty or patent payments for his discovery of penicillin. When American drug companies collected $100,000 for him in recognition of his contribution to science, he passed it on to St Mary's for research.

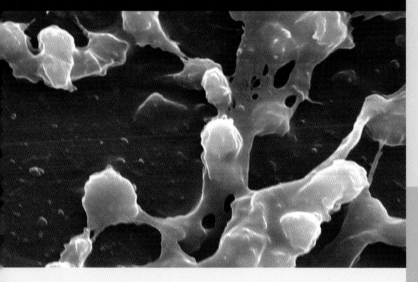

Fleming also discovered that strains of bacteria could develop antibiotic resistance. This magnified view of a bacteria shows a sticky-looking substance known as "biofilm", which protects the bacteria from attacks by antibiotics.

scraped old fruit and smelly shoes, and picked up animal droppings. Soon, he realized that penicillin was a powerful bacteria-killer which seemed to have no side-effects.

Pure penicillin

Fleming adopted the term "antibiotic", meaning "against life", for the penicillin that he tried to purify from mold-juice. But, it took years before other medical scientists began to look again at his discovery, and it was not until 1939, after the invention of vacuum freeze-drying which easily enabled unwanted substances to be removed from a mixture, that the Oxford **biochemists** Howard Florey and Ernst Chain isolated pure penicillin and in tests proved its success in defending against infection.

Key dates

1881	Born in Ayrshire, Scotland.
1901	Receives an inheritance enabling him to go to medical school.
1906	Qualifies as a doctor from St Mary's Medical School, part of London University. Works at St Mary's as a bacteriologist and lecturer.
1914–18	Becomes a battlefield doctor in France during World War I.
1921	Discovers bactericidal substance, which he names lysozyme.
1928	Promoted to professor of bacteriology. Discovers bacteria-killing properties of the mold penicillium.
1929	Fails to communicate importance of his discovery. Continues to study "mold juice" on his own.
1939	Howard Florey and Ernst Chain isolate pure penicillin and conduct successful tests on humans.
1942	Publicly credited with the discovery of penicillin.
1943	Becomes famous worldwide with mass production of the new wonder drug.
1943	Elected a Fellow of the **Royal Society**.
1944	Knighted.
1945	Receives Nobel Prize in Medicine.
1946	Discovers drug-resistant strains of bacteria.
1948	Appointed emeritus professor of bacteriology at the University of London.
1952	He is successfully treated for pneumonia with his own drug.
1955	Dies suddenly in London from heart problems. Buried in St Paul's Cathedral in London.

I preferred to tell the truth that penicillin started as a chance observation. My only merit is that I did not neglect the observation and that I pursued the subject as a bacteriologist.

Nobel Prize Lecture (1945)

"Super bugs"

In 1946 Fleming noted that some strains of bacteria change so quickly that they can become resistant to antibiotics, particularly if the dose is too small or is stopped too soon. With this discovery he predicted today's "super bugs" that show resistance to common drugs.

Niels Bohr

Theoretical physicist Niels Bohr's investigations of atomic structure and the radiation emitted by atoms contributed towards the modern understanding of quantum theory. He was one of a small group whose work revolutionized the scientific understanding of the nature of matter and reality.

At school Niels Bohr's best subject was possibly physical education. Particularly keen on soccer, he played it well, but not quite well enough to represent his country, Denmark, as his brother Harald, his closest friend, did.

Bohr studied physics at the University of Copenhagen, although there was no physics laboratory available. He had to carry out experiments in his father's physiology laboratory at the university. In 1911 he moved to study in England, and in 1912 began to work under Ernest Rutherford (1871–1937), the atomic scientist, in Manchester. From then until 1916 he alternated between Copenhagen and England, developing his theories of **atomic structure**.

In 1921 he had become so acclaimed that a new Institute of Theoretical Physics was created in Copenhagen for him to lead. It was not only a major center for **quantum physics**, but also became a refuge for many German scientists fleeing the Nazis in the 1930s.

During World War II, Germany conquered Denmark in 1940. The following year, Bohr's former friend and protégé, the physicist (and head of the German atomic bomb project) **Werner Heisenberg** paid him a somewhat mysterious visit. Some historians think that Heisenberg was trying to hint that he would not build an atom bomb for the Nazi war-machine, but others think that, since Bohr had some Jewish ancestry and believed in

Essential science

Atomic structure
Electrons were only discovered in 1897, and Ernest Rutherford soon showed that they carry a negative charge while whirling in a cloud around the positively charged **nucleus** of an **atom**. Yet since opposite charges attract, there was no explanation for why electrons did not collapse onto the nucleus, particularly when they lose energy. An additional puzzle was why, since atoms are stable, the light **spectrum** of an atom shows several spectral lines representing colors, or light at different frequencies.

In what is now called the Bohr model, Bohr suggested that electrons lie in set rings or orbits around the nucleus according to how much energy they contain. He also "quantized" the atom, that is he applied Planck's theory of small quanta of energy, and theorized that if an electron receives a quantum of energy, it jumps to an outer orbit, and, when it drops down to a lower orbit, it loses a quantum, but it can never fall beyond the innermost orbit.

Atomic radiation
While they are moving about, Bohr established that electrons themselves radiate a quantum of **electromagnetic** energy in the form of light. His proposal that they start at different orbits or energy levels also explains why the light they emit has different frequencies.

Complementarity
A keen philosopher of science, in 1927 Bohr proposed his principle of complementarity, which argues that things may have

a dual nature, containing complementary or even contradictory properties, but we can only examine one aspect of a pair at a time. Therefore objects have to be analyzed in more than one way, for example **subatomic particles** also act as waves. He applied this principle to the great paradox he helped uncover, the fact that **classical physics** does not describe the behavior of subatomic particles, yet provides a perfectly good explanation of matter and waves at the macroscopic level, where the quanta are too small to have a noticeable effect. According to complementarity, the quantum theory corresponds in its results to classical mechanics at very large levels.

Nuclear fission
In 1937 he proposed the theory that the nucleus of an atom is not one single particle, but is a compound structure, held together only by energy, that might therefore be split or undergo fission. This was the founding theory for nuclear energy and bombs.

Copenhagen Interpretation
In 1927 Bohr and his team at Copenhagen produced an interpretation of quantum physics that accepted Heisenberg's uncertainty principle and effectively suggested that scientists should adopt the complementary principle, and not worry about the quantum effect except in the fields of subatomic particles or light speeds, where the effects do have an impact on experiments.

the peaceful application of science, they simply argued bitterly. Neither spoke much about the meeting, but their relationship was never so close afterwards.

In October 1943 Bohr and his family joined a mass escape from Denmark, when the Resistance helped nearly all the country's Jews flee by boat to Sweden before the Nazis began deportations to the death camps. From Sweden he was brought to Britain: seated in the hastily converted bomb bay of the plane, he passed out when he didn't apply his oxygen mask in time. Fortunately the pilot realized and flew at a lower altitude, and Bohr afterwards reported having enjoyed a good sleep on the flight.

Possibly reluctantly, Bohr joined the Allies' efforts to build an atomic bomb, working on the secret Manhattan Project based at Los Alamos, New Mexico, US. He preferred to investigate the atom's secrets for peaceful – not military – benefits, and argued so vehemently that scientists should share knowledge (in this case between the Allies and the USSR) that British prime minister Winston Churchill warned that he was on the edge of committing a war crime.

When the war ended Bohr campaigned for the peaceful use of nuclear energy. He came to appreciate the Chinese philosophy of Daoism, whose ideas of the flow of energy are surprisingly similar to those of quantum mechanics. He adopted the Chinese yin–yang symbol, mirroring his theories of opposite complements.

Bohr was one of only a handful of Nobel Prize winners whose children have also gone on to win the award: his son Aage Bohr received a Nobel Prize in Physics in 1975.

Legacy, truth, consequence

- Bohr's model of **quantum mechanics** was an enormous step towards understanding atomic structure and the nature of matter.
- Although the orbits of electrons are now known as set energy states, and his model of the atom has been superseded, the image he created of electrons circling the nucleus of an atom remains a symbol of physical science.
- **Albert Einstein**, **Erwin Schrödinger**, and others disagreed with the probabilistic, indeterminate world view suggested by the new quantum theory, but Bohr's Copenhagen Interpretation prevailed.

The great extension of our experience in recent years has brought light to the insufficiency of our simple mechanical conceptions and, as a consequence, has shaken the foundation on which the customary interpretation of observation was based.

Atomic Physics and the Description of Nature (1934)

Humanity will be confronted with dangers of unprecedented character unless, in due time, measures can be taken to forestall a disastrous competition in such formidable armaments and to establish an international control of the manufacture and use of powerful materials.

Open letter to the United Nations (1950)

The Manhattan Project led to the first ever nuclear detonation, known as the Trinity test of July 16, 1945.

Key dates

1885	Born in Copenhagen, Denmark.
1913	Publishes his revolutionary theories on atomic structure.
1921	Becomes director of the new Institute of Theoretical Physics in Copenhagen, which attracts scientists from around the world.
1922	Wins the Nobel Prize in Physics.
1939	Passes on to the US news that German scientists are working on splitting the atom, spurring the US to develop its own atomic bomb through the secret Manhattan Project.
1943	Escapes from Nazi-occupied Denmark, and joins the Manhattan Project.
1955	Helps organize the first Atoms for Peace Conference in Geneva.
1962	Dies in Copenhagen following a stroke.

Erwin Schrödinger

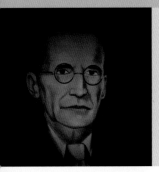

One of the great theoretical physicists of the early part of the twentieth century, Schrödinger founded wave mechanics, a mathematical approach to solving some of the problems of the emerging science of quantum mechanics. He is also known for the thought experiment called "Schrödinger's Cat", which highlights the uncertainty involved in quantum physics.

Schrödinger was 39 – an unusually late age for theoretical physicists to create original work – when he produced his great ideas on **wave mechanics**. His first academic positions were held in experimental physics, not theoretical, though he claimed this gave him a solid practical foundation on which to build hypotheses.

However, by 1917 he was exploring the theories surrounding the new physics, and in 1921 he was given the chair of theoretical physics in Zurich, Switzerland. From then on he began to concentrate on **atomic structure** and **quantum statistics**, and in November 1925 he came across the recently proposed theory of the French physicist Louis de Broglie (1892–1987) that microscopic particles might also have aspects of waves. It took Schrödinger only weeks to develop this revolutionary idea and formulate his ground-breaking mathematical equation, one that satisfactorily described the strange behavior of **subatomic particles** in terms of waves.

Beginning in January 1926, Schrödinger published his ideas in a series of papers that were immediately acclaimed by the scientific community. However, he disliked the **probabilistic** nature of the new physics that he had himself helped create, so after 1926 his main contributions were to different areas of science, such as

biology and the still unexplained **unified field theory**. Like **Niels Bohr**, he loved the philosophy of science and was drawn to Asian mysticism which often paralleled some of the ideas of particle theory. Schrödinger was particularly attracted by the classical Indian philosophy, Vedanta.

Schrödinger left his position in Germany in 1933 in protest at Hitler's treatment of Jews. But he couldn't stay away from his beloved Austria for long, and he was there when the Nazis took over Austria in 1938. His new university rector, a Nazi appointee, persuaded Schrödinger to write a letter saying that he had earlier misjudged his country's "will and destiny". He bitterly regretted this afterwards, and later gave a personal apology to the great physicist **Albert Einstein**, a Jewish refugee from Nazi Germany. In any case, the letter didn't help Schrödinger: he was sacked and repeatedly harassed by the Nazis, so with his wife he fled, eventually settling in Dublin, Ireland, for many years.

Schrödinger was a well-known womanizer, and had an open marriage – both he and his wife had affairs and he had children with several different women, scandalizing academic institutions around the world.

Essential science

Wave mechanics

Schrödinger built on ideas of wave-particle duality and created a complete theory to describe and predict the behavior of subatomic particles in terms of a **wavefunction**, governed by a fundamental differential equation, known as the Schrödinger equation.

Schrödinger's model treated **electrons** as separate three-dimensional waves around the nucleus of the **atom**, instead of particles. He considered that each electron would have its own unique wavefunction, which comprised three separate properties:
1. Its orbital distance from the nucleus, depending on its energy level;
2. The shape of its orbit (not all orbits are elliptical);
3. The orbital magnetic moment, or strength of its magnetic force within its field.

In a complicated equation (but not as complex as **Werner Heisenberg**'s **matrix mechanics**), Schrödinger showed how a wave-function can be calculated, which allows scientists to predict

the probability of a particle, such as an electron, being at a particular spot within its field. He "proved" his own theory by correctly applying the equation to a hydrogen atom to show its energy levels.

Quantum mechanics

Schrödinger's wavefunction was the second form of quantum mechanics, appearing shortly after Heisenberg's matrix mechanics, that also offered a way to explore the behavior of subatomic particles. Although the math was not proven until later, Schrödinger always said that his **differential equation** was mathematically equivalent to Heisenberg's algebraic approach.

Schrödinger's Cat

In this thought experiment, Schrödinger intended to show that the new quantum science, with its discussion of uncertainties and probabilities, was completely bizarre. He imagined an experiment

Legacy, truth, consequence

- Schrödinger confirmed the revolutionary theory of wave-particle duality, that matter and light can behave both like a particle and like a wave, and thereby affected the scientific view of the very basis of reality.
- The concept of many parallel universes rose out of Schrödinger's Cat, since if there can be two possible universes, why should there not be many?
- He contributed to the concept of an observer-related universe, which posits that everything scientists do affects their experiments.
- Schrödinger hoped that his theory of wave mechanics, involving continuous wave-like properties, would do away with the other suggested theory that subatomic particles moved in discontinuous quantum jumps. Like Einstein, he hoped that a classical explanation would be found for the strange behavior of **quantum particles**. However, his work simply contributed to the emerging idea that at the subatomic level, matter and energy intermingle and cannot be described in the same way as at the macroscopic level.
- His 1944 book *What is Life?* on genetic codes and biology is credited with inspiring future research into genes, and particularly influenced **James Watson** and **Francis Crick** in their discovery of **DNA**.

Key dates

1887	Born in Vienna, Austria.
1926	Publishes "The Schrödinger Equation and Wave Mechanics as a Mathematical Formulation for Explaining Quantum Mechanics".
1927	Succeeds **Max Planck** at the University of Berlin, Germany.
1933	Leaves Germany because of his dislike of Nazi anti-semitic policies. Goes to England.
1933	Awarded the Nobel Prize in Physics (shared with **Paul Dirac**).
1935	Introduces the paradox "Schrödinger's Cat".
1936	Moves to the University of Graz, Austria.
1938	Sacked by the Nazis after their takeover of Austria, and flees.
1956	Returns to Vienna.
1961	Dies of tuberculosis.

One can even set up quite ridiculous cases ... If one has left this entire system to itself for an hour, one would say that the cat still lives if meanwhile no atom has decayed. The psi-function [wavefunction] of the entire system would express this by having in it the living and dead cat (pardon the expression) mixed or smeared out in equal parts.

The Present Situation in Quantum Mechanics (1935)

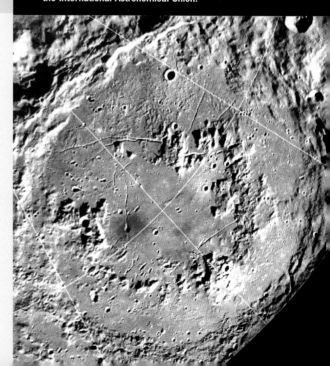

The Schrödinger basin, a large impact crater on the far side of the moon, was posthumously named after the scientist by the International Astronomical Union.

Science cannot tell us a word about why music delights us, or why and how an old song can move us to tears.

Nature and the Greeks (1954)

in which a cat is placed in a sealed box with such a tiny amount of a radioactive material that in an hour one atom has an even chance of decaying. If an atom did decay, this quantum event would trigger a device to kill the cat, but until the box is opened, the observer does not know what has happened. Schrödinger argued that this showed that two universes – with a live and a dead cat respectively – exist in parallel, or in a superposition of states, until the box is opened. Only at the very act of observation does the wavefunction collapse into an actual state of being.

Edwin Hubble

Astronomer, and founder of the science of cosmology, Edwin Hubble changed the world's idea of the universe forever by proving the existence of galaxies beyond the Milky Way. Essentially creating the idea of the universe and our place within it that we recognize today, Hubble discovered dozens of galaxies and provided revolutionary research into the distances between galaxies, as well as proving the theory of the expansion of the universe.

Born in Marshfield, Missouri, US, in 1889, Edwin Powell Hubble was raised in Wheaton, Illinois. Although always a good student, he was noted more for his athletic achievements in his youth than his academic strengths. His first degree at the University of Chicago, in mathematics, **astronomy**, and philosophy, was followed by a masters in jurisprudence at Oxford University in England, though he switched to Spanish before completing the degree. Returning to the US, he took up a teaching position near Louisville, Kentucky. It wasn't until after serving in World War I that he seemed to have made up his mind to study astronomy. This he pursued at the Yerkes Observatory of the University of Chicago, where he obtained a doctorate degree in 1917 after presenting a dissertation entitled "Photographic Investigations of Faint Nebulae".

Hubble started his professional career during an exciting time, when **Albert Einstein** had just published his theory of **relativity**, leading to revolutionary ideas in the field of astronomy. At the beginning of the twentieth century, most astronomers thought that our galaxy, the Milky Way, was the extent of the universe, and that it measured only a few thousand light-years across. In the 1910s Harlow Shapley (1885–1972) was the first to realize that the Milky Way galaxy was much larger than previously believed (about 100,000 light-years), and that the sun was not in the center. Henrietta Leavitt (1868–1921) then noticed that the Large and Small Magellanic Clouds (objects visible close to the edge of the Milky Way, now known to be dwarf galaxies) contained thousands of **variable stars**, which were slightly outside the Milky Way's border. Her observations led to the development of a method for measuring the distances

between stars that would change forever our picture of the universe. But perhaps it was Hubble who contributed most to expanding our view of the universe when he discovered the nature of some of the fuzzy patches of light known as nebulae.

On the faculty of the Mount Wilson Observatory, which housed the 100-inch (254 cm) Hooker telescope, the largest telescope in the world at the time, he focused his research and observations on a part of the sky called the Andromeda nebula. He was led to the conclusion that the stars observed in the Andromeda nebula were too distant to be part of our galaxy and so belonged to an entirely new galaxy (now known as the Andromeda galaxy). Further investigations in the following years led to the discovery of several other galaxies. Hubble was able to compare the galaxies and establish a method for classifying them according to their distance, shape, content, and brightness. This system of galaxy classification is still used today.

One of Hubble's most important discoveries came in 1929, when he published his data on uniform universe expansion. By studying 46 galaxies he found that the farther apart that galaxies are from each other, the faster they move away from each other. This was the basis of his famous Hubble's law.

> *Equipped with his five senses, man explores the universe around him and calls the adventure Science.*
>
> The Nature of Science, and Other Lectures (1954)

Essential science

The Hubble classification

Introduced in 1936, this is the system that Hubble established to distinguish various galaxies by their visual appearance. Sometimes known as the "Hubble tuning fork" because of the style of diagram in which it is traditionally represented, the system classifies galaxies according to whether they appear elliptical, spiral, barred spiral (with a bar across the center), lenticular (barred but not spiral), or irregular (no obvious structure). The system is still the most common form of galaxy classification used today.

Hubble's law and the Hubble constant

For years during observations of the newly discovered galaxies, Hubble and his colleagues encountered repeated problems when observing light from the distant nebulae (the term Hubble used for clusters of stars or galaxies). The colors of light from these nebulae were shifted towards the red end of the color spectrum, and the reason for this "red-shifting" remained unknown. Basing their study on the shared opinion that the cause of red-shifting was because the nebulae were moving away

Legacy, truth, consequence

- Hubble's identification of the Andromeda galaxy was revolutionary; not only had he discovered a new galaxy, but his discovery changed the way we viewed the world: as part of a much larger universe than we had ever imagined.

- Albert Einstein, whose theory of relativity and ideas that the universe was either expanding or contracting, was a great admirer of Hubble. When news of Hubble's evidence of uniform universal expansion became known, Einstein was elated at the effect of the findings on his own work.

- Hubble's law is considered to support the theory of expanding space and the **Big Bang** cosmological model of the universe. Hubble's constant, depicting the rate at which the scale of the universe changes over time, is one of the most important numbers in cosmology and is used to estimate the size and age of the universe.

- Hubble's discoveries not only advanced the field of astronomy, giving birth to the new field of **cosmology**, but also provided the discipline with much-needed recognition. Hubble campaigned for many years to allow astronomy to be considered a part of physics, so astronomers might be considered for the Nobel Prize in this discipline, and although he did not achieve this in his lifetime, it is the case today.

- The Hubble space telescope, which was launched in 1990 and named in honor of the great astronomer, aims to provide further data to confirm and refine the Hubble constant. The telescope has so far helped show that the universe is not only expanding, but is expanding at an accelerated pace, driven by a mysterious force dubbed "dark energy".

Key dates

1889 Born in Marshfield, Missouri, US.

1910 Obtains an undergraduate degree in mathematics, astronomy, and philosophy from the University of Chicago.

1913 Leaves Oxford University, England, with a master's degree.

1917 Obtains a doctorate in astronomy from the University of Chicago.

1919 Accepts a staff position at the Carnegie Institutions' Mount Wilson Observatory, near Pasadena, California, where he remains until his death.

1925 Proves that the nebulae in his observations are separate galaxies and not part of the Milky Way.

1929 Formulates the **empirical** Hubble's law, which is used to measure the distance of galaxies.

1935 Discovers the asteroid "1373 Cincinnati".

1936 Publishes his book *The Realm of Nebulae*.

1953 Dies of cerebral thrombosis in San Marino, California.

1990 The Hubble space telescope is launched.

Hubble concluded that the stars observed in the Andromeda nebula belonged to an entirely new galaxy, now known as the Andromeda galaxy.

from our Milky Way galaxy, they measured the distances to the receding nebulae, which led to what is now known as Hubble's law. This law states that the farther away a galaxy is from the earth, the faster it is moving away – the distances between galaxies are continuously increasing and therefore the universe is expanding (the theory of universal expansion). Hubble then created the equation of universal expansion that is still used today, although in an updated form. With his colleagues he estimated the expanding rate of the universe at 311 miles (500 km)

per second per Megaparsec – a Megaparsec (Mpc) being equal to a distance of approximately 3.26 million light-years; this estimate is known as the Hubble constant.

It is now thought that Hubble underestimated the distances to the galaxies, which makes his calculation of the expansion rate too large. Astronomers today estimate the number at around 44 miles (70 km) per second per Megaparsec, although there still remains a significant amount of uncertainty in the value of the Hubble constant.

Megh Nad Saha

An Indian theoretical astrophysicist, Megh Nad Saha was best known for his work on thermal ionization in the atmospheres of stars, developing the Saha equation which helped with the interpretation of a star's light spectrum. This was the basis for much future work in astrophysics. He was also one of the Indian scientists who worked to establish modern science education in India.

Sometimes referred to as M. N. Saha, he was from a poor rural family, and would have had no further education at all after the elementary village school if it were not for a local doctor who helped arrange for him to go to a secondary school. He later won a scholarship for a school in Dhaka, and although the grant was taken away when he joined protests against the partition of Bengal, he found another school and another scholarship. In 1911 Saha was able to attend Calcutta Presidency College, studying mathematics, only because of a further scholarship. There he was one of a generation of Bengali Indian students (including his classmate and occasional collaborator **Satyendra Nath Bose**) who had inspirational teachers in the scientists **Jagadis Chandra Bose** and Prafulla Chandra Roy (1861–1944).

After graduating, Saha applied to join the civil service, but as he was suspected of having contacts with revolutionaries he was turned down. So he returned to science, which was still in a developmental stage in India. In 1916 he became a lecturer at the recently established department of modern science and mathematics at the University College of Science in Calcutta. There he was instrumental in translating the most recent works from Europe and America, particularly papers in **quantum theory** and **relativity**, and in introducing the new physics to Indian students.

His doctoral thesis in 1919 on radiation pressure in **astrophysics** was the beginning of his involvement in that field, and won him a scholarship that took him to Britain and Germany. There he refined his groundbreaking ideas on thermal **ionization** of gases, publishing his equation and applying this to stellar **spectra** (frequencies of radiations emitted by stars) in 1921. Returning to India he found that a new chair of physics had been created for him in Calcutta, and he began his lifetime work of improving educational facilities in India. With an eye on practical realities, he was keen not only to ensure the best possible teaching facilities, but also to set up lobbying organizations that would keep science and technology at the forefront of national development.

Among his achievements were the establishment of the Institute of Nuclear Physics in Calcutta (later named after him) and of the physics department in Allahabad University. He founded the periodical "*Science and Culture*", the Indian Physical Society, and the Indian National Science Academy. In 1938 his argument for the involvement of science in social planning bore fruit when he was invited to join the national planning committee for a free India run by Pandit Jawaharlal Nehru (1889–1964), the future first prime minister of an independent India.

Saha was a keen supporter of the independence movement, but later in life he found that his thinking diverged from that of the

Essential science

The Saha equation
Also known as "Saha's thermo-ionization equation" or just "the thermal ionization equation", Saha himself called it the "equation of the reaction-isobar for ionization". In this work Saha was the first person to explain the different spectra through means of the **thermodynamic** ionization of stellar matter.

When an element becomes extremely hot its **electrons** can gain so much energy that they escape from the **atom**, creating separated positive and negative **ions**, and producing **radiative energy** of varying frequencies. Saha realized that this chemical process (thermal ionization) could produce differing amounts of ions in the stellar atmosphere, thus explaining the bewildering range of **spectral lines** measured in stars.

His equation links the temperature of the star, its chemical composition, pressure inside the star's atmosphere, and the appearance of its light spectrum, meaning that any of these can be analyzed with more confidence.

Other science
Saha was early on gripped by the possibilities of the new physics, and in 1918 he and Satyendra Nath Bose wrote a paper producing an equation, based on **Albert Einstein**'s theory of relativity, that helped explain aspects of the temperature, pressure, and cubic measurement of gases.

His interest in **spectroscopy** later led him to the study of **nuclear magnetic resonance** (NMR) and its related techniques.

Legacy, truth, consequence

- Saha's thermal ionization theory of stellar atmospheres was widely accepted, and, with further refinement, the Saha equation became one of the standard, basic tools for astrophysics.
- He was a key figure in the establishment of modern physics in India. Among his many achievements were the establishment of the country's first electron microscope and first **cyclotron** (particle accelerator), and he effectively founded the study of **nuclear physics** in India.

The impetus given to astrophysics by Saha's work can scarcely be overestimated, as nearly all later progress in this field has been influenced by it and much of the subsequent work has the character of refinements of Saha's ideas.

S. Rosseland, *Theoretical Astrophysics* (1939)

new Indian government. In 1952 he was elected as an independent member of parliament, where he campaigned for science education and applied his technological expertise to flood prevention. The scheme he developed for the Damodar Valley – a series of dams instead of one large one – is still successfully working. But his health declined in the mid-1950s and he died suddenly. Although he was nominated for a Nobel Prize, he was never awarded one.

Key dates

1893	Born near Dhaka, eastern Bengal, India (modern Bangladesh).
1911	Studies mathematics at Presidency College, Calcutta (Kolkata).
1916	Appointed lecturer in department of new physics and mathematics at University College of Science in Calcutta.
1920	Wins a scholarship and visits Europe to work with leading scientists in Britain and Germany.
1921	Outlines his thermal ionization equation in *Proceedings of the Royal Society*.
1923	Becomes professor at Allahabad University, Uttar Pradesh, India.
1927	Appointed member of the **Royal Society of London**.
1938	Appointed professor and dean of the Faculty of Science at Calcutta University.
1948	Founds separate Institute of Nuclear Physics in Calcutta, later named after him as the Saha Institute of Nuclear Physics.
1952–6	Becomes a member of parliament in order to help develop science in India.
1956	Dies suddenly of a heart attack in New Delhi.

Megh Nad Saha's ionization equation, which opened the door to stellar astrophysics, was one of the top ten achievements of twentieth-century Indian science [and] could be considered in the Nobel Prize class.

Jayant Narlikar, *The Scientific Edge* (2003)

Saha's work was important for astrophysics, a branch of astronomy that studies the physical properties of celestial objects such as stars.

Satyendra Nath Bose

A Bengali Indian physicist, Satyendra Nath Bose contributed key ideas to the development of quantum science, in particular the theoretical foundations of Bose–Einstein statistics and of Bose–Einstein condensates. He also worked tirelessly to improve and advance science education in India.

Known familiarly as Satyen, Bose was the only son in a family of seven children. When he was still a toddler, a fortune-teller is supposed to have predicted that he would face many obstacles throughout his life, but would overcome them and achieve great fame. He started on this path when he was just a schoolboy, achieving a startling 110 per cent in a mathematics exam when he provided different ways of solving some of the questions.

Bose married at the age of 20, the same year he qualified for a master's degree, before going on to further studies in applied mathematics. However he never completed a doctorate, although later in life he received several honorary degrees including that of doctor of science. In 1916 he became one of the first lecturers at the new modern science and mathematics department at the University College of Science in Calcutta, where he first came across European articles on **quantum physics**. Bose and his former classmate and occasional collaborator **Megh Nad Saha** helped translate these from English and German, making them more widely available in India. Bose moved to the University of Dacca in 1921, and it was while lecturing there that he developed his landmark ideas, published in 1924 as "Planck's Law and the Hypothesis of Light Quanta".

Only a few pages long and written in English, this work suggested a new statistical approach to measuring **subatomic particles** and therefore a new way of deriving **Max Planck's** formula for **radiation**. The paper was turned down when Bose first offered it for publication to a science magazine, so, greatly daring for an unknown lecturer, Bose then sent it to **Albert Einstein**, who immediately translated it into German and arranged for its publication in the prestigious German scientific periodical, *Zeitschrift für Physik*. For both his science and for his personal support, Einstein was forever after Bose's hero.

Suddenly Bose was a star. His university funded a tour of Europe, where he worked at the Madame Curie Laboratory in Paris, and in Berlin met the great quantum physicists **Erwin Schrödinger** and **Werner Heisenberg**, as well as Einstein himself. It was Einstein who provided Bose with a reference for the post of professor at Dacca University, to which he was appointed in 1926.

Bose was deeply engaged with science education, since he firmly believed that it would benefit humanity in general. As a result he was a committed and popular teacher. Although he himself was a gifted linguist and could read papers in European languages, he felt that more people would gain an education if classes in all subjects were provided in their own language. So although at the time English was considered to be the language of scholars, he promoted Bengali science classes and wrote textbooks in the vernacular. To his surprise, he had to overcome stiff resistance to this idea.

In 1952 Bose decided to enter the political arena in order to develop and encourage science and intellectual life in India. He campaigned against caste distinctions, and was also a keen musician and composer.

Essential science

Bose–Einstein statistics

In 1900 Max Planck introduced the **quantum theory** – which states that light and other energy moves not continuously but in small packets or quanta – to explain the anomalous spectrum of energy produced by a blackbody (an object that absorbs and reemits all energy radiated upon it). But to produce his mathematical formula describing his **spectrum**, Planck still referred to classical **oscillators**.

In his 1924 paper "Planck's Law and the Hypothesis of Light Quanta", Bose took the radical step of deriving Planck's formula for **radiation** without referring to **classical physics** at all. He achieved this by adapting Einstein's theory to produce the conclusion that the quanta or energy of light behaves like particles (**photons**) as well as like energy waves. Bose treated the energy radiation from the blackbody as a collection of photons similar to a cloud of gas, but instead of considering each photon as an independent, individual particle he proposed that they should be statistically analyzed as groups of particles within defined spaces called cells.

Einstein later successfully applied this new technique to other collections of particles, and as a result the approach became known as the Bose–Einstein statistics. Bose's one, short paper also gave rise to several other features of quantum physics named after him.

■ Bose's work was a major contribution to the emerging science of **quantum statistics**.

■ The rules of the Bose–Einstein statistics only apply to subatomic particles that can exist in the same quantum or energy state at the same time within an **atom**, and can therefore cluster in groups. Light photons belong to this type of particle, which are named "bosons" in honor of Bose's original work on their behavior. (The Fermi–Dirac statistics describe the behavior of particles which cannot share the same quantum state, and which are called "fermions".)

■ Bose–Einstein statistics predicted that at very low temperatures, all the bosons within a dilute gas would cluster together at the lowest energy or quantum state, forming a very dense condensate that would behave as if it were a single atom. Called a Bose–Einstein condensate, this matter was only known theoretically until it was finally produced in a laboratory in 1995.

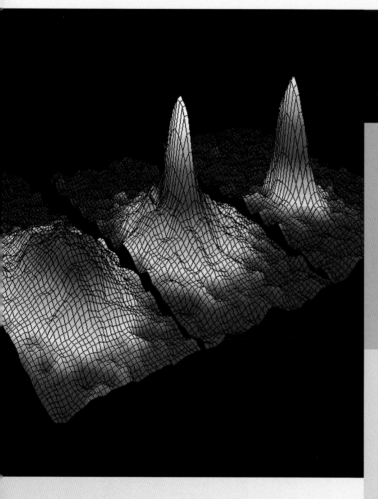

In 1995 researchers created an entirely new state of matter, predicted decades ago by Einstein and Bose, by achieving a temperature far lower than ever produced before. The picture shows three successive states of the Bose–Einstein condensation of rubidium atoms, from less dense (in red, yellow and green) to very dense (blue and white).

Respected Sir, I have ventured to send you the accompanying article for your perusal and opinion. You will see that I have tried to deduce the coefficient ... in Planck's law independent of classical electrodynamics ...

Letter to Albert Einstein (1924)

Key dates

1894	Born in Calcutta (Kolkata), India.
1916	Becomes a lecturer in physics at the University of Calcutta.
1924	Writes paper on quantum physics, entitled "Planck's Law and the Hypothesis of Light Quanta", and becomes internationally acclaimed.
1926	Appointed professor of physics at the University of Dacca.
1945	Moves back to the University of Calcutta as professor of physics.
1948	Founds an institution to promote science education in the Bengali language.
1949	Appointed president of the National Institute of Science of India.
1952	Elected a member of the upper house of parliament of India.
1956	Retires and is appointed emeritus professor at Calcutta University. Becomes vice-chancellor of Viswa-Bharati University, West Bengal.
1958	Elected a member of the **Royal Society of London**; appointed a national professor by the Indian government.
1974	Dies following a heart attack.

Other science

Even before his groundbreaking work on energy quanta, Bose had been interested in the new physics. In 1918 he and Megh Nad Saha produced an equation, based on Einstein's theory of relativity, that helped explain aspects of the temperature, pressure, and cubic measurement of gases.

In later years Bose studied a range of physical topics including the structure of crystals, **thermoluminescence** and **fluorescence**, the reflective properties of the ionosphere (the uppermost part of the earth's atmosphere where ionization caused by solar radiation affects radio waves), and **general relativity**.

Leo Szilard

Leo Szilard, the Hungarian-born nuclear physicist, is best known for initiating the American atomic energy program during World War II and helping develop the atom bomb. Horrified at what he had helped unleash, after the war he switched to biology and campaigned against nuclear weapons. He was also an inventor, anticipating and patenting a range of technical equipment.

Szilard followed in the footsteps of his father, a successful engineer, by studying engineering. His education was interrupted by World War I, when he was drafted, but – probably fortunately for him – he escaped from active duty because of influenza. In 1919 he moved to Berlin in Germany to study, where, coming in contact with physicists such as **Albert Einstein** and **Max Planck**, he switched to physics, joining the University's Institute of Theoretical Physics in 1923.

While in Berlin he filed several patents on technical inventions such as the cyclotron, the **linear accelerator**, the electron microscope, and, with Einstein, different designs for refrigerators. He later filed patents for a **nuclear reactor** and the neutron chain reaction (see below).

Szilard was Jewish and soon after the Nazis came to power in 1933, he read the writing on the wall and left Germany, settling in 1934 at the medical college of St Bartholomew's Hospital, London, where he worked on artificial **radioactive isotopes** and helped other refugees.

His interest in atomic energy was spurred when he read an article arguing that it was an impossible idea. Immediately he searched for an answer, finding one in 1939 after he had moved to Columbia University, US.

That year Szilard was the prime mover in persuading Einstein to write to President Franklin D. Roosevelt urging that the US should immediately begin work on an atomic bomb. He had realized that German scientists themselves would soon be on the track of a bomb, so he cast aside his pacifist principles, though, like Einstein, he later regretted that he had initiated a globally destructive weapon.

Einstein's letter was eventually responsible for the Manhattan Project, America's secret nuclear weapons research during World War II. Szilard was coopted onto the project in 1942, working with **Enrico Fermi** at the University of Chicago. There they built the first nuclear reactor and witnessed the first controlled nuclear reaction at the end of 1942.

Szilard soon became worried by the military's grip on nuclear science, and began to argue that the tide of war had turned and a nuclear bomb would not be needed. Later he organized an unsuccessful petition from scientists asking for the bomb to be dropped only on an uninhabited area of Japan. At one point the military head of the project wanted to lock him up.

After the war Szilard switched to **microbiology** and campaigned vigorously against the arms race.

For most of his life Szilard lived in hotels, with his scientific papers packed in suitcases ready to move at a moment's notice. Even when he married another German refugee, Dr Gertrude Weiss in 1951, he usually lived apart from her.

Essential science

Nuclear power

In 1933 Szilard began to consider how to achieve sustained nuclear power when atomic pioneer Ernest Rutherford (1871–1937) claimed it would not be possible. Within a year he conceived of a neutron chain reaction, whereby if the nucleus of an **atom** of a certain type of element could be brought to decay, or release a **neutron** particle, it might start a self-sustaining chain of reactions, with the atom breaking apart to form simpler atoms; these would in turn be forced to decay and release more than one neutron because of the particles already set loose. Every step would involve the release of energy, quickly reaching a massive reaction. As Szilard realized, the military would be interested in the explosive implications of this sort of power.

At first, most of his scientific colleagues thought the idea of a nuclear chain reaction was ridiculous. Also, no one agreed with him that such avenues were in any case too dangerous to be explored. Szilard thus continued his investigations, suspecting that the elements beryllium or indium might be suitable.

He found that a gamma ray directed at a beryllium atom did indeed cause a neutron to be emitted from the nucleus, but the chain reaction quickly fell apart.

Fission

In 1939, while at Columbia, Szilard heard about European experiments in **nuclear fission**, when a uranium atom was bombarded with one single neutron, causing the nucleus to split into two. Szilard realized this might be the source of his neutron chain reaction. Uranium has a large nucleus containing many neutrons, and if after fission it was left with at least one loose

Legacy, truth, consequence

- Szilard's ideas helped bring about the atom or nuclear bomb, as well as nuclear energy in general.

- He was proven right when he feared that a nuclear arms race between the post-war superpowers of the US and the USSR would threaten the destruction of all life on earth. Fortunately the world survived the Cold War, but the very existence of modern nuclear weapons is a matter of continuing concern.

- His campaigns for peaceful use of nuclear energy and scientific collaboration led to a series of international conferences, as well as the American Physical Society's Leo Szilard Lectureship Award, given for "*accomplishments by physicists in promoting the use of physics for the benefit of society*".

The second atomic bomb dropped on Nagasaki, Japan, on August 6, 1945. Szilard opposed the use of the nuclear bomb on civilian populations.

> *We turned the switch, saw the flashes, watched for ten minutes, then switched everything off and went home. That night I knew the world was headed for sorrow.*
>
> Szilard's comment on 1939 experiment at Columbia University proving nuclear fission of uranium

Key dates

1898	Born in Budapest, Austro-Hungary, now Hungary.
1920	Studies engineering in Berlin and switches to physics.
1923	Joins staff of the Institute of Theoretical Physics in Berlin.
1929	Writes paper foreshadowing cybernetics theory.
1933–4	Flees the Nazi regime for Vienna, then London.
1935	Joins the Clarendon Laboratory, Oxford.
1937	Moves to US to teach at Columbia University, New York.
1942	Joins the Manhattan Project.
1942	With Enrico Fermi sets up first controlled nuclear reaction.
1944	Argues against using the atom bomb.
1947	Abandons physics for molecular biology.
1956	Become professor of biophysics at University of Chicago.
1959–60	Diagnosed with bladder cancer. Designs own successful radiation therapy.
1962	Founds Council for Abolishing War (renamed Council for a Livable World) against use of nuclear weapons. During Cuban Missile Crisis takes on diplomatic mission to avoid war between US and USSR.
1963	Appointed fellow of the Salk Institute for Biological Studies, La Jolla, California.
1964	Dies in his sleep from a heart attack in La Jolla.

neutron, this could carry on bombarding the nuclei, causing them to fission again and again.

With Walter Zinn (1906–2000), Szilard reproduced the fission experiment on uranium, confirming that neutrons were emitted, and that they could produce more than one reaction.

Worried that German scientists were following the same path, he initiated Einstein's letter urging the President to develop atomic energy before Germany did. Although the American government agreed, the formal Manhattan Project was only set up in 1942, and before then Szilard had to make do with tiny grants.

He experimented with uranium-water and uranium-carbon graphite systems for sustaining a chain reaction and, working with Enrico Fermi, designed an atomic pile or "neutronic reactor" – the laboratory in which the experiments would take place – together with a reactor cooling system. In 1942, now part of the Manhattan Project, they set up and witnessed the first nuclear chain reaction.

In the laboratory Szilard carefully controlled the reaction, whereas the bombs developed and used by America in World War II allowed the chain reaction to occur without any controls.

Other work

In his 1929 paper on entropy in a thermodynamic system, Szilard introduced the information unit of a "bit", which is now the standard of computing and all its applications.

While in London, he developed the Szilard-Chalmers process with British physicist T. A. Chalmers, a method of chemically separating artificially produced radioactive isotopes, and, later in life, he studied the biology of anti-aging.

Enrico Fermi

Enrico Fermi was an Italian-born American physicist who won a Nobel Prize for his work on radioactivity. Known as both a great theorist and practitioner in physics, he was involved in the construction of the first atomic bomb while working on the US-led Manhattan Project during World War II. "Fermium", a synthetic element, is named after him.

Born and raised in Rome, Enrico Fermi was the son of a government official and he received a traditional education in the capital's public schools. From the Real Scuola Normale Superior in Pisa, he went on to the University of Pisa, where he gained his doctorate in 1922 with a thesis on **X-rays**.

Intending to broaden his horizons and develop his knowledge of theoretical physics, he spent brief spells at Göttingen University, Germany, working with physicist and mathematician Max Born (1882–1970), and at Leiden University, Holland, working with physicist and mathematician Paul Ehrenfest (1880–1933), before returning to Italy to gain his first academic position as professor of mathematical physics at the University of Florence.

In 1928, having accepting a position at Rome University the year before, Fermi initially made his name with the publication of *Introduzione Alla Fisica Atomica* (*Introduction to Atomic Physics*), the first modern physics textbook to be published in Italy. The same year he married Laura Capon, who after his death wrote *Atoms in the Family*, a heartfelt biography of their life together. Because of her Jewish ancestry, the couple became increasingly alarmed at the tide of anti-Semitism sweeping across Mussolini's fascist Italy throughout the 1930s, and so after the Nobel Prize ceremony in 1938, at which

Fermi won the physics prize for his pioneering work on **radioactivity**, the couple and their two children did not return to Italy but made their way instead to the US, where Fermi had accepted a position at Columbia University, New York.

Although settled in the United States, the family moved again in 1941, to Chicago, where projects Fermi had begun in New York resulted in the construction of a **nuclear reactor**; and from 1943 to 1945, Fermi was stationed in Los Alamos, New Mexico, where he was involved in the Manhattan Project, the organization to design and build the first atomic bomb during World War II led by the United States, the United Kingdom, and Canada.

Though known to be a charming man with a great joy for life, Fermi was very dedicated to his research and had few interests outside physics. He was not interested in music or art and read fiction rarely, although he did make an exception for the novels of Aldous Huxley and H. G. Wells.

After the war, Fermi adopted US citizenship and returned to Chicago to continue his research, completing important work in particle physics. However, his health began to deteriorate and, despite his determination to carry on with his professional commitments, he eventually died of cancer in 1954.

Essential science

Radioactivity

Fermi's research on radioactivity throughout the 1920s, which eventually earned him the Nobel Prize for Physics, provided further evidence of the existence of the "neutrino", meaning "little neutron" in Italian, and referring to elementary particles created as a result of radioactive decay. First hypothesized by Wolfgang Pauli (1900–58), neutrinos are hard to detect because their mass is extremely low and they possess no electric charge. Fermi found that the "beta decay" (radioactive decay caused by a weak nuclear force), which occurs in the unstable nuclei of radioactive elements results from the conversion of a **neutron** into a **proton**, an **electron** (beta particle), and an antineutrino (the antiparticle equivalent of the neutrino, but with opposite charge and magnetism).

Another of Fermi's major contributions was to produce new radioactive **isotopes** by neutron bombardment, and the discovery that a block of paraffin wax or a jacket of water surrounding the neutron source produces thermal neutrons that are more effective at producing such elements.

The nuclear reactor and the atomic bomb

Once he arrived in the US, Fermi continued his work on radioactivity by constructing the first nuclear reactor. At that time this was known as an "atomic pile" because it involved a moderator (a substance used in a nuclear reactor to decrease the speed of fast neutrons and increase the likelihood of fission) consisting of a pile of purified granite blocks with drilled holes for the control rods of enriched uranium.

Legacy, truth, consequence

- The nuclear reactor that Fermi and his team built in 1942 is now considered to be the ancestor of all nuclear weapons and all nuclear power plants.
- Element number 100, "Fermium" ("Fm"), discovered in 1955, was named in his honor.
- In his later years he considered a problem now known as the "Fermi paradox". In conversation on the topic, around 1950, he is supposed to have quipped "*But where is everybody?*" since, according to the Fermi paradox, the size and age of the universe suggest that somewhere there must be other life forms that could have developed the technology to make contact with humans and yet we have no **empirical** evidence of their existence. Thus the question: where are they?
- See pages 164–5 for the Fermi–Dirac statistics.

A sketch of the Chicago Pile-1 (CP-1), the first artificial nuclear reactor.

Fermi's team built the first artificial nuclear reactor, known as Chicago Pile-1 (CP-1), on the squash courts under the stadium at Chicago University. On December 2, 1942, the control rods were removed for the first time and the reactor became operational. This was the first manmade, controlled, self-sustaining nuclear reactor, which means that it could be stopped and started according to the constructors' wishes. Fermi's famously meticulous attention to detail was indispensable in bringing this hugely significant task to fruition.

The work done by Fermi and his colleagues, including physicists Arthur Compton (1892–1962) and **Leo Szilard**, was eventually incorporated into the Los Alamos branch of the Manhattan Project, directed by Robert Oppenheimer (1904–67). Fermi was involved in this operation during the latter stages of its development, mainly as a general consultant, and he oversaw the

The fact that no limit exists to the destructiveness of this weapon [i.e. the atomic bomb] makes its very existence and the knowledge of its construction a danger to humanity as a whole ... For these reasons, we believe it important for the President of the United States to tell the American public and the world what we think is wrong on fundamental ethical principles to initiate the development of such a weapon.

Fermi and I. I. Rabi, "Minority Report of the General Advisory Committee", United States Atomic Energy Commission: In the Matter of J. Robert Oppenheimer: Transcript of Hearing before Personnel Security Board, Washington, D. C. (1954)

construction of an atomic bomb employing similar principles to those used in Chicago.

In later years, in particular in 1949, he opposed the development of the hydrogen bomb on moral and technical grounds, most notably co-penning a report for the General Advisory Committee of the Atomic Energy Commission with physicist Isidor Rabi (1898–1988), despite the fact that he had worked on the hydrogen bomb at Los Alamos for the Manhattan Project.

Werner Heisenberg

One of the pioneers of the field of quantum mechanics, Heisenberg was recognized in his own lifetime as one of the world's greatest scientific minds. He was also a philosopher of science and developed the uncertainty principle, which is named after him and is one of the central concepts in modern physics.

Like some other great scientists, as a young student Heisenberg excelled in mathematics and theoretical physics, but he did so poorly in other subjects that he nearly failed to gain his doctorate: during the oral exam he was unable to explain to his experimental or laboratory physics lecturer how a simple battery works.

He wavered between subjects before finally concentrating on **atomic structure**, a field where experimental results did not match with the theory. He was only in his early twenties when, on holiday recovering from hay fever, he invented **matrix mechanics**, the first mathematical formulation to describe some of the concepts of **quantum mechanics**, later developing the concept with Max Born (1882–1970) and Pascual Jordan (1902–80).

Two years later, after further observations of the atom, Heisenberg proposed his revolutionary **uncertainty principle**, arguing that measurements of certain aspects of quantum mechanics contain inbuilt uncertainties, and that it is not possible to predict the path of a particle. He claimed that: *"The path comes into existence only when we observe it."*

At first Heisenberg's mentor **Niels Bohr** did not accept the principle, and according to one report they argued so intensely that the young Heisenberg burst into tears. Later, however, Bohr and his colleagues did adopt uncertainty into their mainstream Copenhagen Interpretation of quantum physics.

Almost from the moment the Nazi party took power in Germany, Heisenberg was embroiled in political controversy. In 1935 he was accused of being a "white Jew" for his work on quantum physics, which the Nazis saw as "Jewish science" instead of "Aryan science". At the request of Heisenberg's family, the police chief, Heinrich Himmler, intervened to stop personal attacks on Heisenberg, but his appointment as professor of theoretical physics in the University of Munich was blocked. Later, however, Heisenberg was brought into the German atomic research group during World War II.

He was the one man that the Allies' team of great physicists feared would be able to build an atomic bomb all on his own. He was seen as so important that the Allies plotted to kidnap or even kill him. However, in 1942 Heisenberg convinced the Nazi government that he would not be able to make a bomb. Many of his biographers believe that he deliberately misled the Nazis, and had already made the correct calculations necessary for a bomb. In support of this they quote cryptic comments by scientists smuggled out of Germany, and his 1941 visit to Niels Bohr in German-occupied Denmark: although he should not have been discussing top-secret projects at all, he may have tried to drop hints that he would not complete a bomb.

After the war Heisenberg worked to reestablish relations between German and international scientists. He supported peaceful uses of atomic energy, and was one of the founders of CERN, the pan-European nuclear research council, in Geneva, Switzerland.

> **The more precisely the position is determined, the less precisely the momentum is known in this instant, and vice versa.**
>
> Paper on Uncertainty (1927)

Essential science

Matrix mechanics

Heisenberg invented matrix mechanics when he was trying to discover a mathematical calculation to explain the spectral lines (or frequencies of light) given off by the movement of particles within an atom. He decided to ignore the electron orbits, which could not be observed, and instead to work backwards from the only quantities that could be observed – the frequencies and intensities of light absorbed and emitted by a particle. His resultant mathematics represented the position and momentum of the particle as a matrix of coefficients, indexed by the start and end energy levels, and used established mathematical matrix rules covering arrays of numbers, which could then be applied to produce an equation.

The Heisenberg uncertainty principle

Also known as the principle of inexactness or indeterminacy, this states that it is possible to measure a particle's position or its momentum, but not both at the same time, and therefore it is not possible to exactly predict the particle's path, or where it is going to go. Heisenberg's explanation is that since an observation involves the exchange of a **photon** to create the observed data, the

Legacy, truth, consequence

■ **Quantum mechanics:** Matrix mechanics was soon joined by **wave mechanics** as a rival approach to understanding the behavior of sub-atomic particles. At first most physicists queried matrix mechanics, preferring the newer wave mechanics since the math was simpler and less abstract, and it offered a clear visual model. In fact the systems were later shown to be mathematically equivalent, and they are both useful methods for exploring quantum mechanics.

■ **Observer-related universe:** Heisenberg's uncertainty principle contributed to theories of a non-causal or indeterministic universe, which states that, at the sub-atomic level, science can only suggest probabilities, not certainties. Based on this, and on other concepts such as Schrödinger's Cat (see pages 146–7), new physics includes the concept that when a scientist observes or measures an event, that act actually fixes the event, which, up to that point, had not been determined. In this respect, reality is considered to be "chance", determined by the behavior of the observer – a completely new view of the universe compared to the classical view that reality is fixed and determined. Heisenberg's work has given rise to many philosophical questions to do with the basic nature of the universe.

■ **Copenhagen Interpretation:** His theories were a part of the 1927 Copenhagen Interpretation of quantum physics, proposed by Niels Bohr and his colleagues, which came to be accepted as the paradigm. This interpretation accepted indeterminacy in the form of the uncertainty principle along with the fact that wave mechanics only gives a probability of locating where the particle is.

The smallest units of matter are not physical objects in the ordinary sense; they are forms, ideas which can be expressed unambiguously only in mathematical language.

Comment after conversations with Rabindranath Tagore, as quoted in *Uncommon Wisdom: Conversations With Remarkable People* (1988) by Fritjof Capra

The Fifth Solvay International Conference, October 1927, where the world's most notable physicists met to discuss the newly formulated quantum theory. Werner Heisenberg is the third from the right, back row.

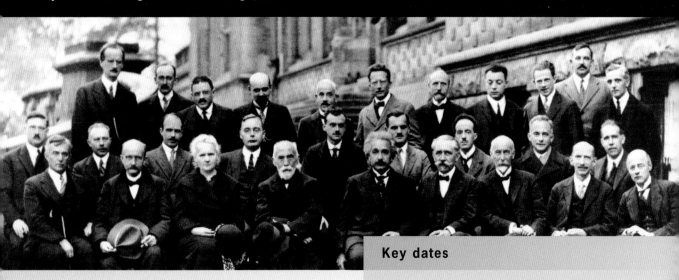

energy or wave state of the observed object changes when observed, collapsing into an unpredictable and random position.

Since **subatomic particles** are the basis for all matter, this principle is extended to say that there is an inbuilt uncertainty in examining the basic physics of matter and radiation. Such indeterminism clashes with the causal or **deterministic** nature of **classical Newtonian physics**. However, at a macroscopic or everyday level, Heisenberg argued that the errors are so small they do not have an impact; it is only at the atomic or sub-atomic level that the indeterminacy becomes critical.

Key dates

1901	Born in Würzburg, Germany.
1922	Begins long-term collaboration with Niels Bohr in Copenhagen, Denmark.
1925	Invents matrix mechanics as a formulation of quantum mechanics.
1926	Works with Bohr and others to formulate the Copenhagen Interpretation of quantum physics.
1927	Formulates his uncertainty principle.
1932	Wins the Nobel Prize in Physics.
1939	Coopted into the Nazi atomic bomb/energy program.
1957	Opposes nuclear armament in Germany.
1976	Dies of cancer in Munich, Germany.

Linus Pauling

Regarded as one of the most important chemists of the twentieth century, Pauling was awarded the Nobel Prize for Chemistry in 1954. He helped found the field of molecular biology, and completed important research into the properties and use of vitamin C to aid human health. A peace activist as well as a scientist, and a respected public speaker and writer, he received the Nobel Prize for Peace in 1962, becoming the first person to win two unshared Nobel prizes.

Pauling's fascination with chemistry began when he was 14 years old and saw his friend's chemistry set. After witnessing impressive chemical reactions, he straightaway set up his own laboratory in the basement of his Oregon home. In 1917 he began a course on **chemical engineering** at Oregon Agricultural College (now Oregon State University). Before long, and still an undergraduate, he was asked to teach chemistry to other undergraduates, such was his advanced knowledge of the field.

At college Pauling turned his attention to chemical theory. He wanted to discover how **atoms** bond to form **molecules**. In 1922 he entered the California Institute of Technology (Caltech), where Professor of Chemistry Roscoe Dickinson (1894–1945) taught him how to study the structures of crystals using **X-rays**. After receiving his doctorate in 1925, he went to Europe on a Guggenheim Fellowship. Over the next 18 months he learnt a great deal from a group of scientists working in the field of **quantum mechanics**, and would later apply quantum mechanics to his work on the chemical bond, resulting in his influential and now famous publication, *The Nature of the Chemical Bond and the Structure of Molecules and Crystals*, completed in 1939.

From the age of 38, Pauling was head of chemistry at Caltech. In the 1930s his research focused on the structure of large **biomolecules**. Breaking off from this to help develop explosives

and shells for the National Defense Research Commission during World War II, he returned to Caltech after the war and in 1949 his own team's research led to the discovery of the molecular basis of sickle-cell anemia. In the 1950s, using X-ray diffraction (which involves the scattering of X-rays by the lattice of atoms or molecules in a crystal, and measuring the diffraction pattern yielded to gather information about the structure of the crystal), he was the first to recognize the three-dimensional helical structure – the alpha helix – in many **proteins**. However, he was less successful in producing an accurate model for the structure of **DNA**.

Influenced by his pacifist wife, Ava Helen Pauling, a great deal of his time was spent in the 1960s and 1970s on peace activism and on encouraging the scientific world to call for a ban on nuclear bomb testing.

During the 1970s and 1980s his scientific work focused on orthomolecular medicine (see below), in which he thought that vitamin C was an essential molecule for optimal health. He encouraged people to take large daily amounts. This, and his publication of a popular science book on the subject, brought great criticism from other scientists. However, as an effective communicator of scientific concepts, and now focused on human health, he found success and popularity in writing more science books for the public. *How to Live Longer and Feel Better* became a best-seller in 1986.

Essential science

Protein research

In the early 1930s the study of protein structures faced many difficulties – the structures appeared to be huge, complex, and fragile. Pauling used his model-building approach, which involved focusing first on the structures of the building blocks of the molecules, and then on how they are linked, and finally building a model to test his findings. In his study of protein molecules, the team began by examining the structures of some of the amino acids, which are the building blocks of protein molecules, after which they established a theory that the amino acids were bonded, end-to-end, to form a rigid structure. The work resulted in the team's groundbreaking papers on the structures of several proteins and on the structure of the common component of most proteins – the alpha helix.

Molecular disease

After learning from a medical expert that sickle-cell anemia was caused by the fact that the sufferers' red blood cells were twisted from normal disc shapes to sickle shapes, Pauling set about examining the contents of red blood cells – hemoglobin. After a year, he and his team made an astonishing discovery when they used a method to separate hemoglobin molecules by their electric charge. Sickle-cell hemoglobin molecules had more electric charge than normal hemoglobin molecules. That a potentially deadly illness could be caused by such a slight difference in the molecules gained great attention, and led the way for important research into "molecular diseases". Pauling himself extended his research on molecular diseases by studying mental illnesses that appeared to be hereditary, to see if they could be caused by

- In finding the structural defect in blood cells that caused sickle-cell anemia, Pauling and his team discovered the first "molecular disease", and so began an important branch of research. Work by Pauling's colleagues later showed that the disease was inherited, which cemented a vital link between the fields of molecular medicine and **genetics**.

- Pauling is often described as the "founding father of **molecular biology**" because of his methods and discoveries in the field of molecular structures.

- His attempts to find the structure of DNA were limited by poor X-ray evidence. **James Watson** and **Francis Crick**, who identified the double helix of DNA in 1953, proved his theory of a three-chain helix was incorrect.

- Since Pauling's book on the benefits of vitamin C, studies on the effects of dietary supplements have been ongoing. Dietary supplements are now a major industry and commonly purchased by millions. Pauling co-founded The Linus Pauling Institute of Science and Medicine to conduct research on orthomolecular medicine. The Institute today works in part to "*determine the function and role of vitamins and essential minerals (micronutrients) and chemicals from plants (phytochemicals) in promoting optimum health and preventing and treating disease*".

1901	Born in Portland, Oregon, US.
1925	Receives his doctorate in chemistry at the California Institute of Technology.
1931	Becomes professor of chemistry at the California Institute of Technology.
1939	Publishes the results of ten years' research in *The Nature of the Chemical Bond and the Structure of Molecules and Crystals*.
1942	Begins work in various departments of the National Defense Research Commission.
1950	Builds the first accurate model of a protein molecule.
1954	Receives the Nobel Prize in Chemistry for "*his research into the nature of the chemical bond and its application to the elucidation of the structure of complex substances*".
1956	Begins research on mental illnesses.
1958	Presents petition to end nuclear testing to UN Secretary. The petition has over 9,000 signatures of scientists from around the world.
1963	Receives the Nobel Peace Prize (for 1962).
1970	Pauling's ideas on the benefits of vitamin C are published in *Vitamin C and the Common Cold*.
1986	*How to Live Longer and Feel Better* is published, a best-seller outlining Pauling's views on nutrition and health.
1994	Dies in Big Sur, California.

> *I believe medicine is just now entering into a new era ... when scientists will have discovered the molecular basis of diseases, and will have discovered why molecules of certain drugs are effective in treatment, and others are not effective.*

Interview with Portland *Oregonian* (February 13, 1952)

A computer-generated model of the molecular structure of L-ascorbate (vitamin C).

molecular defects. However, his five years' work did not provide specific results, although his studies were an opportunity to learn more about psychiatry and general health.

Orthomolecular medicine

Pauling gave the term "orthomolecular medicine" to the concept that optimal physical and mental health can be acquired by making sure "*the right molecules in the right amounts*" are in the body. He believed that if the body had the right balance of chemicals, the necessary chemical reactions for health could be optimized. In 1967 he first used the phrase "orthomolecular" in relation to his belief that conditions like schizophrenia could be treated with nutrients such as niacin. On biochemist Irwin Stone's (1907–84) recommendation, he tested the effects of large doses of vitamin C on himself, and found he had fewer colds. After further studies, and in spite of criticisms from other scientists, his ideas were printed in what came to be a bestseller, *Vitamin C and the Common Cold*.

Kurt Alder

A joint recipient of the Nobel Prize for Chemistry, Kurt Alder was a German chemist who managed to overcome the obstacles that hindered original research in the chaotic Germany of his time. He is most well known for co-discovering a process for the synthesis of complex organic compounds called the Diels–Alder reaction and for his pioneering work in the rubber industry.

Kurt Alder, the son of a schoolteacher, was born and brought up in an industrial region of pre-World War I Prussia, during a time of great political upheaval and uncertainty. He studied chemistry at the University of Berlin and later at Kiel, where he was taught by the German chemist Otto Diels (1876–1954). Under the guidance of his mentor, Alder produced a doctoral thesis entitled "On the Causes of the Azoester Reaction". A little more than a year later, their working collaboration led to pioneering work as developers of the Diels–Alder reaction (also called "diene synthesis") – a method for synthesizing types of organic compounds.

In the 1930s, still working alongside Diels, Alder attained a professorship at Kiel, but in 1936 he took advantage of his expertise in plastics development by becoming a research director at the world's largest chemical establishment, I. G. Farbenindustrie. There he directed the development of synthetic rubber – a worldwide target of organic chemists at the time. After the outbreak of World War II Alder moved to the University of Cologne to take up an appointment as professor of chemistry. Despite the difficult war-torn circumstances surrounding him at that time, he was able to stay focused on his research.

In his lifetime Alder produced over 150 papers on the synthesis of organic compounds. He was the recipient of many awards and honorary degrees. His diligent and systematic approach to work was ultimately rewarded eight years before his death when he and Diels jointly received the Nobel Prize for Chemistry. In 1955 he joined other Nobel laureates in their appeal to world governments to end war.

Essential science

Synthesis of organic compounds

Inorganic compounds are compounds without the structure or characteristics of matter from living organisms: in chemical terms, they are compounds that do not contain hydrocarbon groups (groups of carbon and hydrogen atoms). By contrast, organic compounds (derived from organic matter, or living organisms), contain groups of carbon and hydrogen atoms with chemical bonds of a "covalent character", i.e. the carbon–hydrogen (C–H) bonds share electrons between the two atoms.

Chemists at the beginning of the nineteenth century were interested in forming compounds from simpler compounds or elements, a process known as synthesis. At the time it was believed that the structure of organic compounds (derived from living organisms), was too complicated to be synthesized artificially from non-living matter, a belief generated from the entrenched view of scientists of this era that life processes arise from a non-material "vital force", which cannot be explained in terms of physical or chemical phenomena. As a consequence, investigations from this period were primarily focused towards the seemingly more promising synthesis of "inorganic" compounds.

Organic chemistry began in earnest when scientists realized that organic compounds could be created in a laboratory, without the need for any kind of "vital force", and thus treated in ways similar to inorganic compounds. So, for example, the French chemist Michel Eugène Chevreul (1786–1889) began to study fatty acids from organic sources (animal fats), and showed that it was possible to make a chemical change in fats and produce new compounds without needing anything like a "vital force". He is credited with discovering margarine. The possibility of new products beckoned and chemists such as William Henry Perkin (1838–1907), who manufactured an organic dye ("Perkin's mauve"), made large amounts of money, helping to fuel the interest in organic chemistry.

Crucially, in 1858, the work of German organic chemist Friedrich August Kekulé (1829–96) and Scottish chemist Archibald Scott Couper (1831–92) helped to develop the concept of chemical structure. Both men, working independently, suggested that certain carbon atoms could link to each other to form a carbon lattice, and the bonds between atoms within a substance may be revealed through appropriate chemical reactions. This led to the discovery of petroleum and the birth of the petrochemical industry, with the success story of artificial rubbers and plastics that followed.

From the beginning of the twentieth century, organic chemistry went from strength to strength as chemists searched for new compounds to synthesize.

Legacy, truth, consequence

- The Diels–Alder reaction has been used to amalgamate steroids such as cortisone and reserpine, morphine, insecticides, and many other polymers (substances with a structure composed of small, identical bonded molecules) and alkaloids.
- The synthetic rubber, "Buna", that Adler had a pivotal role in developing, became commercially available during World War II and, though its manufacture was not without difficulties, it was a benefit to the German war effort when supplies of natural rubber ran out. The Americans also made use of it.
- At the ceremony for the Nobel prize awarded to Alder and Diels, they were praised for their "*discovery of great theoretical importance and of enormous practical consequences*". The speaker went on to say that their contribution showed that "*German scholarship is emerging from the wreck of the last world war*".

Key dates

1902	Born in Königshütte, Prussia (now Chorzów in Poland).
1922	Begins studies in chemistry in Berlin.
1926	Receives a chemistry doctorate from the German University of Kiel.
1928	Publishes a paper, with Otto Diels, following their work on diene synthesis.
1934–6	Teaches at the University of Kiel as professor of chemistry.
1936	Works as a research director for I. G. Farbenindustrie in Leverkusen in western Germany. Instigates crucial core research into the development of plastics.
1940	Becomes director of the chemical institute at the University of Cologne, as well as the university's professor of chemistry.
1943	Discovers the ene reaction.
1950	Receives the Nobel Prize for Chemistry jointly with Diels, for their discovery and development of the diene synthesis.
1958	Dies in Cologne, in former West Germany.

It appears to us that the possibility of synthesis of complex compounds related to or identical with natural products such as terpenes, sesquiterpenes, perhaps also alkaloids, has been moved to the near prospect.

Article in *Justus Liebig's Annalen der Chemie*
(1928, with Otto Diels)

A synthetic rubber production plant, 1941.

The Diels–Alder reaction

The Diels–Alder reaction or diene synthesis, which became Alder's greatest contribution and most important work, proved to be especially important in the production of synthetic rubber and plastics in the twentieth century.

The method developed in 1928 by Alder and Diels involved combining organic compounds with two carbon-to-carbon double bonds (strong chemical bonds characterized by the sharing of electrons between atoms), known as "dienes", with "dienophiles" (named after their affinity to react with dienes). Dienes and dienophiles are often gases, and the reaction therefore usually took place inside a sealed container at elevated pressure and temperature. The process not only produced syntheses of many cyclic organic substances (compounds with atoms arranged in a ring or closed-chain structure), but also revealed the molecular structure of the products obtained.

Paul Dirac

Considered by some scientists to be the unsung hero of physics, Dirac is not a household name, but he was a brilliant theoretical mathematician whose ideas of quantum theory had a significant impact on many branches of modern physics.

The British theoretician Paul Dirac had a stern and socially isolated upbringing. His Swiss father, a French teacher, insisted the family should speak French at dinner, and since the young Paul could not comfortably express himself in that language, he became taciturn and silent. He remained quiet, terse, modest, and shy throughout his life, but, on the other hand, he was known for expressing mathematical concepts in a clear, elegant, and eloquent way.

Initially Dirac studied electrical engineering, but switched to mathematics, settling at the University of Cambridge in 1923. Two years later he began to investigate the statistical behavior of **quantum particles**, and over the next few years he produced some fundamental contributions to **quantum mechanics** and the emerging **quantum electrodynamics**.

From 1926 onwards he made many research trips or lecture tours, working with **Niels Bohr** in Copenhagen; with Robert Oppenheimer (1904–67), Max Born (1882–1970), and others in Göttingen; visiting Japan with **Werner Heisenberg**, and making several trips to the USSR. During World War II he worked on identifying fissionable isotopes, the foundation work for building an atom bomb, but he refused to join the next stage of the bomb project in America. Even so, the British government worried that he might discuss atomic secrets with Russian scientists, and

banned him from returning to the USSR until 1957. Because of his suspected Marxist leanings he was also refused an entrance visa by the US in 1954, although one was granted later on.

After the war Dirac worked on a wide range of physical and mathematical problems, always arguing that the mathematics should be elegant as well as logical.

Dirac produced about 200 scientific papers containing an astonishing number of original ideas. He relied on intuition and an inbuilt sense of elegant mathematical logic, and his theories were always recognized as important by the world's great physicists, even though he shrank from publicity and was not known outside his own field. He even wanted to turn down the award of a Noble Prize, until he was warned that he would get a great deal more media attention if he did actually refuse it. Niels Bohr said of him: "*Of all physicists, Dirac had the purest soul.*"

> ## A theory with mathematical beauty is more likely to be correct than an ugly one that fits some experimental data.
>
> Scientific American interview (1963)

Essential science

General theory of quantum mechanics

In 1926 Dirac supplied the first general theory of quantum mechanics combining the two mathematical approaches of **matrix** and **wave mechanics** into one logically satisfying algebraic solution. He "proved" this formula by using it to correctly predict the energy levels of a hydrogen **atom**.

Fermi–Dirac statistics

One of the many puzzles of quantum physics was that the fundamental particles whose spin or angular momentum is half an integer (whole number) followed different statistical rules from those particles whose spin was a full integer. Dirac applied his own formulation to this issue, and worked out the statistical rules. The Italian physicist **Enrico Fermi** had independently begun to investigate the statistical behavior of quantum particles just before

Dirac began his work, so the rules are named in honor of both the scientists. Always retiring, Dirac insisted that the particles that follow these rules were called "fermions".

Quantum electrodynamics

In 1927 Dirac launched the science of quantum electro-dynamics or quantum field theory by applying quantum theory to the electromagnetic field. In particular, he introduced the idea of "second quanti-zation", treating a particle itself as an oscillator producing a wave-function.

> ## Physical laws should have mathematical beauty.
>
> Answer to a question about his philosophy of science, Moscow (1933)

Legacy, truth, consequence

■ In what is praised as a triumphant example of logical scientific prediction, Dirac's theory of antiparticles was confirmed in 1932 when Carl D. Anderson (1905–91) experimentally identified the antielectron, an electron with a positive charge, later called the **positron**. Dirac therefore laid the foundations for the study of **antimatter**.

■ Fermi–Dirac statistics are widely used to determine the distribution of electrons at different energy levels.

■ The Dirac equation can be used to correctly predict the spin of an electron, an intrinsic property that had only recently been identified, as well as other aspects of the particle's behavior, such as its magnetic moment, or turning point.

■ His book *The Principles of Quantum Mechanics* (1930) soon became a classic textbook that is still used today, a testimony to its clear logic and simple though abstract style. It offered the first overall view of quantum mechanics, bringing together Werner Heisenberg's matrix mechanics and **Erwin Schrödinger**'s wave mechanics into a cohesive description of events at the **subatomic** level.

■ His work exploring other potentials of science was a foundation for modern concepts such as **string theory** and **M-theory**.

■ His clear and precise logic inspired younger scientists, particularly **Richard Feynman**.

■ Some of Dirac's other theories have not yet been proved, such as his prediction that there should be a single **magnetic monopole** (although there is some evidence hinting at its existence).

Key dates

1902	Born in Bristol, England.
1926	Gives the first complete mathematical solution of quantum mechanics.
1928	Introduces his equation that finds a connection between relativity and quantum mechanics.
1930	Publishes the textbook *The Principles of Quantum Mechanics*, the first full overall view of quantum mechanics.
1931	Announces his theory of antiparticles, introducing the concept of antimatter.
1932	Only 31, becomes Lucasian Professor of Mathematics at the University of Cambridge.
1933	Wins the Nobel Prize for Physics, shared with Erwin Schrödinger.
1971	Moves to the US to become research professor of physics at Florida State University.
1984	Dies in Tallahassee, Florida, US.

The Dirac equation

Published in 1928, this equation incorporated both the quantum theory and the theory of **special relativity** and therefore described the behavior of electrons at any speed up to the speed of light. Dirac provided an interpretation that allowed scientists to relate the processes of the quantum world to the observations of the laboratory, where **classical physics** still provided the explanation.

Antiparticles

Dirac continued to explore the implications of his equation, particularly the paradoxical conclusion that energy could have a negative value. He eventually realized that particles had to have opposites or antiparticles, so that the negatively charged electron should have a positively charged opposite. At first most scientists rejected or laughed at this theory.

> *People had got used to the determinism of the last century, where the present determines the future completely, and they now have to get used to a different situation in which the present only gives one information of a statistical nature about the future … it is certainly the best that we can do with our present knowledge.*

The Development of Quantum Mechanics (1972)

St John's College, Cambridge, where Paul Dirac studied and later became Lucasian Professor of Mathematics.

Zhang Yuzhe

Considered to be the father of modern Chinese astronomy, Zhang Yuzhe (also known as Y. C. Chang) was one of the promising Chinese students who were sent abroad to bring advanced science and technology back to China. His work helped lay the foundation for China' to become a spacefaring nation.

Zhang Yuzhe is the modern Pinyin spelling, but the old Wade-Giles transliteration of his name was Chang Yu-Che, so while studying in America he used the name Y. C. Chang.

Zhang was able to go to America solely because of the Boxer Rebellion of 1899–1901. This uprising by poor, dissatisfied Chinese against foreign missionaries and traders – who were blamed for China's humiliation and poverty – was supported by the Imperial government. But a small military force of foreign allies easily defeated the technologically backward Chinese army and peasant Boxers, and China had to pay a massive compensation of US $330 million, of which America's share was seven per cent – more than the US had originally claimed. Eventually the US government agreed to put aside the extra in an Indemnity Fund to allow Chinese students to go abroad and learn the modern science and technology China so desperately needed.

This Indemnity Fund not only paid the travel and fees for students but also set up Qinghua (Tsinghua) University, which Zhang attended, as a preparatory institute before the scholars went to America.

Back in China, Zhang applied his expertise to help make the Purple Mountain Observatory a major astronomical institute. In 1955 he received the country's greatest scientific honor with election to the Chinese Academy of Science, the state's central research organization.

Essential science

Asteroids

Zhang early on began to specialize in following and calculating the orbits of comets and minor planets. While at the University of Chicago in 1928 he discovered a new asteroid, assigned the number 1125, which he named China. Unfortunately, it dropped out of sight before its orbit could be traced, so it was declared "lost" and the name was made available again.

While he was director of the Purple Mountain Observatory, China's first modern observatory, his team discovered another new asteroid in 1957, which they gave the same designation as Zhang's lost one, 1125 China. However, in 1986 his original asteroid was rediscovered, so it was renamed 3789 Zhongguo (the Chinese name for their country, literally "Middle Kingdom").

During his directorship the Observatory discovered several other asteroids as well as three new comets.

Legacy, truth, consequence

■ Zhang was one of a generation who rebuilt science in China after its long decay at the end of the Imperial era. Their knowledge also provided the foundation of a scientific structure in the new People's Republic after 1949.

■ His work contributed to the launches of China's first space satellite in 1970 and first manned space flight in 2003.

■ The asteroid 2015 Chang was named after him.

In transforming a backward agricultural China into an advanced industrial country, we are confronted with arduous tasks ...

Chairman Mao on Chinese students such as Zhang Yuzhe (speech in 1955, quoted in *The Private Life of Chairman Mao* by Li Zhisui, 1994).

Key dates

1902	Born in Minhou, Fujian province, China.
1919	Attends Qinghua (Tsinghua) University, Beijing.
1925	Goes to the University of Chicago, US, for postgraduate studies.
1928	Discovers his first new asteroid.
1941–50	Leads astronomy research institute at National Central University, China.
1946–8	Returns to US to study aspects of stars.
1950–84	Directs Purple Mountain Observatory, near Nanjing.
1955	Appointed to Chinese Academy of Sciences.
1986	Dies in China.

Other work

Zhang contributed to calculating the past orbit of Halley's Comet, tracking it back to 1057 BCE.

He studied the rotation of asteroids and, in 1957, wrote a paper with Zhang Jiaxiang on the orbits of artificial satellites.

Gregory Goodwin Pincus

Gregory Goodwin Pincus was an American endocrinologist best known for his work on the development of the contraceptive pill and for his important contributions to knowledge of the effects of steroid hormones and of mammalian reproduction. His research transformed family planning throughout the world.

Pincus studied science at both Cornell University and Harvard University. In 1927 he gained his doctorate from Harvard, under the tutelage of geneticist William E. Castle and animal psychologist W. J. Crozier. From 1929 to 1930 he was at Cambridge University, UK, working with biologists F. H. A. Marshall and John Hammond, before moving on to the Kaiser Wilhelm Institute where he worked with geneticist Richard Goldschmidt.

During World War II, Pincus conducted research on stress for the US Navy and Air Force, and in 1944 he and Hudson Hoagland co-founded the Worcester Foundation for Experimental Biology. After the war he worked at Tufts Medical School before moving to Boston University. In 1967 he died in Boston from myeloid metaplasia, a rare blood disease thought to be a result of his work with organic solvents.

> **Objective appraisal is surely but slowly replacing heated partisanship.**
>
> On birth control, in *The Control of Fertility* (1965)

Essential science

The oral contraceptive pill

Throughout the 1930s Pincus had been interested in mammalian fertilization and development. In the 1940s he focused specifically on the effects of steroid hormones in general **physiology**, especially in reproduction. This all came to fruition in 1951 when Pincus and Min Chueh Chang (1908–91) studied the effects of various newly synthesized hormones on reproduction in laboratory animals. They found that several progestational compounds when administered orally would prevent pregnancy by inhibiting ovulation. With John Rock (1890–1984), Pincus made studies on humans and produced the oral contraceptive pill.

Experiments began on hundreds of women in Puerto Rico and Haiti in 1956 with great success. The following year, the Food and Drug Administration authorized the sale of the pill for miscarriages and menstrual disorders. In 1960 the Federal Agency licenced Enovid, a contraceptive pill. Tests showed that the pill was 100 per cent effective. Pincus and Chang also worked on another pill, known colloquially as "the morning after pill", which affected the female egg after ovulation.

Legacy, truth, consequence

- The invention of the contraceptive pill constituted a revolution in family planning and helped to curb the problem of overpopulation throughout the world.
- Together with his associates, Pincus published around 350 papers on tropism in rats; genetics of mice; fertilization and transplantation of eggs; diabetes; cancer; schizophrenia; adrenal hormones; and ageing.
- He was a major figure in mammalian reproductive physiology and **endocrinology** for more than 30 years.
- Once it was founded, the Worcester Foundation for Experimental Biology became an internationally recognized center for the study of steroid hormones and mammalian reproduction.

Key dates

1903	Born in Woodbine, New Jersey, US.
1927	Receives a doctorate in science from Harvard University.
1929–30	Studies at Cambridge University, England, and the Kaiser Wilhelm Institute of Biology in Germany.
1936	Publishes *The Eggs of Mammals*.
1940	Publishes "The Comparative Behavior of Mammalian Eggs *in vivo* and *in vitro*" in *Proceedings of the American Philosophical Society*, coauthored with H. Shapiro.
1951	Appointed Professor at Boston University. Co-invents the oral contraceptive pill.
1956	Experiments with the pill in Puerto Rico and Haiti. Publishes "Studies of the Physiological Activity of Certain 19-Nor Steroids in Female Animals" in *Endocrinology*, coauthored with M. C. Chang et al.
1957	Food and Drug Administration authorize the sale of the pill.
1960	The Federal Agency licences Enovid.
1965	Publishes *The Control of Fertility*.
1966	Publishes "Control of Conception by Hormonal Steroids" in *Science*.
1967	Dies in Boston, Massachusetts, US.

Barbara McClintock

Barbara McClintock was a pioneering geneticist. Her research led to the discovery of transposable genetic elements (genetic material, able to move from one place to another on a chromosome or between chromosomes). Initially ignored by many scientists, her research was finally recognized as advances in genetics proved her findings correct. In 1983 she received the Nobel Prize for her ground-breaking work.

Barbara McClintock was born in Connecticut, US, at a time when her father, a physician, was struggling to establish his medical practice, so she was sent to stay with an aunt and uncle for a couple of years before starting school. Always very close to her father, she reportedly had a difficult relationship with her mother.

McClintock discovered her love of science during her time in high school and wanted to go on to university. Although her family had little money at the time, and her mother believed that higher education would make her daughter "unmarriagable", with the intervention of her father she was able to study **botany** at Cornell University. After taking her first course in **genetics**, she was invited by her professor to follow a graduate genetics course, and by that time she was convinced she wanted to continue her studies in the subject. She remained at the university after obtaining her doctorate and became an influential member of a small group of researchers who studied maize **cytogenetics**.

McClintock continued her work in genetics at the California Institute of Technology, the University of Missouri, and Cornell University, and in 1933 she also spent a few months in Germany. But as a result of increasing political tensions, she soon returned to Cornell University. At that time, few positions commensurate with her experience were open to women, but in 1936 she was offered an assistant professorship at the University of Missouri, where she examined the effects of **X-rays** on maize chromosomes. She left in 1940, realizing that she would never be offered a permanent position there, and the following year took a research post at the Carnegie Institute at Cold Spring Harbor. This turned into a long-term placement.

In 1948 McClintock discovered that certain genetic elements, which she had previously identified, were able to change their position (transpose) on chromosomes. She further argued that these transposable elements inhibited or changed the effects of genes around them. When presenting her findings in the early 1950s she was met with scepticism and even hostility, and in 1953 she stopped publishing her research.

The importance of her discovery was only recognized in the 1960s and 1970s, when techniques that could isolate transposable elements were available and these elements were also discovered in other organisms.

During the later stages of her career, McClintock spent many years studying different varieties of maize from Central and South America.

Essential science

Genetic recombination

In 1931 McClintock and her graduate student Harriet Creighton (1909–2004) showed that during reproduction the exchange between genes (genetic recombination) is accompanied by the exchange of physical parts of chromosomes. During cell division the complementary chromosomes (one from each parent) pair and cross over (exchange chromosome parts), giving rise to new genetic traits in the offspring.

Ring chromosomes and telomeres

In the 1930s McClintock started examining the effects of **irradiation** (by X-rays) on maize chromosomes. Chromosomes can be broken by irradiation and McClintock discovered that the ends of newly broken chromosomes often fuse with each other to form a ring, giving rise to a so-called "ring chromosome". She hypothesized that, at the chromosome tip a special structure exists that normally prevents this from happening. She called this structure "the telomere".

Telomeres are sequences at the ends of chromosomes that protect them from damage, and prevent the chromosomes from fusing into rings. When a cell divides, chromosomes are copied by **enzymes**, but these enzymes are unable to completely reproduce the tips of the chromosomes. This means that the duplicate chromosome lacks a small amount of the original telomere sequence. This does not particularly affect cellular functioning until enough cell divisions have occurred, and the telomeres become extremely short.

Cells with these short telomeres do not, under normal circumstances, replicate themselves anymore.

Breakage-fusion-bridge cycle

McClintock expanded her research on the effects of X-rays on maize chromosomes, and she observed a repeating pattern of

Legacy, truth, consequence

■ McClintock's work explains how changes affecting the structure of chromosomes occur. These changes result in genetic variability, but can also cause problems by affecting the development and function of a cell or an organism.

■ Ring chromosomes, described by McClintock, have now also been demonstrated in other species, including humans, where they can be associated with a number of diseases.

■ McClintock's hypothesis, that telomeres protect the ends of chromosomes, has turned out to be correct. In the majority of cells the progressive shortening of telomeres with each cell division results in a finite life span of the cells. Cancer cells, on the other hand, are able to divide indefinitely. Many cancer cells are known to be able to prevent shortening of the telomere.

■ **Transposons** can "turn off" genes around them and are now used by scientists studying genetics to determine the function of those genes. The changes induced by transposition may result in the development of resistance to antibiotics in bacteria. Because transposons affect the function of genes, they can also cause a number of diseases, such as haemophilia and muscular dystrophy, and they can confer a predisposition to cancer.

> *Over the many years, I truly enjoyed not being required to defend my interpretations. I could just work with the greatest of pleasure … If I turned out to be wrong, I just forgot that I ever held such a view …*
>
> Barbara McClintock (1983)

chromosome behavior. Sometimes triggered by an initial chromosome breakage, the broken chromosome then fuses to the other chromosome of the pair, forming a bridge, which is ripped apart again during cell division and this restarts the cycle. This cycle demonstrates the source of large-scale changes in genetic material.

Transposable elements

Transposable genetic elements are segments of genetic material that can move to a different position on a chromosome. McClintock was the first person to describe this and in 1983 she eventually received the Nobel Prize in recognition of her discovery. The changes she observed were insertions (addition of a genetic sequence to a chromosome), deletions (loss of genetic material), and translocations (genetic material moves from one chromosome to another). These transposable elements have since been described in many organisms, including humans.

Key dates

1902	Born in Hartford, Connecticut, US.
1921	Studies botany at Cornell University.
1927	Completes her doctorate and starts working in the Department of Botany, studying the genetics of maize at a cellular level.
1933	Receives a Guggenheim Fellowship enabling her to spend part of her postdoctoral training in Freiburg, Germany. Returns to Cornell after six months.
1936	Accepts position as assistant professor at the University of Missouri at Columbia, where she studies the effects of X-rays on maize chromosomes.
1941	Takes up a research position at the Carnegie Institution of Washington's Department of Genetics at Cold Spring Harbor. She remains affiliated to the Cold Spring Harbor laboratory for the rest of her life.
1944	Elected to the prestigious Academy of Sciences.
1945	Becomes the first woman to be elected president of the Genetics Society of America.
1948	Discovers for the first time the phenomenon of transposing genes.
1957	Begins studying different varieties of maize from Central and South America.
1983	Receives the Nobel Prize in Physiology or Medicine for her work on genetic transposition.
1992	Dies in Huntington, New York, US.

Barbara McClintock's microscope and some of the sample materials she analyzed.

Rachel Carson

A biologist, ecologist, and science writer, Rachel Carson almost single-handedly started the modern worldwide environmental movement with her powerful book *Silent Spring*, which eloquently showed the devastating effects of pesticide pollution on the natural world.

Growing up in the small riverside town of Springdale, Rachel Carson was encouraged by her mother to appreciate the natural world. Throughout her career, as a marine biologist and later as a general ecologist, her work was just as much her lifetime interest and hobby as a job. She was particularly absorbed by the seas, sea life, and the ecology of the seashore, a love she later expressed in her book *The Sea Around Us*.

After studying English then zoology, she spent several years working for the US Bureau of Fisheries, later the Fish and Wildlife Service, being only the second woman to secure a full-time professional job within the service. She eventually rose to become editor-in-chief of publications, with her enthusiasm for biology and nature shining through her writing.

After the success of *The Sea Around Us* she left her job to become a full-time writer on environmental issues, and began to research the material that would be *Silent Spring*, the book that woke the world to the dangers of industrial and agricultural pollution caused by indiscriminate crop-spraying. The title *Silent Spring* refers to an example she quoted of how all the wildlife in an area could be destroyed by widespread use of artificial chemical pesticides, resulting in an unnaturally silent region devoid of all nature whether animal or vegetable.

In 1957 she found herself in charge of an orphaned grandnephew as well as her aging mother, and had to rearrange her life in order to care for them.

Carson worked at a time when Americans believed that science could only be a force for good, and her proof that scientific progress was damaging the environment came as a double shock, particularly since she wrote in scientific terms and was already known as a science writer for her work on the seas and marine biology. Her books were recognized by other scientists for their scholarship, and were popular with ordinary people not just because of the topical subject matter, but also because of her eloquent and poetical language that brought science alive.

Silent Spring, however, was a call to arms, and was less lyrical and more angry and combative than her earlier books.

Essential science

Anti-pesticides

Carson was the first scientist to point out on a broad platform that pesticides created to kill just one weed or one insect or animal pest had a much wider impact by poisoning the food supply of other species, sometimes killing all insects, birds, fish, and wildlife in the area, and lingering on in the soil to have a lasting effect. She called these chemicals "biocides", and identified more than 200 chemicals that had been developed since the 1940s to kill pests or weeds, and which then became widely available for public use in the US.

Pioneering later holistic concepts, she stressed that human beings are also part of nature, and our health is harmed by destructive environmental practices just as much as any other species. She showed that pesticides can go on to contaminate the human food chain.

Research

Carson was not automatically opposed to the use of chemicals in agriculture, but she argued that while the long-term effects of newly developed pesticides was not known, it was scientifically and morally wrong to use them indiscriminately and on a large scale. As with the dumping of nuclear waste at sea, she pointed out that there was a lack of research on the long-term effects.

Public awareness

Carson believed passionately that ordinary people should be informed of the environmental and health impacts of modern agricultural practices, feeling that only if people were informed could they make choices about their environment. She was also keen to warn people that humanity was recklessly squandering natural resources.

Scientific integrity

The big multi-national chemical companies were quick to attack or even deride her views, but Carson remained calm and dignified, and maintained confidence in her scientific integrity and impartiality in the face of vicious personal attacks. Although she was quiet and shy, she felt so strongly about the issues that she was prepared to fight back against attempts by industrial organizations to suppress her work.

Legacy, truth, consequence

- Carson's book *Silent Spring* was an environmental "wake-up" call that rocked not just the American public but also most of the world. It brought existing conservation and wildlife organizations together and inspired a new generation of environmental activists.
- Her work on pesticides sparked an immediate public debate that forced the government to examine the issue. As a result, a federal government advisory committee in 1963 called for research into the potential health hazards of indiscriminate pesticide use. Eventually several artificial pesticides, including DDT, were banned in the US because of her arguments, and many other countries also introduced restrictions on pesticide use.
- The chemical industry fought back, but it failed to disprove her science, and resorted to personal attacks – even calling her a "hysterical woman" – which swung public sympathy towards Carson's cause.
- She was the inspiration behind the establishment of the US's Environmental Protection Agency.
- She introduced into the public consciousness several concepts such as "**ecosystem**" that are now part of everyday language, and her ability to bring biology to life helped publicize and popularize science.
- Carson inspired scientific studies of the relationship between environmental pollution and human health; in a horrible irony, breast cancer, the disease she died from, has since been identified as a cancer which can sometimes be caused by environmental contamination.

Rachel Carson pointed out the danger of intensive crop-spraying of new chemical pesticides, sparking an important public debate.

I truly believe that we in this generation must come to terms with nature, and I think we're challenged as mankind has never been challenged before to prove our maturity and our mastery, not of nature, but of ourselves.

CBS television program, *The Silent Spring of Rachel Carson* (1963)

Key dates

1907   Born in Pennsylvania, US.
1936–52 Works for the federal government as a researcher and scientific editor.
1945   First notices the widespread effect of new chemical pesticides such as DDT.
1952   Publishes the award-winning book on the sea and marine life, *The Sea Around Us*, which becomes a bestseller and makes her a household name. Becomes a full-time author and begins to investigate effects of pesticides.
1962   Publishes *Silent Spring*, alerting the world to the environmental damage caused by pesticides and industrialized agriculture.
1963   Testifies before a Congressional enquiry into the labeling of pesticides.
1964   Dies in Maryland, US, of breast cancer.

To dispose first and investigate later is an invitation to disaster, for once radioactive elements have been deposited at sea they are irretrievable. The mistakes that are made now are made for all time.

Preface to the 1961 edition, *The Sea Around Us*

Subrahmanyan Chandrasekhar

A physicist, astrophysicist, and mathematician, Chandrasekhar made contributions to a wide range of fields. He is particularly known for his work on the structure, origins, and dynamics of stars, such as his important discovery that there is an upper limit to the mass of a star that could become a white dwarf, and his prediction of black holes.

Chandrasekhar, usually known simply as Chandra, was born in Lahore when it was part of British India; it is now in Pakistan. The oldest boy in a family of ten children, he was perhaps inspired by his scientist uncle, Sir C. V. Ramen, who won the Nobel Prize for physics in 1930. Chandra was a bright student and his first paper was published in Britain in 1928. That same year he was introduced to the new **quantum mechanical** theories by a visiting German physicist, Arnold Sommerfeld (1868–1951), a meeting that Chandra called the single most important encounter in his scientific career.

After graduating in 1930, Chandra was awarded a special scholarship by the Indian government to carry out postgraduate work at the University of Cambridge, England. On the long sea journey to Britain he first came up with one of his most significant theories – that **white dwarf** stars have an upper mass limit – when he applied his new interest in **special relativity** and quantum statistics to known observations of white dwarfs.

He proposed this theory in a presentation at the Royal Astronomical Society in 1935. To his dismay, the preeminent **astrophysicist** Sir Arthur Eddington (1882–1944) publicly ridiculed the whole idea of applying relativity to star structure. Although no one could fault Chandra's logic, no European scientist supported him, and Chandra was glad to accept the offer of a post in Chicago, US. After this setback he moved onto a different research topic, setting the pattern for his life of exploring one subject intensively, writing about it, then moving on. These periods, as he called them, covered nearly all branches of theoretical astrophysics, involving more than 400 publications.

Chandra and Eddington remained friendly, but the disagreement delayed acceptance of Chandra's theory and his Nobel Prize.

Unusually for Indians of his social class at the time, Chandra did not have an arranged marriage, but married for love – his university girlfriend. They were married in India before Chandra took up his position in America: to get round immigration controls, his entrance visa to the US was for a missionary. The couple faced some racial prejudice in America, but committed themselves to living there.

Although he was an atheist, Chandra agreed with Mahatma Gandhi's non-violence stance, but during World War II he felt that Nazi Germany had to be defeated, so he worked on ballistics problems for the American government.

In 1952 he took on the editorship of the university's *Astrophysical Journal* and, although keen to build it up and to share scientific knowledge through it, found that the job was eating into his research time. His solution was to enforce strict working hours for the journal and simply refuse to discuss editorial issues outside those hours.

Essential science

The Chandrasekhar limit

Chandra's most famous theory was also his earliest. This states that when the nuclear energy source at the heart of a star (such as our sun) runs out and the star approaches its last stage of evolution, it does not necessarily end up as a small, stable, slowly cooling fragment known as a white dwarf. Instead, if its mass is above a certain limit, now called the Chandrasekhar limit, he proposed that it will explode in a **supernova**, then carry on collapsing into itself to form an infinitely dense, infinitely small point, later called a **black hole**.

As with all his work, Chandra followed his physical intuition but applied rigid mathematics, in this case employing the new ideas of quantum mechanics and the special theory of relativity to the known properties of white dwarf stars.

He surmised that at large masses the **Pauli exclusion principle** (formulated by Wolfgang Pauli in 1925) – also known as the electron degeneracy principle – would apply. This states that no two electrons can occupy exactly the same quantum space. A consequence of this, according to Chandra, is that the contracting pressure of a massive collapsing star will force **electrons** to move outwards to higher energy levels at near light speeds. This will bring about an explosion, blowing away the envelope of electron gases surrounding the dying star and leaving the remnant as a dense, still-collapsing fragment. The mathematics for this can only be formulated by using the Fermi-Dirac statistics that apply to the relevant **subatomic** particles (see pages 156–7 and 164–5), not by **classical physics**.

> *My scientific work has followed a certain pattern motivated, principally, by a quest after perspectives ... this quest has consisted in my choosing a certain area which appears amenable ... And when after some years of study, I feel that I have accumulated a sufficient body of knowledge and achieved a view of my own, I have the urge to present my point of view ...*

Autobiography, Nobel Prize lecture (1983)

A supernova remnant. The ring marks the outer limits of the shock wave produced as the material ejected in the explosion collides with interstellar gas. This picture was taken by the Chandra X-ray Observatory, a NASA satellite named after Chandra.

After retiring in 1980 Chandra continued to give occasional lectures, especially on science and art, and he spent the last years of his life studying the works of Sir Isaac Newton. His last book, *Newton's Principia for the Common Reader*, was published just months before he died.

Combining this relativistic and quantum mechanical approach with fundamental constants such as gravity and the mass of the hydrogen **atom**, he was able to deduce that the mass limit for a white dwarf is 1.44 solar masses, or 1.44 times the mass of the sun.

Other ideas

Chandra's white dwarf/black hole theories were so spectacular that they overshadowed his many other important contributions. These include: a new statistical approach to studying changes in the orbit of a star; new mathematics to explain observed variations in the energy absorbed and emitted in stellar atmospheres; and an explanation of polarization of light by the earth's atmosphere.

Legacy, truth, consequence

■ Chandra's theory that an object larger than our sun could eventually collapse to infinity was revolutionary, and roused initial opposition and scepticism. However, his mathematics could not be faulted and observations eventually showed that white dwarf stars did, indeed, have a limited range of masses. In the 1960s **neutron stars** were first observed, and the black holes that he had predicted were also later verified.

■ The Chandrasekhar limit is now a fundamental concept in modern astrophysics.

■ Chandra was much admired for his intellectual and lucid writings, as well as his original ideas. Many of his books are considered to be classics, and his 1983 *Mathematical Theory of Black Holes* is the definitive textbook on the subject.

■ He was also a committed teacher, and supervised about 50 postgraduates, including future Nobel Prize winners.

Key dates

1910	Born in Lahore, Punjab.
1929–39	Researches the structure of stars, including white dwarfs.
1933	Receives a doctorate from the University of Cambridge, England, and becomes research fellow of Trinity College, Cambridge.
1937	Moves to University of Chicago, US.
1938–43	Researches **stellar dynamics**.
1943–50	Researches **radiative transfer** in stellar atmospheres.
1952–71	Edits the university's *Astrophysical Journal*, turning it into the world's leading publication on the subject.
1953	He and his wife become US citizens.
1962–7	After studying other astrophysics topics, returns to research combining the general theory of relativity and astrophysics.
1974–83	Researches the mathematical theory of black holes.
1983	Wins Nobel Prize for Physics, shared with William Fowler (1911–95). Publishes his *Mathematical Theory of Black Holes*.
1995	Dies in Chicago, US, of heart failure.

Alan Turing

Alan Turing was an English mathematician whose work focused on issues in mathematical logic and paved the way for many later developments in computer science. He is best known for pioneering the Turing test and the Turing machine, which raised many important questions about the possibility of artificial intelligence, and for his code-breaking work at Bletchley Park during World War II.

Born into a London-based middle-class family in 1912, Turing was an awkward yet prodigious child who, by the age of 16 had mastered **Albert Einstein**'s account of **relativity**. An atheist from an early age, Turing nurtured a fascination for science and technology throughout his schooldays.

He graduated with a degree in mathematics from Kings College, Cambridge, and was elected a fellow of the college in 1935. Two years later he published the paper "On Computable Numbers …" which made a major contribution to mathematical logic. He then attended Princeton University, New Jersey, US, where he received his doctorate in 1938, under the tutelage of the **logician** Alonzo Church (1903–95).

Back at Cambridge in 1939, delivering undergraduate lectures, Turing attended the philosopher Ludwig Wittgenstein's characteristically informal lectures on the foundations of mathematics. For a few weeks the class was treated to an impassioned dialogue between the two great thinkers, although ultimately Turing stopped attending, undoubtedly frustrated by Wittgenstein's radically unconventional approach to the issues.

Turing spent World War II at the code-breaking establishment Bletchley Park, Buckinghamshire, working on the computational machinery required to break the German cypher encrypted by the German enciphering machine Enigma, and thus played an important role in the Allied victory. For security reasons, the nature of his work remained secret for many years. After the war, he worked on electronic digital computers and theoretical issues related to **artificial intelligence**, most notably publishing "Computing Machinery and Intelligence" in 1950.

Away from his work, Turing was a talented amateur marathon runner, so much so that he narrowly missed out on qualifying for the British Olympic team for the 1948 summer games based in London.

In February 1952 Turing was arrested in connection with his homosexuality. To avoid a jail sentence he was subjected to estrogen injections intended to curb his libido. Ironically, given his contributions during the war, he was now deemed a national security risk. In 1954, at the age of 41, Turing committed suicide by eating a cyanide-poisoned apple.

Essential science

Wartime cryptanalysis

At Bletchley Park, with the mathematician Gordon Welchman, Turing designed a **cryptanalytic** machine known as the "Bombe", for breaking the Enigma cipher. It was an improvement on the original, Polish-designed "Bomba". Eventually over 200 Turing–Welchman Bombes were in operation, deciphering the codes of the naval version of Enigma and reading U-boat communications. For a period Turing was also head of Hut-8, the section in charge of reading the German naval signals.

During short stays at Bell Laboratories, New York, and the Special Communications Unit at Hanslope Park, Buckinghamshire, UK, Turing conducted research on speech-encipherment, producing a portable machine, code-named *Delilah*, for secure voice communications.

Machine intelligence and human intelligence

In "Computer Machinery and Intelligence" Turing considered whether a machine can think and he hypothesized that if human intelligence is explained in terms of processes in the brain then this could be simulated by a machine. His view was that if a machine cannot be distinguished from a human then it passes the Turing test and, therefore, possesses human intelligence. Turing conceived of electronic digital computers *simulating* human intelligence, where the successful simulation of human intelligence counts as human intelligence.

To illustrate his point, he devised a test called "the imitation game". This involved three previously unfamiliar participants: a human, a machine, and an interrogator, in different rooms but in contact via teleprinter communication. Modeling the human mind as a physical machine, he (1) produced a series of considerations purporting to show that the responses of the computer are indistinguishable from the human participant, since the notion of "human thinking" can be equated with successful participation in the imitation game, and (2) presented a series of arguments *against* the claim that machines *cannot* exhibit human intelligence.

Legacy, truth, consequence

- Turing's work in breaking the code of the Enigma machine not only demonstrated great advances in cryptanalysis but also, more significantly, played a vital part in the outcome of World War II. The theory of information and statistics that he advanced during this period is credited with making cryptanalysis a science.
- He had a pivotal role in the invention of the modern computer. Bell Laboratories put the ideas embodied in the Turing machine into practice in 1939 by developing the first relay computer. Alongside the American mathematician John von Neumann (1903–57), Turing is a founding father of **computer science**, producing a plan for the construction of the electronic computer.
- His research on machine intelligence raised many important philosophical questions – about artificial intelligence, human consciousness, and the mind-body problem – that still provoke discussion to the present day, particularly in relation to **functionalism** in the philosophy of mind. Though critical of Turing's conclusions, John Searle's celebrated **Chinese Room Argument** (1980), concerned with highlighting the importance of human **intentionality**, owes a debt to the Turing test.
- Turing's work in mathematical logic charted new territory regarding **decidability questions** in pure mathematics.

> *I believe that at the end of the century the use of words and general educated opinion will have altered so much that one will be able to speak of machines thinking without expecting to be contradicted.*
>
> "Computer Machinery and Intelligence" (1950)

The Decision Problem and Computability

Turing's article "On Computable Numbers ..." solved a question posed by the mathematician David Hilbert (1862–1943) and attempted to set limits for computation. Coming in the aftermath of mathematician and philosopher Kurt Gödel's 1931 proof that no system of arithmetic can be both consistent and complete, Turing set out to address Hilbert's *Entscheidungsproblem* (*The Decision Problem*). He considered whether there is a general methodology for *deciding*, for any mathematical proposition, whether it is *provable*, and he framed his solution in what are now known as Turing machines: universal computers designed to solve mathematical problems by reducing them to coding in a given set of commands. Defining "computable" as that which a Turing machine can perform when acting alone and identifying computability with the general methodology in question, Turing proved that there could be no general method for addressing Hilbert's decision problem and that, therefore, there is no definite method for addressing all mathematical questions.

Detail of the German Enigma machine. Turing helped to crack the Enigma's code.

Key dates

1912	Born in London.
1926	Attends Sherborne School, Dorset.
1931–4	Reads mathematics at King's College, Cambridge.
1935	Elected fellow of King's.
1936–7	"On Computable Numbers with an application to the *Entscheidungsproblem*" published in *Proceedings of the London Mathematical Society*. Attends Princeton and wins the Procter Fellowship.
1938	Gains doctorate from Princeton. Returns to Britain and attends the Government Code and Cypher School.
1939	Begins lecturing at Cambridge and publishes 'Systems of Logic based on Ordinals' in *Proceedings of the London Mathematical Society*. Attends Wittgenstein's Cambridge lectures on the foundations of mathematics. Begins working on the German naval Enigma encoding machine at Bletchley Park.
1942	Becomes chief research consultant at the Government Code and Cypher School.
1943	Spends from January to March at Bell Laboratories, New York, researching speech-encipherment.
1944	Works on speech-encipherment project *Delilah* at the Special Communications Unit at Hanslope Park, Buckinghamshire, UK.
1945	Awarded OBE for wartime services. Works on ACE computer at the National Physical Laboratory, Teddington.
1947	Returns to Cambridge.
1948	Works on a prototype computer at Manchester University.
1950	"Computing Machinery and Intelligence" published in *Mind*.
1952	Arrested for his homosexuality and deprived of his security clearance. "The Chemical Basis of Morphogenesis" is published in *Philosophical Transactions of the Royal Society*.
1954	Dies in Wimslow, Cheshire.

Jonas Salk

An American microbiologist, Salk developed the first vaccine against polio, making him a popular scientific hero almost overnight. To some extent, his fame alienated the medical scientific community, and perhaps he never received the academic honors he deserved.

Born into an Orthodox Jewish–Polish immigrant family, Salk was encouraged by his parents to study and work hard. Talented and motivated, he excelled at school, graduating when only 15 and going on to law school. However, he attended some science classes and was gripped, switching courses and receiving his science degree at the age of 19.

He then attended medical school, which was only possible because his parents scrimped and saved and borrowed money for him. But after his first year he found scholarships and grants that made it possible for him to continue his studies. It was there that he met Dr Thomas Francis, Jr (1900–69), an **epidemiologist** who was to become a mentor, colleague, and friend.

After a two-year internship at Mount Sinai Hospital, New York, Salk received funding from the National Research Council to work with Francis on developing a flu vaccine for the US army. The flu virus had only recently been identified, and the authorities were frantic to prevent a repeat of the flu epidemic that had killed millions after the end of World War I. Francis and his team successfully created a vaccine that was used by the armed forces.

In 1947 Salk gained a position at Pittsburgh. He was horrified to find an ancient, ill-equipped laboratory, and the first thing he did was apply his energies to finding funds and converting the lab into a state-of-the-art research center. He was soon drawn by the polio problem, which he thought must be similar to flu.

American parents in the 1940s and 1950s were terrified by the growing incidence of poliomyelitis or polio (sometimes called

infantile paralysis). This virus attacks the nervous system and crippled or killed about one child in 5,000 in annual summer epidemics in America, so it seemed to many people like a plague. After Salk had discussed the problem in a couple of articles, the prestigious National Foundation for Infantile Paralysis (now the March of Dimes Foundation), attracted by his obvious energy and enthusiasm, offered him nearly all its research funds to find a cure.

This was the beginning of the scientific community's dislike of Salk. Researchers such as Albert Sabin (1906–93) had spent many years in careful research, and suddenly a newcomer to the field had been given seemingly limitless money.

Sabin and others also scorned Salk's research ideas. Although saddened by their lack of confidence, Salk was supported by Francis, and persevered. As a result, he came up with the first successful inoculation against polio.

Normally, scientific results are first published in academic journals before being announced to the world at large. However, under pressure in 1955, Salk agreed to a press conference before publishing his test results, and although he did not claim personal credit, he found himself overnight a media and public darling. Unfortunately it also meant that he was a villain to many other scientists who felt that he had not given enough credit to other researchers in the field.

In 1970 Salk became notorious again when he married for a second time, to Françoise Gilot, who had been the mistress of the artist Pablo Picasso.

Essential science

"Killed" polio

At the time, most vaccine-hunters were working with "live" but weakened polio viruses, thinking that with this they might be able to create a strain that would produce a mild infection in a patient, who would be able to recover and be immune to future infection. Many scientists thought that previous mild exposure to the disease was, in fact, the only way to bring about immunity.

From working on a flu vaccine, Salk knew that a "killed" or deactivated virus could sometimes work as an antigen, triggering the body's immune system to generate antibodies that would attack and destroy any future invasion by that virus. This sort of vaccine would work without actually infecting the patient, with all those attendant

risks. His most important insight was to apply the same principle to polio, and try to find a vaccine based on a "killed" virus.

He entered the polio field at the right time. A few years earlier, John Enders (1897–1985) at Harvard had succeeded in growing polio in test tubes, so that researchers now had a plentiful supply of the substance at their fingertips all the time. Enders would go on to win the Noble Prize for Medicine in 1954, as some scientists argued that the real breakthrough against polio was his, not Salk's.

Other researchers had come up with the idea of using formaldehyde to kill the virus, so Salk focused his experiments on using this as a vehicle that would kill the virus, but leave it intact enough to spark the immune system into action. Once he had

1963 poster from the US Department of Health promoting the polio vaccine.

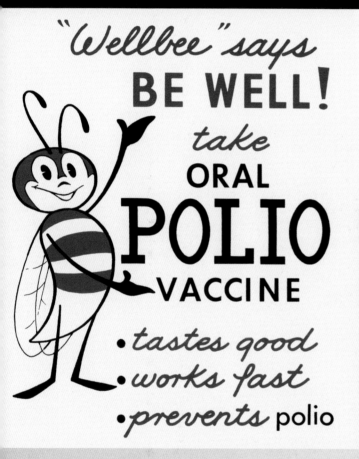

"Wellbee" says
BE WELL!
take
ORAL
POLIO
VACCINE
• tastes good
• works fast
• prevents polio

Legacy, truth, consequence

■ In the minds of ordinary people, Salk will forever be the man who beat polio, removing a dreadful fear and relegating it to the ranks of the many "conquered" diseases that no longer affect the modern world. He particularly endeared himself to the public by refusing to patent the vaccine or profit from it personally.

■ In 1958 Sabin's vaccine based on a "live" virus was introduced. This could be taken orally, unlike the Salk vaccine which had to be injected, and it needed fewer booster shots, so it began to replace Salk's "killed" virus vaccine. Today, the two are usually used in tandem.

■ The scientific medical community never really forgave Salk for what they considered to be his self-seeking publicity stunt in announcing his results publicly before presenting them to his peers. He did not receive any of the major awards such as the Nobel Prize. There was also resentment because of the research funds that he was able to raise in the future.

■ The Salk Institute for Biological Studies is now a famous and prestigious institution for molecular biology and genetics.

Key dates

1914	Born in New York, US.
1942	Works on flu vaccine for the US army.
1947	Becomes head of the Virus Research Lab at the University of Pittsburgh. Begins work on a polio vaccine.
1952	Begins tests on humans, including himself, his wife, and their three sons.
1954	Begins national testing of his vaccine with a study involving two million children.
1955	Announces the success of his polio vaccine to immediate public acclaim.
1963	Founds the research center, Salk Institute for Biological Sciences, in La Jolla, California.
1995	Starts research on an AIDS virus.
1995	Dies of congestive heart failure in La Jolla.

secured the necessary funding, he charged ahead efficiently and effectively, achieving in a few years what other scientists had worked for slowly and steadily over the course of decades.

He first tested his vaccine on monkeys, then on a small group of humans. All showed the successful production of antibodies with no unwanted ill effects. The next stage was a large-scale test program on children, launched in 1954, which by April 1955 showed that his vaccine was effective and safe.

Double-blind studies

Salk's was one of the first large-scale studies incorporating the **placebo** effect and the **double-blind** approach. Some children were not treated, forming a natural control group, and of the children who were given an injection, half of them received nothing but a placebo. Out of his two million subjects, the few deaths were caused by a contaminated sample provided by one drug company.

World campaigns

Salk worked tirelessly to promote the advantages of vaccination and disease prevention in general, and he also campaigned for world peace, saying that war was "the cancer of the world".

Francis Crick   James D. Watson

The English physicist Francis Crick and the American geneticist James Dewey Watson are best known for their role in the discovery of the structure of DNA (deoxyribonucleic acid), for which they were awarded the Nobel Prize in 1962. Both Crick and Watson subsequently continued to make significant contributions to scientific research.

Francis Crick developed an interest in science during his childhood and, to help answer his many questions, his parents bought him a children's encyclopedia.

Once he'd finished school, he studied physics at University College London. His studies for a doctorate were interrupted by World War II, and after the war he found himself unsure of his future career. Then he read *What is life? The physical aspects of the living cell* (1944) by the physicist **Erwin Schrödinger**, which sparked his interest in biology, especially in the notion that biological processes within living organisms could be explained by physics and chemistry. In 1949 he joined the Medical Research Unit at the Cavendish Laboratory in Cambridge, where his research focused on determining the structure of **proteins**. Crick met James Watson in 1951, when Watson arrived at the Cavendish Laboratory.

James Watson grew up in Chicago. He first became interested in science as a child, when, like his father, he enjoyed bird-watching. He was a precocious student and entered the University of Chicago at the age of 15 to study **zoology**.

In 1951, when Watson was a postdoctoral student, he met the **biophysicist** Maurice Wilkins (1916–2004) and saw for the first time the **X-ray** images of **DNA**, generated at King's College in London by Wilkins and his colleague Rosalind Franklin (1920–58). Watson then decided to change the direction of his research towards discovering the structure of proteins and DNA and started a new research project at the Cavendish Laboratory. When Crick and Watson met, they soon discovered their shared interest in solving the DNA structure. In April 1953, after combining biochemical evidence, recently published experimental

Essential science

DNA

The DNA molecule stores the hereditary information of living organisms. When Crick and Watson started their research on the molecule's structure, DNA had recently been shown to control heredity. Without knowing its structure, however, no additional conclusions could be drawn about its function.

DNA is made up of chains of subunits, known as nucleotides, and the four different DNA-nucleotides are called adenine, cytosine, guanine, and thymine. Several important pieces of information for the determination of the correct structure of DNA were made available to Crick and Watson. These were the experimental evidence for the most likely structure of nucleotides, and the evidence that the amount of guanine is equal to cytosine and the amount of adenine is equal to thymine. Using cardboard models of each nucleotide, Watson tried to establish how these could interact and fit together. He soon realized that they could only be paired in a certain way: adenine with thymine and cytosine with guanine.

Crick and Watson obtained another important clue in the form of Rosalind Franklin's X-ray images of DNA (shown to them without her knowledge), which suggested a helical structure and offered additional information about the dimensions of the structure.

Armed with this data, Crick and Watson came up with the correct structure of two parallel strands, gently twisted to give the

appearance of a double-helix. The nucleotide pairs, just described, connect the two chains like the rungs of a ladder.

Replication

The DNA model by Crick and Watson immediately suggested a possible mechanism for replication and the transmission of hereditary information from one generation to the next: because the nucleotides can only be paired in a certain way, the nucleotide sequence on one strand can serve as a template for the assembly of a new complimentary strand during cell division.

Genetic code

Proteins are fundamental components of living cells and are necessary for the functioning of an organism. They are made up of long chains of subunits called amino acids. After discovering the structure of DNA, Crick went on to study how the DNA's nucleotide sequence could be translated to the amino acid sequence in proteins, and thus how the instructions encoded by DNA could be translated into living things. By 1961 Crick and his colleagues had shown that this translation involves a three-nucleotide code, which means that a sequence of three nucleotides encodes one particular amino acid. The amino acid sequence then determines the type of protein and its function.

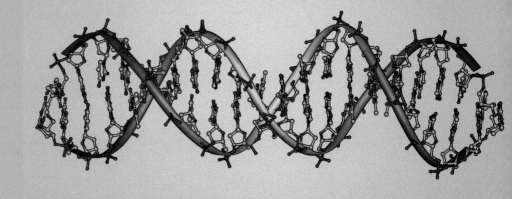

No one may have the guts to say this, but if we could make better human beings by knowing how to add genes, why shouldn't we?

James D. Watson,
"Risky Genetic Fantasies",
The Los Angeles Times
(July 29, 2001)

The double-helix structure of a DNA molecule.

data, and evidence from the X-ray images by Wilkins and Franklin, they were able to write their famous paper, proposing the double-helical structure of DNA. Experimental evidence later proved that their proposal was correct and their findings greatly stimulated further research in the newly emerging field of **molecular biology**. In 1962 Crick, Watson, and Wilkins received the Nobel Prize for their discovery. Rosalind Franklin had already died by that time and was, therefore, not eligible for the prize.

After their discovery, Crick went on to study the genetic code of DNA. In the 1970s he moved to the Salk Institute in California, where his research focused on **neurobiology**, including the nature of consciousness and the phenomenon of human dreams.

Watson, whose personality has sometimes been described as abrasive, has occasionally attracted controversy by his remarks on a variety of topics. He has, however, continued a distinguished career in science and his work has included research on tumor **virology**, the study of oncogenes (genes that cause the transformation of normal cells into cancerous tumor cells), and the launch of the Human Genome Project (for more on this see **Craig Venter**, page 193).

Legacy, truth, consequence

■ The Human **Genome** Project, led by James Watson for four years, had the aim of deciphering the entire human genetic code (genome). A "rough draft" of the genome sequence was announced in 2000, and the complete sequence was published in 2003, revealing the existence of over 3 billion nucleotide pairs and 20,500 human genes. The project provides information about the structure, organization, and function of the complete set of human genes.

■ The uses of DNA technology in medicine include the diagnosis and prediction of diseases and of disease susceptibility, drug design, and potentially also gene therapy.

■ DNA-typing is an identification system examining differences or similarities in the DNA nucleotide sequence, either between two individuals or between a DNA sample and the known sequence of a reference sample. This can, for example, be used in criminology or when trying to establish paternity.

■ DNA science is now commonly applied to veterinary medicine, and to plant and animal breeding.

Key dates

1916	Crick is born in Nothampton, UK.
1928	Watson is born in Chicago, Illinois, US.
1949	Crick joins the Medical Research Unit, then located at the Cavendish Laboratory, Cambridge, UK.
1951	Watson starts working at Cavendish Laboratory and meets Crick. They share a common interest in solving the DNA structure and begin work on their first DNA model.
1953	Crick and Watson publish their proposal of the double helical structure of DNA in "Molecular Structure of Nucleic Acid: A Structure for Deoxyribose Nucleic Acid".
1955	Watson is a faculty member at the University of Harvard, US.
1961	Crick and Sydney Brenner (b. 1927) discover that the DNA code is a triple code.
1962	Crick, Watson, and Maurice Wilkins are awarded the Nobel Prize in Physiology or Medicine.
1968–2007	Watson serves as director, president, and eventually as chancellor of the Cold Spring Harbor Laboratory, US. successfully steering the laboratory into the field of tumor virology.
1976	Crick becomes Kieckhefer Professor at Salk Institute for Biological Studies, US, where he begins his research in neurobiology.
1988–1992	Watson leads the Human Genome Project at the National Institutes of Health, US.
2004	Crick dies in San Diego, California.
2008	Watson becomes the Institute Advisor for the Allen Institute for Brain Science, US.

Both of us had decided, quite independently of each other, that the central problem in molecular biology was the chemical structure of the gene.

Francis Crick, *What Mad Pursuit* (1988)

Richard Feynman

Although he made important contributions to quantum electrodynamics, Richard Phillips Feynman is best known as a communicator whose books and audio or video recordings made the intricacies of science available to everyone. An eccentric, bongo-playing practical joker, he became one of the world's best-known scientists because of his ability to communicate his enthusiasm for physics.

Feynman became a precocious child who studied math and science on his own. Although in his entrance exam for graduate school he excelled in physics and mathematics, he nearly missed getting into Princeton University because his scores in arts subjects were so poor.

During World War II he was the youngest member of the team building the atom bomb. There he mingled with some of the world's greatest theoretical physicists, including **Niels Bohr**, who loved debating with him since Feynman would lose his awe of the grand old man of atomic science in his love of abstract discussion.

After World War II Feynman became depressed about the presence of nuclear weapons, and at one point thought there was no point in starting new projects since the world would probably be blown up. Despite this, his extensive writings on science are light-hearted and full of entertaining anecdote. Throughout his teachings and writings, he was concerned to show the integrity of science and to condemn "pseudo-scientists".

> ## I think I can safely say that nobody understands quantum mechanics.
>
> *The Character of Physical Law (1965)*

Essential science

Quantum electrodynamics and Feynman diagrams

His new approach to exploring **quantum mechanics** applied the principle of least action (that changes are the least that are possible) and therefore involved analyzing all the possible paths a particle could take in moving from one position to another.

To show the result, he produced a graphic representation mapping the possible movements in space-time, with the final path being a sum over all the possible ways the particle could move. His pictorial diagrams offered a simple way of representing the complicated mathematical expressions that describe the interactions of **subatomic** particles.

Challenger investigation

Feynman correctly identified o-ring seals in one of the two solid rocket boosters as the cause of the shuttle explosion, and raged about the scientific incompetence of **NASA**'s management. His views – calling for a complete overhaul of NASA – were included as an appendix to the official report.

Other areas

He worked in several other fields and is credited with pioneering quantum computing and **parallel computing**, and introducing the idea of **nanotechnology**.

Legacy, truth, consequence

■ As an approach to formulating quantum mechanics, Feynman diagrams led to an improved theory of **quantum electrodynamics**, enabling scientists to more accurately predict the effect of electrically charged particles in a radiation field.

■ Feynman diagrams are fundamental for new concepts such as **string theory**.

■ His books, particularly *The Feynman Lectures*, are classic introductions to physics and helped popularize science.

Key dates

1918	Born in New York, US; goes on to study physics.
1941–5	Works on the atom bomb during World War II.
1950	Becomes professor of theoretical physics at the California Institute of Technology.
1961–3	Delivers introductory course in physics, later published as *The Feynman Lectures*.
1965	Wins the Nobel Prize in Physics – shared with Sin-Itiro Tomonaga (1906–79) and Julian Schwinger (1918–94).
1986	Appointed to the committee investigating the Challenger space shuttle disaster.
1988	Dies in Los Angeles, US, from cancer.

Paul Berg

An American biochemist and molecular biologist, Paul Berg made a major technological contribution to the science of genetics by developing genetic engineering. He worked out a technique of splicing DNA from two different organisms, called recombinant DNA technology, which has many possible applications, sometimes controversial.

Paul Berg grew up in Brooklyn in New York, and discovered an enthusiasm for scientific discovery when he was still at school.

His studies of **biochemistry** at Pennsylvania State University were interrupted by World War II, during which he volunteered for the navy and served on a submarine chaser. He chose to go to Western Reserve University for his doctorate, a pioneering center for biochemistry, and became one of the first people to show the role of folic acid and vitamin B12 in metabolism.

During the following years he became increasingly interested in the nature and structure of the **gene**, and also in cell biology. His work on recombinant **DNA** came about because he thought it would be easier to study a gene in a new environment, when there would be no confusing interactions between it and its natural neighbor genes.

> *There is serious concern that some of these artificial recombinant DNA molecules could prove biologically hazardous.*
>
> "The Berg Letter", published in *Science* journal (July 26, 1974)

Essential science

Cancer cells

While studying why some cells spontaneously turn cancerous, Berg thought that interactions between their genes and their cellular biochemistries were responsible, so he decided he could examine this if he could introduce a cancer gene into a simple single-celled organism like a bacteria. He felt he could smuggle the gene into the **bacteria** if he could somehow combine it with genetic material that can normally enter bacteria, such as a bacteriophage, a **virus** that is known to infect bacteria. He chose to work with a virus that causes cancer in monkeys, SV40, and the bacterium *E. coli* which is found everywhere and is often used in labs.

Gene splicing

First Berg used a particular **enzyme** to cut the double strands of a DNA **molecule** at exactly the spot that he wanted. Then he used a different enzyme to add sections to just one strand, creating a long "sticky end" ready to be linked to another similarly treated

Legacy, truth, consequence

■ Berg is recognized not just for his discoveries but also for his stance on responsible science.

■ Gene therapy and the controversial genetically modified food crops are some of the results of recombinant DNA. Insulin, human growth hormone (the hormone that regulates growth in an individual), and some antibiotics are now made using his technique, inserting the genes that trigger the required protein growth into fast-multiplying bacteria. There are many other potential applications.

Key dates

1926	Born in New York, US.
1959	After further researches, joins the medical school of Stanford University, becoming Willson Professor of biochemistry in 1970.
1974	Stops his work when realizes potential dangers.
1975	Helps organize international conference drawing up guidelines for genetic engineering.
1980	Wins Nobel Prize for Chemistry (shared with Walter Gilbert and Frederick Sanger).

piece of DNA. He did this to the SV40 gene and to the bacteriophage, and successfully recombined the DNA piece.

Dangers

At this point Berg voluntarily put a stop on the research. *E. coli* can sometimes exchange genetic material with other types of bacteria, including some that cause human disease. He realized that if he inserted his hybrid DNA into the bacteria, and any escaped and spread, he could not predict the circumstances, but it might cause a medical disaster.

Guidelines

In 1974 Berg called for a moratorium on genetic engineering until the dangers could be assessed. The next year a conference of 100 scientists from around the world agreed guidelines and a ban on any experiments where genetically engineering organisms might be able to survive in humans if they escaped the lab.

Elias James Corey

The American organic chemist and Nobel Prize winner Elias James Corey has completed pioneering work in synthetic chemistry; in particular he developed the principles of so-called retrosynthetic analysis, a technique that simplifies the synthesis of large, complex organic molecules. With his methods Corey's research group has synthesized over 100 natural products, resulting, among other things, in the synthesis of many commercially available pharmaceuticals.

William James Corey was born to Christian Lebanese immigrants in Methuen, Massachusetts, US. His birth was followed, 18 months later, by the death of his father Elias, and his mother subsequently changed his name from William to Elias. Despite growing up during the Depression and World War II, Corey grew up in a happy and loving household, which included his aunt and uncle, both functioning as second parents.

After attending school in nearby Lawrence, he entered the Massachusetts Institute of Technology, planning to study electrical engineering. During his first course in basic sciences, however, he soon discovered his interest in chemistry, especially in the field of **organic chemistry**. By 1950 he had completed both his undergraduate and postgraduate degrees in chemistry, and a few months later he took up a position as an instructor at the University of Illinois. Upon promotion to assistant professor in 1954 he established a research group working on a range of projects involving the structure and the synthesis of complex, naturally occurring organic compounds.

He was offered a professorship at Harvard University in 1959, where he was able to start a number of new scientific projects and to teach an advanced graduate course in synthetic chemistry. By this time he had developed several ideas for a definitive approach to organic synthesis, and by the 1960s he had developed and described the strategy of what is called retrosynthetic analysis. Using this strategy, Corey's research group synthesized more than 100 natural products, a number of which can be used for medicinal purposes. They have developed numerous new methods and Corey has demonstrated the use of computer analysis to generate potential new synthetic pathways. His methods are now taught and practiced worldwide.

Corey became Sheldon Emery professor of Chemistry at Harvard University in 1968, where his research group has continued to synthesize a wide range of complex structures, including many rare substances, and to study disease mechanisms at a molecular level.

In 1990 Corey was awarded the Nobel Prize in Chemistry for *"his development of the theory and methodology of organic synthesis"*.

Essential science

Organic synthesis

Organic chemistry is the study of carbon–containing **molecules**, including their structure and the reactions they undergo. Organic synthesis is the production, via chemical processes, of complicated organic compounds using simple starting materials. The compounds that are generated are used to make plastics, rubber, paints and dyes, pesticides, many medicinal products, and synthetic fibers such as nylon.

The traditional way of designing syntheses of complicated organic molecules was to begin with simple or readily available materials that could be assembled, by a sequence of chemical reactions, to form the desired end-product, the so-called target molecule. This was often done rather intuitively and chemists found it difficult to explain how exactly they came up with the starting materials and reactions. Corey recognized the need for a planned and structured approach and so he developed the principles of retrosynthetic analysis.

Retrosynthetic analysis

Corey developed retrosynthetic analysis, a simpler, faster, and more efficient method of synthesis, in the 1960s. This planned and logical approach involves starting with the target molecule and then analyzing how it can be broken down into smaller subunits. These subunits are then further disassembled to eventually end up with simple starting materials, thus simplifying the structure step by step, while ensuring that all the steps can be reversed at each stage. After working backwards in this way, it is then possible to build the target molecule. Corey has subsequently shown that retrosynthetic analysis is amenable to computer programming, which means that potential synthetic pathways can be generated with the help of computers.

With this widely applicable method Corey and his research group have been able to complete a large number of total syntheses (complete chemical syntheses of complex molecules from simple starting materials), thereby making it possible to produce many biologically active, complicated natural products.

Legacy, truth, consequence

■ In order to synthesize large numbers of complex molecules, Corey has had to develop numerous new, or considerably improved, methods. These can be applied by chemists all over the world. Many synthetic reactions now carry Corey's name.

■ Thanks to Corey's contributions, many pharmaceuticals can be synthesized and are commercially available. He is probably most famous for the total syntheses of a group of molecules called eicosanoids, which includes prostaglandins, thromboxanes, and leucotrines. These molecules occur naturally only in small quantities and are frequently very unstable. They are hormone-like compounds that control many bodily systems and are very important medically because they are involved in the induction of labor, in blood pressure control, blood clotting, and the treatment of infections and allergies.

■ Corey is also well known for the synthesis of ginkgolide B, a substance found in very small amounts in the roots of the ginko tree, used in Chinese folk medicine. This compound can now be synthesized and is used to treat circulatory problems and asthma.

■ The Corey research group at Harvard University continues to study the total synthesis of a variety of complex molecules and to develop new methods and strategies for the construction of these molecules. In addition, the group participates in collaborative studies on diseases such as arthritis, asthma, cardiovascular diseases, and AIDS, with the aim of understanding the disease mechanisms at a molecular level. The application of computers in organic synthesis also continues to be studied.

Key dates

1928	Born in Methuen, Massachusetts, US.
1945	Enters Massachusetts Institute of Technology.
1948	Obtains undergraduate degree in chemistry. Stays on as a graduate student, working on John C. Sheehan's (1915–92) pioneering program on synthetic penicillins.
1950	Obtains a doctorate in chemistry.
1951	Becomes an instructor at the University of Illinois, Urbana-Champaign.
1957–8	Receives a Guggenheim fellowship and spends sabbatical leave at Harvard University, Massachusetts, as well as in Switzerland, England, and Sweden. Starts to develop his ideas on a logical strategy for chemical synthesis.
1959	Becomes professor of chemistry at Harvard University.
1960s	Develops the concept of retrosynthetic analysis. Synthesizes prostaglandins.
1961	Marries Claire Higham, with whom he has three children.
1968	Becomes Sheldon Emery professor at Harvard University.
1988	Achieves the synthesis ginkgolide B.
1990	Awarded the Nobel Prize for Chemistry.

My special fascination has been to understand better the world of chemistry and its complexities ... The naturally occurring organic substances are the basis of all life on earth, and their science at the molecular level defines a fundamental language of that life ... Chemical synthesis is uniquely positioned at the heart of chemistry, the central science. Its impact on our lives and society is all pervasive.

Elias James Corey (1990)

This diagram shows the synthesis for oseltamivir (an antiviral drug marketed under the trade name Tamiflu) designed by E. J. Corey following his method of retrosynthetic analysis.

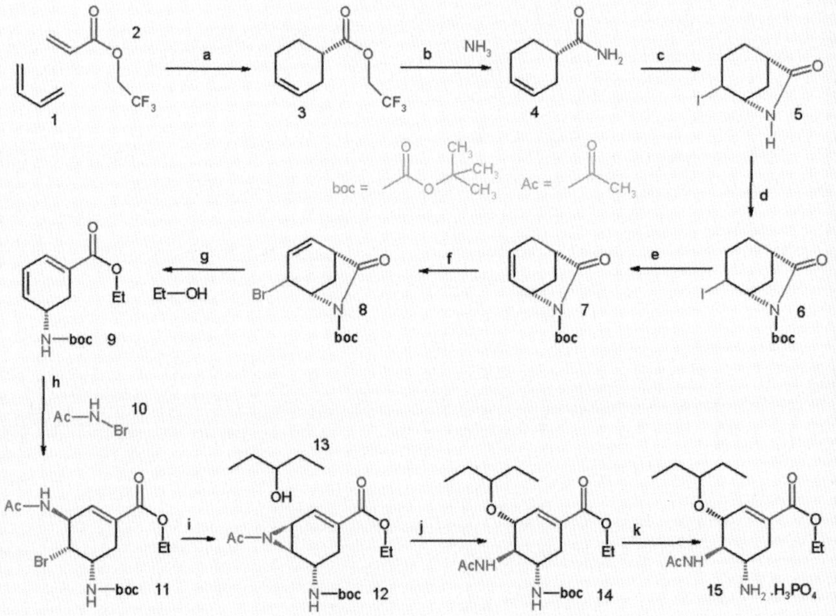

Sylvia Earle

An oceanographer and environmentalist, Sylvia Earle made many contributions to our knowledge of marine life, particularly ocean algae and humpback whales. She was one of the first marine biologists to carry out research at first hand, underwater, and was also one of the first women scientists to break into active research areas.

A born naturalist, as a child Sylvia Earle enjoyed exploring the wildlife and botany found in the woods around her home on a small farm. She became fascinated with marine life and the ocean through trips to the seaside, and when she was 13, her family moved to the west coast of Florida, on the Gulf of Mexico, putting the habitats of sea and shore on her doorstep.

Although her parents could not afford to send her to university, Sylvia won a scholarship and specialized in marine botany, learning scuba diving and beginning active marine exploration. In 1968 – when four months pregnant – she became the first woman scientist to look through the porthole of a submersible (a craft operating under water).

By 1969 she had spent more research hours undersea than any other American scientist, but she was still turned down by the Tektite Project, a new scheme run by NASA, the Smithsonian Institution, the US Department of the Interior, and the US Navy to place people in underwater habitats for weeks at a time. The managers argued that they did not want a mixed group of women and men, so in 1970

Earle organized a separate group of women aquanauts for the Tektite II Project. They spent two weeks in an underwater craft submerged 50 feet (15 meters) below the ocean surface near the Virgin Islands, spending up to 12 hours a day in the water.

It was during this project that Earle first noticed the damage that pollution was causing to coral reefs. She was also concerned about how impure the ocean waters were becoming, and from then on she became a leading light in the conservation and environmental movements.

Her group in Tektite II was the first ever all-women submarine expedition, and led to a surge of interest from the media, giving Earle a publicity platform that she used to inform the public about the science of the oceans and about the dangers of pollution.

I want to get out in the water. I want to see fish, real fish, not fish in a laboratory.

Sylvia Earle

Essential science

Active research

Instead of simply studying specimens in a laboratory, Earle was one of the pioneers of active marine research. She believed that biologists needed to explore marine habitats and observe species – whether plants or fish – within their own **ecosystems**. To this end she swam with whales, walked on the ocean floor, explored the ocean depths in submersibles, and spent two weeks living underwater.

Scuba equipment had only recently been invented when Earle made her first scuba dive at the age of 17, and since then she has led more than 60 research expeditions, spending more than 7,000 hours underwater. In these trips she was able to make discoveries such as luminescent coral that pulses when touched, and make observations on the behavior of fish and whales that could not be seen in dead laboratory specimens.

During her record-breaking depth dive in 1979, Earle was only connected to her support submersible. Unlike most deep dives, she was not tethered to the surface. After this exploit she was

nicknamed "Her Deepness" by the media. In 1985 she broke another record, by making the deepest ever solo descent in a submersible, to 3,280 feet (1,000 meters). Earle also designed and tested submersibles, technology which might be applicable for space missions.

Algae

Earle's first and central interest was the algae of the Gulf of Mexico, and her 1966 doctoral thesis on this is still a valuable resource. Over time, she collected more than 20,000 samples of algae from the Gulf waters.

Publicity

Earle thought that ignorance is one of the main threats to ocean life, so she was concerned to inform ordinary people, other scientists, and the government about marine life and ecosystems. In her many books, lectures, and television programs, her enthusiasm and sense of wonder about the ocean helped to

> *... planetary health, planetary wealth, is very directly linked to the health of the oceans. To the extent that we take care of the sea, we will help insure our ultimate survival and well being. To the extent that we ignore this ... we are in trouble.*

American Academy of Achievement interview (1991)

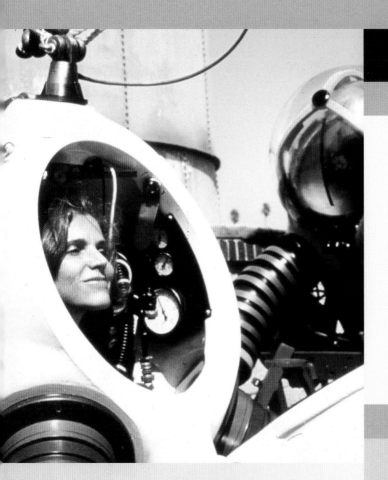

Dr Sylvia Earle prepares to dive in a JIM suit, an atmospheric diving suit designed for deep diving.

Legacy, truth, consequence

- Earle was one of the first marine biologists to intensively study her subject from the inside, by spending long stretches of time underwater.
- She was instrumental in forcing the scientific community to accept the presence of women on long scientific expeditions, and she is an inspiration to other women who want to carry out active research.
- Other biologists have marked their respect for Earle by naming species of marine life after her: the sea urchin *Diademai sylvie* and the alga *Pilinia earleae*.
- Earle has received a host of awards recognizing her pioneering work and her environmental campaigns, including the Living Legend award by the Library of Congress. In 1998 *Time* magazine named her its first "hero for the planet".

Key dates

1935	Born in New Jersey, US.
1966	Her doctoral dissertation on aquatic plants is the first extensive and detailed study of marine botany. Goes on to be a researcher at several academic institutes.
1970	Leads a team of women living below sea for two weeks.
1976–86	Becomes curator of **phycology** at the California Academy of Sciences.
1979	In pressurized diving suit, makes a depth record by walking 1,250 feet (381 meters) below sea level.
1980–4	Serves on the President's Advisory Committee on Oceans and Atmosphere.
1990–2	Becomes chief scientist of America's National Oceanic and Atmospheric Administration (NOAA), responsible for monitoring oceanic pollution.
1992	Co-founds two companies to design and build sub-sea equipment.
1998–2002	Leads Sustainable Seas Expeditions, a National Geographic-sponsored program to study the United States National Marine Sanctuaries.

popularize marine science and also alerted ordinary people to the dangers faced by pollution. She has written more than 150 publications on marine science ranging from introductory children's books to academic papers, and in 1980 she presented her groundbreaking research on the lives of humpback whales in a documentary film, *Gentle Giants of the Pacific*.

Environmentalism

The more time Earle spent in the water, the more she realized how much damage had been done by human activities. She played a leading role in raising public awareness of the need to protect the oceans and sealife, making the compelling argument that 97 per cent of the world's water is ocean, and human survival could depend on making sure it is still habitable.

Earle also campaigned against over-fishing, and was particularly enraged by the widespread deaths caused by discarded fishing traps and huge trawler nets floating around the seas, that go on needlessly catching and killing sealife for years.

Günter Blobel

A pioneering cellular and molecular biologist, Günter Blobel discovered the key mechanisms of protein transportation and localization within the cell, also known as "protein targeting".

Günter Blobel was born in 1936 in the Silesian village of Waltersdorf, in eastern Germany (now in Poland). At the age of nine, Blobel and his family were forced to flee ahead of the advancing Russian Red Army to relatives close to Dresden. Blobel managed to continue his education in Freiberg but discovered that his further education would be hindered under the East German regime due to his perceived "capitalist" origins, so he crossed the border into West Germany to study medicine in Frankfurt and then at the University of Tübingen.

After graduating in 1960, Blobel moved to the US to take a doctorate in **oncology** at the University of Wisconsin. It was in 1967 that he joined the laboratory of the pioneering scientist George Palade (*b*. 1912), at the Rockefeller University (formerly the Rockefeller Institute). Blobel's collaboration with Palade's laboratory led to revolutionary discoveries in **protein** transportation and signaling, for which he earned the Nobel Prize for Physiology or Medicine in 1999. By this time Blobel had settled in the United States and assumed citizenship, marrying artist Laura Maioglio.

Essential science

Protein targeting

When Günter Blobel joined the laboratory at the Rockefeller University, Palade and his colleagues had already made a significant breakthrough in protein science, having established that secretory proteins are able to permeate the membrane of special functional structures (known as organelles) found within certain cells, and to move between cells. The **biochemical** mechanisms involved in these processes, however, such as how the proteins moved around and established their desired location within the cell, were not known.

After conducting a series of experiments spanning approximately 20 years, Blobel and his colleagues showed that proteins contain a code of amino acids, an "address tag", within their structure that guides them to their desired destination. Depending on their function, the proteins are enabled by their destination code to permeate an organelle membrane, integrate alongside the membrane, or be exported outside the cell. This ability of proteins to locate their destination by integrated programming, which is subsequently recognized by a receptor protein at the desired destination, is called "protein targeting".

Legacy, truth, consequence

- When Blobel embarked on his career in the field of **cellular** and **molecular biology**, little was known about the cell and its components, largely because of the limitations of a light microscope, which was the only available technology at the time.
- Blobel's findings not only established how proteins are transported and localized within the cell, but also presented the principles for cellular membrane formation and overall cellular organization. This formed an essential foundation for the development of bioengineered drugs – pre-programmed therapeutics targeting specific cell types – such as insulin. His research also had a significant effect on the development of the treatment of the congenital disease, cystic fibrosis.

What began as an inquiry ... proceeded into an exciting voyage that revealed the principles by which cells organize themselves ...

Nobel Prize Lecture (1999)

Key dates

1936 Born in Waltersdorf (now Niegoslawice), Silesia, in eastern Germany, now in Poland.

1945 Forced to take refuge near Dresden, a city that is to remain dear to him for the rest of his life.

1960 Graduates with a medical degree from the University of Tübingen, West Germany.

1967 Earns a doctorate in oncology from the University of Wisconsin, US, and joins George Palade's Laboratory of Cell Biology at the Rockefeller University.

1976 Appointed professor at the Rockefeller University.

1999 Continuing the work of Palade and colleagues and discovering that protein transportation and localization in the cell is regulated by signals earns him the Nobel Prize for Physiology or Medicine.

Sidney Altman

A Canadian–American molecular biologist, Sidney Altman shared the 1989 Nobel Prize for Chemistry with Thomas R. Cech for their independent discoveries of the catalytic properties of RNA, or ribonucleic acid, in genetics. This discovery has far-reaching implications for chemistry, biochemistry, and medicine.

Altman did not have an academic background; his father was a grocer and his mother had been a mill-worker. But he was drawn towards science, moving from Canada to the US to study physics at the Massachusetts Institute of Technology, then taking a doctorate in **biophysics** in Colorado.

He worked as a teaching assistant, switching to the relatively new discipline of **molecular biology**, then became a researcher. While working at the University of Cambridge in England from 1969 to 1970, he first began to intensively examine ribonucleic acid (**RNA**), which is involved in the translation of the genetic material deoxyribonucleic acid (**DNA**) into proteins.

Although they worked independently, Altman at Yale and Thomas R. Cech (*b.* 1947) at the University of Colorado both discovered at much the same time that RNA, far from being just a carrier of genetic codes to different parts of a cell, could also take on an active **catalytic** role.

> *Major progress in the study of RNase P has resulted from crystallography of bacterial catalytic subunits ... Our current world should be called the "RNA-protein world" rather than the "protein world".*
>
> A View of RNase P (2007)

Essential science

When Altman started his research it was not clear how DNA, or **genetic** information, is conveyed into living cells to instruct the process of growth. The standard view was that nucleic acids such as RNA simply carried the genetic codes from DNA, which created **enzymes** made of proteins, which in turn triggered or catalyzed living cells into vital chemical and biological reactions.

The actual catalyst that causes molecular change was ribonuclease-P (RNase P), an enzyme made up of RNA and a protein. Altman and his team (and Cech and his research group) discovered conclusively that it is only the RNA component that is the catalyst for biochemical development, and therefore RNA can take on active enzymatic functions.

Altman also identified intermediary stages. DNA first undergoes a change into a long strand of what is called precursor RNA, which contains extra genetic sequences at each end of its strand. The enzyme RNase P works at this stage by removing these additional sequences, and converting the strand into the small component of RNA called transfer RNA or tRNA. It is this tRNA that actually helps in the development of proteins.

Legacy, truth, consequence

- The work of Altman and Cech contributed to our understanding of how life originates and develops. They showed that nucleic acids are some of the basic building blocks of life, acting as both genetic codes and enzymes.

- The discovery of RNA's catalytic activities opened up completely new areas and directions of scientific research. Many different types of RNA have since been discovered, all with different roles to play in genetic expression.

- There are important and optimistic medical implications from the discovery of RNA's actions. Since the RNase P enzyme cuts out specific sections of RNA, it is possible that the enzymes could be applied to cut out infectious or abnormal sequences from the genetic material of a patient with certain diseases such as cancer or AIDS.

Key dates

1939	Born in Montreal, Canada.
1960	Graduates with a physics degree from Massachusetts Institute of Technology.
1967	Awarded a doctorate in biophysics from the University of Colorado.
1967–9	Becomes a Fellow in molecular biology at Harvard University.
1969–70	Carries out research at the University of Cambridge, England.
1971	Joins biology faculty at Yale University.
1980	Appointed full professor at Yale.
1983–5	Appointed department chairman.
1985–9	Serves as dean at Yale.
1984	Takes out US citizenship while keeping Canadian citizenship.
1989	Receives Nobel Prize for Chemistry (shared with Thomas R. Cech).

Richard Dawkins

The British zoologist, ethologist, and evolutionary biologist Richard Dawkins popularized the gene-centered view of evolution which suggests that evolutionary change is driven by the survival needs of the gene. He coined the term "meme" as a unit of cultural transmission, and in his many popular science books he argues vehemently for rationalism and against religious superstition.

Dawkins' father was sent to Kenya from Britain as a soldier during World War II, but the family returned to England when Dawkins was a young child. As a boy he wavered between religious belief and atheism, then settled into atheism when he was introduced to the theory of **evolution** and natural selection and realized that this explains all the complexity of life without the need to introduce a "designer" or creator.

As an undergraduate studying **zoology** at Oxford University he was heavily influenced by one of his tutors, Nikolaas Tinbergen (1907–88), the Nobel Prize-winning Danish **ethologist** who was a pioneer in the study of instinct and learning in animal behavior. Dawkins began to focus more and more on animal behavior at the same time as working out his own theories of the primacy of the **gene** in evolution. His postgraduate research was in the field of animal decision-making.

Dawkins was always keen to communicate scientific ideas to the general public. As a result, his first book, *The Selfish Gene* (1976), was written in a clear way that everyone could understand. It became a best-seller, proving his contention that the public wants to explore advanced scientific ideas as long as they are not presented in too dry an academic style.

His several other books were also successful, with his 2006 *The God Delusion* being the most popular yet. Never afraid of controversy or challenge, in this book he argues that there is no supernatural creator of the world, and that religious faith is practically self-delusion. He is also dismissive of "New Age" superstitions, and is actively associated with sceptical and **humanist** associations in the UK and the US. Dawkins has been called by journalists "Darwin's Rottweiler", as a reference to the nineteenth-century biologist Thomas Huxley who was called "Darwin's Bulldog" for his defense of **Charles Darwin**'s original theory of evolution.

As well as books, Dawkins has published many scientific papers, and has explained the life sciences through lectures, articles, and up-to-date methods such as DVDs and a CD-ROM, which allows users to interactively explore evolutionary stages.

Although many scientists disagree with some of his premises, he is well respected within the scientific community, and has received many honors (both literary and scientific). His numerous public roles include being on the judging panel for scientific awards, and serving as president of the Biological Sciences section of the British Association for the Advancement of Science.

Essential science

Gene-centered evolution

In *The Selfish Gene* Dawkins presented his argument that the driving force of evolution is not a species or a population or an individual organism, but the gene. He theorized that natural selection takes place at this much reduced level, and the bodies of living organisms are nothing more than the carriers of genes, used by genes to ensure their own survival. Our impetus to survive and reproduce is driven by our genes' needs, not our own. "*We are survival machines – robot vehicles blindly programmed to preserve the selfish molecules known as genes,*" he wrote.

He developed this conclusion partly by applying the study of animal behavior to genes, looking at how they compete and survive, and how they behave as individuals or in groups. He also took ideas from the emerging field of information technology: with an understanding that animal behavior can often appear machine-like, he added the theory that life is an information process; evolution is basically a transfer of binary information from gene to gene, and an individual gene is nothing but a coded information system or replicator.

These ideas are explored more thoroughly in his 1982 book, *The Extended Phenotype*, in which he said that natural selection is "*the process whereby replicators out-propagate each other*". But, while genes compete, they also cooperate in order to survive and reproduce, and they might especially be prepared to assist copies of themselves that are carried in an individual's relatives. Dawkins also argued that the **phenotype**, or the outward appearance determined by genes and environment, extends beyond the body to include the wider setting, such as the nests of birds or the dams of beavers, and in modern humans our entire technology.

Memes

Dawkins did not originate the theory of cultural evolution as a parallel to genetic evolution but he provided it with a complete

Legacy, truth, consequence

- Dawkins galvanized the field of evolutionary biology.
- His many ideas and his ability to communicate them clearly have brought him millions of fans from within science and from the general public.
- Partly inspired by his work, a whole new scientific area of study, memetics, has grown up to study memes as units of culture.
- *Time* magazine considered him to be one of the world's 100 most influential people in 2007.

In *The Selfish Gene*, Dawkins discusses the effect that a gene can have on an organism's environment, citing as an example how beaver dams have kept waterways healthy and in good repair for hundreds of thousands of years.

framework, coining the term "meme" for the unit of culture which is transmitted in human society. He described the meme as similar to the gene – a basic replicator or information code that tries to survive and self-replicate.

Ideas or memes compete, cooperate, stagnate, or mutate, just like genetic replicators. Everything from temporary fashion trends to complex philosophical concepts can be explained by memes. Ideas that are only adopted by small numbers of people are unsuccessful memes, whereas the concept of god is a meme with an extremely high survival value, that has successfully transmitted itself over generations and in most human societies.

According to Dawkins, memes can infect individuals or societies. In his 1991 article "Viruses of the Mind", he wrote: "*Like computer viruses, successful mind viruses will tend to be hard for their victims to detect. If you are the victim of one, the chances are that you won't know it, and may even vigorously deny it.*"

Today the theory of evolution is about as much open to doubt as the theory that the earth goes round the sun.

The Selfish Gene (1976)

Key dates

1941	Born in Nairobi, Kenya.
1949	Moves to England with his family.
1959	Studies zoology at Balliol College, University of Oxford.
1967–9	Moves to University of California, Berkeley, US, as assistant professor of zoology.
1970	Returns to University of Oxford to lecture in zoology.
1976	Publishes his first book, *The Selfish Gene*, arguing for gene-centered evolution and introducing his concept of cultural memes.
1982	His more academic book, *The Extended Phenotype*, expands the gene-centered evolution hypothesis.
1986	In his book *The Blind Watchmaker* he explains modern theories of evolutionary processes and natural selection.
1995	Appointed Simonyi Professor for the Public Understanding of Science at the University of Oxford.
1996	Produces interactive CD-ROM, *The Evolution of Life*.
2006	In his ninth book, *The God Delusion*, he argues that a creator god almost certainly does not exist.
2006	Sets up an educational charity: the Richard Dawkins Foundation for Reason and Science.

The argument of this book is that we, and all other animals, are machines created by our genes.

The Selfish Gene (1976)

Rationalism

Dawkins' support for the scientific method as a way of establishing facts led him to criticize religion and advocate **rationalism** and atheism instead. He has argued that religion is a mind virus, that faith without evidence is "*lethally dangerous nonsense*", and that followers of astrology, spiritualism, and other "New Age" practices have abandoned critical thought. He is particularly scathing about creationism, the fundamentalist religious argument that the universe and life on earth were created by a deity exactly as described in several holy books. In *The Blind Watchmaker* (1986) he argued that the complexity of life can adequately be explained by natural selection; there is no need to posit an intelligent designer of the universe.

Stephen Hawking

Stephen Hawking is an English physicist who has pioneered research in the field of black holes and furthered human understanding of the creation, evolution, and present structure of the universe. Having done much to bring theoretical physics to a public audience, he is indisputably the world's most famous and recognizable living scientist.

Stephen Hawking was born in 1942; his birthplace in Oxford, England, was a strategic decision by his parents because the British and German governments had a pact that Oxford and Cambridge would not be bombed during World War II. He was brought up in Highgate, London, and later in nearby St Albans, and throughout this time was a studious, intellectual child.

Inspired in part by his father, a specialist in tropical diseases, Hawking was interested in fundamental scientific questions from his early teens. Already with ambitions to carry out research in science, he began a physics degree at Oxford University.

After completing his degree, he applied to be a doctoral candidate in **cosmology** at Cambridge University, under the impression that he would be supervised by the British astronomer Fred Hoyle (1915–2001). However, to Hawking's disapointment, when he arrived at Cambridge his supervisor was not Hoyle but Denis Sciama (1926–99), with whom he was unfamiliar. But what appeared to be an initial setback turned out to be a blessing because Sciama proved to be a stimulating and supportive mentor to Hawking.

Soon after his arrival at Cambridge, Hawking was diagnosed with amyotrophic lateral sclerosis (ALS), the motor neurone disease, otherwise known as Lou Gehrig's disease, an ailment that produces weakness and muscle wastage. Doctors gave him only years to live. Rather than surrendering to his apparent fate, this news instilled in Hawking the drive to capitalize on his ability and realize his ambitions to discover the secrets of the universe.

In 1965 he married Jane Wilde, who supported him through the trauma of ALS, and they later had three children. He completed his doctorate, and went on to begin a Fellowship at Caius College, Cambridge.

Physical deterioration generated by his ALS eventually confined Hawking to a wheelchair. Over the years his speech became slurred and research students would often have to read his lectures on his behalf. In 1985, after an operation which removed his ability to speak altogether, he was fitted with a computer system and speech synthesizer enabling him to deliver public lectures with an electronically-generated voice.

In the early nineties he and Jane separated and he married his nurse, Elaine Mason, a few years later. However, they separated in 2006 amid unsubstantiated tabloid rumors of a Hawking affair and physical mistreatment on her part, allegations they both denied.

Alongside his technical work, Hawking continues to be a public figure. In 2007 he became the first quadriplegic to experience microgravity flight in order to prepare for a proposed Virgin Galactic sub-orbital spaceflight in 2009.

Essential science

Singularities

Working on **general relativity** between 1965 and 1970, Hawking and Roger Penrose (*b.* 1931) developed new mathematical techniques designed to show that in the past there must have been a state of infinite density, or **Big Bang** singularity, where all the galaxies are on top of each other and the density of the universe is infinite. Hawking and Penrose also made progress with general relativity's prediction that massive stars collapse in on themselves when they have exhausted their nuclear fuel, showing that this collapsing would continue until they reached a **singularity** of infinite density, the gravitational field of which would be so strong that not even light could escape from the region but would instead be pulled back inwards. This is the region known as a "**black hole**". Its boundary is an "event horizon".

Black holes

From his cosmological research of the late 1960s, Hawking realized that many of the techniques he had developed on singularities could be applied to black holes. Thus black holes became his central research interest for the next few years.

Prior to Hawking, scientists believed that nothing at all could escape a black hole. Hawking discovered that under certain conditions, where the event horizon interacts with pairs of **virtual particles** and attracts one of the particles inside, a black hole could emit certain **subatomic particles**, known as "Hawking Radiation".

Importantly, Hawking discovered that this radiation has a perfect thermal **radiation spectrum**, just like any other body in the universe. He showed that black holes have **entropy**, they have

Legacy, truth, consequence

■ Hawking's book, *A Brief History of Time*, in which he attempts to summarize his theories about black holes and the nature of the universe for a general readership, is undoubtedly one of the most famous popular scientific texts of all time, selling millions of copies all over the world, with translations into numerous languages. Although Hawking has acknowledged that the majority of readers will not understand everything in the book, he believes that most people have an abiding interest in the workings of the universe in which they live and will learn something from the work.

■ Hawking is a great believer in making his views available to the general public and has frequently courted celebrity, with appearances on numerous television shows, including *The Simpsons* and *Star Trek: The Next Generation*. The popularization of science has also extended to a children's book series, coauthored with his novelist daughter Lucy. The first book in the series, entitled *George's Secret Key to the Universe*, has George and a supremely intelligent computer called Cosmos embark upon an adventure through space, time, and the universe.

A simulated view of a black hole.

Key dates

1942	Born in Oxford, England.
1950	Moves from Highgate to St Albans.
1962	Gains a degree in physics from Oxford University.
1965	Marries Jane Wilde. Awarded a doctorate from Cambridge University.
1973	Publishes *The Large Scale Structure of Space-Time*, coauthored with G. F. R. Ellis (*b.* 1939).
1974	Discovers that black holes are not completely black and can emit radiation. Publishes "Black Hole Explosions?" in *Nature*. Elected Fellow of the **Royal Society**.
1979	Appointed Lucasian Professor of Mathematics at Cambridge University.
1981	Publishes *Superspace and Supergravity*.
1983	Publishes *The Very Early Universe*.
1988	Publishes *A Brief History of Time*.
1991	Separates from Jane Wilde.
1993	Publishes *Black Holes and Baby Universes*.
1995	Marries Elaine Mason (divorces 2006).
1996	Publishes *The Nature of Space and Time*, coauthored with Roger Penrose.
2002	Edits *On the Shoulders of Giants*, a collection of the great works of physics and astronomy.
2005	Edits *God Created The Integers: The Mathematical Breakthroughs That Changed History*, a collection of great works in the history of mathematics.
2006	Awarded Copley Medal from the Royal Society.
2007	Experiences zero-gravity flight.

I picture the origin of the universe, as like the formation of bubbles of steam in boiling water. Quantum fluctuations lead to the spontaneous creation of tiny universes, out of nothing. Most ... collapse to nothing, but a few ... will expand ... and will form galaxies and stars, and maybe beings like us.

Millennium Mathematics Project, University of Cambridge (Issue Jan 18, 2002)

a temperature, and they are not entirely black. They also obey the laws of **thermodynamics**, and eventually they evaporate.

Big Bang theory

In the late 1970s Hawking called the Big Bang theory into question, suggesting instead the hypothesis that the universe has no distinct beginning and, equally, no end. He developed the "no boundary principle", which claims that space-time is finite but does not have any boundaries, thus removing the singularity at the beginning of the universe.

A Theory of Everything

Since 1974 Hawking has worked on combining general relativity and quantum mechanics into a consistent theory, most famously popularizing his view and, with it, theoretical physics itself, in his 1988 best-seller, *A Brief History of Time*. He continues to investigate the possibility of such a combination, in order to achieve what is known as a "Grand Unified Theory" or a "Theory of Everything": an explanation, manifested in one single equation, of the origins of the universe; one unifying picture.

Ahmed H. Zewail

An Egyptian–American chemist, Ahmed Zewail developed a technique using rapid lasers to actually see what was happening to atoms and molecules during chemical reactions. His pioneering work resulted in a new field of physical chemistry called femtochemistry.

Ahmed Zewail grew up near the ancient city of Alexandria in Egypt. He was always keen on science, but in 1960s Egypt would-be university students applied to a central bureau who assigned them a university and a faculty. Zewail was lucky enough to be given the subject of his inclination, science, at the University of Alexandria.

For further research he wanted to go to America, and although he received a grant offer from the University of Pennsylvania, he had to overcome bureaucratic difficulties to get there, since Egypt at that time perceived America as being a supporter of its then enemy, Israel.

America proved to be a culture shock, but Zewail's researches and publications on a range of subjects were successful, and in 1976 he settled at Caltech, the California Institute of Technology. It was there that he began to work on his ground-breaking examinations of molecules using laser technology.

In this race against time, femtosecond resolution is the ultimate achievement ...

Extract from Zewail's Nobel Lecture (1999)

Essential science

Femtoseconds

Chemical reactions happen so quickly that they can only be described in tiny divisions of a second, femtoseconds. A femtosecond is just 0.000000000000001 of a second, or 10^{-15}, and during the transition state of a chemical reaction the **atoms** of a **molecule** move extremely rapidly, taking less than 100 femtoseconds to rearrange themselves. Until Zewail's discovery, most scientists thought that it would never be possible to actually see what happens during such a fast reaction.

Fast lasers

Zewail realized that the new technology of fast lasers might supply the sort of super-fast "camera" that chemistry needed. Fast lasers are able to produce flashes of light that last just a few femtoseconds in duration. Zewail used the series of flashes to initiate a chemical reaction and record the changes.

After many experiments, he developed a process involving mixing molecules in a vacuum tube, then beaming pulses at the mixture from a fast laser. The first flash of light energizes the

Legacy, truth, consequence

■ Zewail's discovery changed chemists' views of chemical reactions. Instead of imagining what was happening, they were able to see them in action, and more easily predict and control their experiments.

■ **Femtochemistry** is expected to have applications in **bioscience** and **electronics**, as well as within chemistry. The Nobel Prize Foundation said: "*Scientists the world over are studying processes with femtosecond spectroscopy in gases, in fluids, and in solids, on surfaces and in polymers. Applications range from how catalysts function and how molecular electronic components must be designed, to the most delicate mechanisms in life processes and how the medicines of the future should be produced.*"

Key dates

1946	Born in Damanhur, near Alexandria, Egypt.
1967	Graduates with a science degree from the University of Alexandria.
1969	Awarded a master's degree; moves to the University of Pennsylvania, US.
1974	Gains a doctorate.
1976	Moves to Caltech, California.
1980s	Uses laser technology to study molecular reactions.
1990	Appointed first Linus Pauling professor of chemical physics at Caltech.
1982	Becomes an American citizen.
1999	Awarded the Nobel Prize for Chemistry.

chemicals, starting the reaction, and the subsequent beams record the resulting light patterns or spectra from the molecules. These can then be analyzed to determine how the molecules are structurally changing. To a chemist this is the equivalent of watching as the chemical bonds are broken and formed.

Zewail's technique became known as femtosecond **spectroscopy** or femtospectroscopy.

Craig Venter

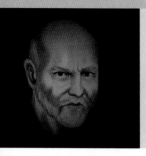

J. Craig Venter is an American geneticist and high-profile businessman who was instrumental in mapping the human genome. His visionary contribution to genomic research has seen him heralded as one of the most important scientists of the twenty-first century.

Born into a working-class neighborhood of Salt Lake City, US, John Craig Venter showed little application for education as a teenager, preferring instead to spend his time surfing. However, after military hospital work in Vietnam – which drove Venter to the brink of suicide – he returned a changed man, motivated to study.

He was awarded a doctorate in **physiology** and **pharmacology**, and gained an academic position at the State University of New York, Buffalo.

Having parted from his wife Barbara Rae, Venter married one of his then students, microbiologist Claire Fraser. He took up a position at the National Institutes of Health, which includes the Human Genome Project. But, following disagreements with his colleague **James D. Watson**, he left and has since founded several companies – The Institute for Genomic Research, Celera, and Synthetic Genomics – all designed to investigate the field of **genomics**.

Legacy, truth, consequence

■ With the collective achievement of Celera and the Human Genome Project in 2000, for the first time the three billion DNA letters of the human genome had been read and put in their correct order. A small percentage of the human genome still remains un-sequenced, as mapping of these few regions is difficult using current technology.

■ The data published by the HGP does not represent the exact sequence of each and every individual's genome (all humans have unique gene sequences): it is the combined genome of a group of anonymous donors.

> *The future of life depends ... perhaps in creating new synthetic life ... not forged by Darwinian evolution but created by human intelligence.*
>
> The Richard Dimbleby Lecture (2007)

Essential science

Genomic research and the human genome

The Human Genome Project began in the 1980s as an international scientific research project to determine the complete genome sequence of a human being. This required the determination of the sequence of chemical base pairs, which make up **DNA**, for each of the **chromosomes** in a human.

Backed by biotech firm PerkinElmer, Venter founded Celera in 1998. He challenged the publicly-funded Human Genome Project (HGP) to a race, promising to decipher the genetic code, using shotgun sequencing (a fast system), for a tenth of the cost in a fraction of the time, thus provoking one of the most exciting races in the history of science. Venter's aim caused great controversy because of his expressed wish to use the discovery for commercial purposes. However, Celera eventually caved in to industry campaigns claiming that the data should be shared.

In the end, Venter and Francis Collins (head of the HGP) made a joint announcement on the mapping of the human genome in

Key dates

1946	Born in Salt Lake City, Utah, US.
1984	Joins the National Institutes of Health (NIH).
1991	Pioneers fast new approach to gene discovery using Expressed Sequence Tags (ESTs).
1992	Founds The Institute for Genomic Research (TIGR).
1998	Founds Celera Genomics.
2000	Co-announces the mapping of the human genome (complete draft in 2003).
2002	Terminated from Celera Genomics.
2005	Founds Synthetic Genomics.
2007	Announces the development of *Mycoplasma laboratorium*, potentially the first synthetic bacteria. Delivers the Richard Dimbleby lecture.

2000. President Clinton declared their contest officially a draw and claimed their achievement marked "*a day for the ages*".

Artificial life to combat climate change

He is presently working on reducing the earth's dependence on fossil fuels by developing an organism that is capable of producing energy. Believing that every era is defined by its technologies, Venter claims that the twenty-first century will be "*fundamentally shaped by advances in biology, particularly the field of genomics*".

Tim Berners-Lee

Sir Tim Berners-Lee is an English computer scientist who revolutionized communications and the flow of information by inventing the World Wide Web. Often dubbed "the father of the web", he is the founder and director of the World Wide Web Consortium (W3C), an organization to promote global web standards and development.

Tim Berners-Lee was born in London, England. As the son of two mathematicians, Conway Berners-Lee and Mary Lee Woods, who both worked on the Manchester Mark I computer, he entered a household that coveted complex mathematical problems. The family frequently puzzled over them at the dinner table.

He attended the Sheen Mount Primary School and the Emmanuel School, Battersea, London, and then went to Queen's College, Oxford, to study physics, graduating from there in 1976. During his time at Oxford, he fostered his interest in **computer science**, assembling a makeshift computer from assorted remnants, including an old processor and a television set. Unfortunately, on one occasion he was banned from using the university computers after an incident involving computer hacking during rag week.

After leaving university, Berners-Lee married his first wife Jane, and worked for several companies, such as Plessey Telecommunications Ltd, D. G. Nash Ltd, and Image Computer Systems Ltd, in the field of communications and computer technology. Jane, who he'd met at Oxford, also worked for Plessey and the couple settled in Poole on the south coast of England, where the company was located.

He also briefly worked for CERN (The European Organization for Nuclear Research), the world's largest particle physics laboratory, in Geneva, Switzerland, as an independent contractor in 1980. Four years later he returned there on a Fellowship, and it was at CERN, in early 1989, that he wrote his initial proposal for the **World Wide Web**. Before the year ended, he had brought his idea to fruition, and throughout the early 1990s he continued to work for CERN on its design.

In 1994 he left CERN and founded the World Wide Web Consortium (W3C), the main international standards organization for the Web. The organization was based at the Massachusetts Institute of Technology (MIT) and was supported by the Defense Advanced Research Projects Agency (DARPA) and the European Commission. Five years later he was appointed to the 3Com (Computer Communication Compatibility) Chair at the laboratory for computer science, MIT.

In 2004 Berners-Lee accepted a chair in computer science at the School of Electronics and Computer Science, University of Southampton, UK, to work on his new project named "the Semantic Web". He and his second wife Nancy divide their time between Southampton and their base in Lexington, Massachusetts, US.

Essential science

Invention of the World Wide Web

Although Berners-Lee invented the World Wide Web in 1989, he laid the foundations for this innovation in 1980 when he was employed as a temporary software consultant at CERN. It was then that he wrote a computer program called "Enquire", which he called a "memory substitute" because it allowed him to record the connections between various people and projects at CERN through **Hypertext** links. These links allowed a computer user to "jump" from one document to another related document.

When he returned to CERN in 1984 he had a vision of a global information space where computers were linked in a vast network and data was freely available to all. He immediately set to work. In 1989 Berners-Lee combined his use of Hypertext with the **Internet**, at this time a basic communications and computer-to-computer networking tool for scientists and the military, which had been around for approximately 20 years. This combination resulted in the World Wide Web, an invention that allowed computer users to share their information by combining their data in a web of Hypertext documents. The basic idea was that Hypertext would allow documents to be linked to one another and the Internet would be the vehicle through which they were transmitted.

Although CERN was initially unresponsive to his innovation, Berners-Lee continued to make progress. In 1990 he wrote the Hypertext Transfer Protocol (HTTP), the language computers use to transmit Hypertext documents across the Internet, and he developed the idea of giving the documents on the Web an address, which he called a Universal Resource Identifier (URI – later known as URL, or Universal Resource Locator). In the same year he wrote a client program (or browser) to enable the

Legacy, truth, consequence

■ Berners-Lee declined the chance to patent his invention because he thought this would hinder its chances of spreading, and he has campaigned in favor of keeping it out of the hands of private companies. He believes that the web should be free and open to all.

■ He has campaigned against the use of exclusive domain names like ".xxx" and ".mobi", claiming that all users should be able to access the same Web, and he has argued that no one should own domain names like ".com".

■ Writing recently in the *Observer*, Internet historian John Naughton described him as the *"man who invented the future"*. There are currently more than 100 million active websites on the World Wide Web.

■ In 2006 MIT and the University of Southampton launched the Web Science Research Initiative (WSRI) as a multidisciplinary project to study the social and technological implications of the growing Web adoption. Berners-Lee sees this as an opportunity to study the impact on society of the Web and to deal with problems resulting from misuse of the technology, such as viruses and spam email.

Berners-Lee's own computer was used as the first web server hosting the first ever website, which was put online in 1991.

Hypertext documents to be viewed, which he called "WorldWideWeb", he wrote the Hypertext Markup Language (HTML) with which to format the Hypertext pages, and he wrote the first web server, known as "info.cern.ch".

CERN was still slow to realize his achievements, so Berners-Lee used the Internet community to spread the word about his invention. In 1991 he made his WorldWideWeb browser and web server publicly available. Within five years the number of Web users jumped from 600,000 to 40 million.

World Wide Web Consortium

Concerned that private companies would seek to promote projects to destroy the open nature of the Web, in 1994 Berners-Lee created the World Wide Web Consortium (W3C), an organization to "*develop common protocols to enhance the interoperability and evolution*

Key dates

1955	Born in London, England.
1973–6	Studies physics at Oxford University.
1978	Works for D. G. Nash Ltd.
1980	Software consultant at CERN.
1981–4	Works at Image Computer Systems Ltd.
1984	Begins Fellowship at CERN.
1989	Invents the World Wide Web.
1990	Writes the first web client and server.
1991	Puts the first website online.
1991–3	Works on the design of the web.
1994	Founds the World Wide Web Consortium (W3C).
1999	Appointed to the 3Com (Computer Communication Compatibility) Chair at the laboratory for computer science, MIT. Publishes *Weaving the Web*. Included by *Time* magazine in their list of the 100 most influential people of the twentieth century.
2001	Appointed Fellow of the **Royal Society**.
2002	Awarded the Japan Prize. Named among the 100 Greatest Britons by the British public in a BBC poll.
2004	Knighted by Queen Elizabeth II. Accepts a chair in computer science at the School of Electronics and Computer Science, University of Southampton, UK. Awarded the Millenium Technology Prize.
2006	Publishes "A Framework for Web Science" in *Foundations and Trends in Web Science*, coauthored with several colleagues.
2007	Awarded the Order of Merit by Queen Elizabeth II. Inducted into the National Academy of Engineering. Awarded the Charles Stark Draper Prize.

> *The Web is a tremendous grassroots revolution. All these people coming from very different directions achieved a change. There's a tremendous message of hope for humanity in that.*
>
> Interview in the *Independent* (May 17, 1999)

of the Web" on royalty-free standards. He has remained director of the consortium ever since.

The Semantic Web

Since the mid-1990s Berners-Lee has been working on a project called the Semantic Web, a system of technology designed to aid the organization and correlation of data regardless of whether it is on a website, in a database, or contained within a piece of software.

Select Bibliography and Further Reading

Listed here is a selection of sources that the reader may wish to consult in addition to the sources noted in the individual entries on scientists.

Ackerknecht, Erwin H., *Rudolf Virchow: Doctor, Statesman, Anthropologist* (Madison, 1953)

Adler, Robert, *Science Firsts: From the Creation of Science to the Science of Creation* (John Wiley & Sons, 2002)

Appleyard, Rollo, *Pioneers of Electrical Communication* (Books for Libraries Press, 1930)

Appleyard, Rollo, *Pioneers of Electrical Communication* (Books for Libraries Press, 1930)

Arnold, Lois Barber, *Four Lives in Science: Women's Education in the Nineteenth Century* (Schocken Books, 1984)

Babkin, B. P., *Pavlov, A Biography* (University of Chicago Press, 1949)

Baxter, Stephen, *Ages in Chaos: James Hutton and the Discovery of Deep Time* (Forge, 2003)

Berners-Lee, Tim, *Weaving the Web: The Past, Present and Future of the World Wide Web by its Inventor* (Texere, 1999)

Blunt, Wilfrid, *Linnaeus The Compleat Naturalist* (Frances Lincoln, 2004)

Bowers, Brian, *Michael Faraday and Electricity* (Priory Press Limited, 1974)

Brown, G. I., *Invisible Rays: The History of Radioactivity* (Sutton Publishing, 2002)

Browne, Janet, *Darwin's Origin of Species: A Biography* (Atlantic Books, 2006)

Calle, Carlos I., *Einstein for Dummies* (Wiley, 2005)

Carson, Rachel, *The Sea Around Us* (Signet, 1961)

Chorley, Richard J., Beckinsale, Robert P., & Dunn, Antony J., *The History of the Study of Landforms: Vol. 2, The Life and Work of William Morris Davis* (London: Methuen, 1973)

Christianson, G., *Edwin Hubble: Mariner of the Nebulae* (Straus & Giroux, 1995)

Claxton, K. T., *Wilhelm Roentgen* (Heron Books, 1970)

Clegg, Brian, *The first scientist: A life of Roger Bacon* (London: Constable, 2003)

Connor, James A., *Pascal's Wager: The Man Who Played Dice with God* (HarperOne, 2007)

Coveney, Peter, & Highfield, Roger, *The Arrow of Time: The Quest to Solve Science's Greatest Mystery* (Flamingo, 1991)

Coveney, Peter, & Highfield, Roger, *The Arrow of Time: The Quest to Solve Science's Greatest Mystery* (Flamingo, 1991)

Creese, Mary R. S., *Ladies in the Laboratory? American and British Women in Science, 1800–1900: A Survey of their Contribution to Research* (The Scarecrow Press, 1998)

Crowther, J. G., *British Scientists of the Nineteenth Century* (Kegan Paul, 1935)

Dickinson, Alice, *Carl Linnaeus, Pioneer of Modern Botany* (Franklin Watts, 1967)

Everitt, C. W. F., *James Clerk Maxwell: Physicist and Natural Philosopher* (Charles Scribner's Sons, 1974)

Fedoroff, N.V., *Barbara McClintock in Genetics 136(1)* (1994)

Fermi, Laura, *Atoms in the Family: My Life with Enrico Fermi* (George Allen & Unwin, 1955)

Francis, Keith A., *Charles Darwin and The Origin of Species* (Greenwood Press, 2007)

Frayn, Michael, *The Human Touch* (Faber & Faber, 2006)

Gardiner, C. I., *An Introduction to Geology* (G. Bell & Sons, 1914)

Gascoigne, John, *Joseph Banks and the English Enlightenment: Useful Knowledge and Culture* (Cambridge University Press, 1994).

Geikie, Archibald, *Life of Sir Roderick I. Murchison* (John Murray, 1875)

Geison, Gerald L., *The Private Science of Louis Pasteur* (Princeton University Press, 1995)

Gilman, Daniel C., *Life of James Dwight Dana* (Ayer Co Pub, 1977)

Glasser, Otto, *Wilhelm Conrad Röntgen and the Early History of the Roentgen Rays* (John Bale, Sons & Danielson, 1933)

Grey, Vivian, *Secret of the Mysterious Rays: The Discovery of Nuclear Energy* (Constable Young Books, 1966)

Guevellou, Jean-Marie Le, *Louis Pasteur* (Hart-Davis, 1981)

Hager, Thomas, *Force of nature: The Life of Linus Pauling* (Simon & Schuster, 1995)

Hamilton, James, *Faraday: The Life* (HarperCollins, 2002)

Hayes, J.R. (ed.), *The Genius of Arab Civilization* (Eurabia, 2nd edn, 1983)

Henig, Robin Marantz, *A Monk and Two Peas* (Weidenfeld & Nicolson, 2000)

Hodges, Andrew, *Alan Turing: The Enigma of Intelligence* (Unwin Paperbacks, 1986)

Hodgson Mazumdar, P.M., "The Linnaeans: Ferdinand Cohn and Robert Koch" in *Species and Specificity: An Interpretation of the History of Immunology* (Cambridge University Press, 1995)

Hoskin, Michael (ed.) *Cambridge Illustrated History of Astronomy* (Cambridge University Press, 1997)

Hunter, Michael (ed.) *Robert Boyle by Himself and His Friends* (William Pickering, 1994)

Isaacson, Walter, *Einstein: his Life and The Universe* (Pocket Books, 2008)

Jaffe, Bernard, *Michelson and the Speed of Light* (Heinemann, 1961)

Lodge, Oliver, *The Work of Hertz and Some of his Successors* (The Electrician, 1894)

Mahon, Basil, *The Man Who Changed Everything: The Life of James Clerk Maxwell* (Wiley, 2003)

McIntyre, Donald B., & McKirdy, Alan, *James Hutton: The Founder of Modern Geology* (National Museum of Scotland Publishing Ltd, 2001)

McMurray, Emily J. (ed.), *Notable Twentieth-Century Scientists* (Gale Research, 1995)

Menand, Louis, *The Metaphysical Club* (Flamingo, 2001)

Meyer, Herbert W., *A History of Electricity and Magnetism* (The MIT Press, 1971)

Michelson Livingston, Dorothy, *The Master of Light: A Biography of Albert A. Michelson* (University of Chicago Press, 1973)

Miller, Dayton Clarence, *Sparks, Lightening, Cosmic Rays: An Anecdotal History of Electricity* (MacMillan, 1939)

Morgan, Michael Hamilton, *Lost History* (National Geographic Society, 2007)

Murchison, Roderick Impey, *Murchison's Wanderings in Russia: His Geological*

Murdin, Paul (ed.), *Encyclopedia of Astronomy and Astrophysics* (Bristol: Institute of Physics Publishing, 2001)

Needham, Joseph, *Science and Civilisation in China* (Cambridge, 1954)

Pancaldi, Giuliano, *Volta* (Princeton University Press, 2003)

Pincus, Gregory, *The Control of Fertility* (Academic Press, 1965)

Proffitt, Pamela (ed.), *Notable Women Scientists* (Gale Group, 1999)

Repcheck, Jack, *The Man Who Found Time: James Hutton and the Discovery of the Earth's Antiquity* (Simon & Schuster, 2003)

Romer, Alfred, *The Restless Atom: The Awakening of Nuclear Physics* (Dover Publications, 1982)

Schuck, H. & Sohlman, R. *The Life of Alfred Nobel* (William Heinemann, 1929)

Segrè, Emilio, *Enrico Fermi: Physicist* (University of Chicago Press, 1970)

Segrè, Emilio, *From X-Rays to Quarks:*

Modern Physicists and Their Discoveries (University of California, 1980)

Shapin, Steven, & Schaffer, Simon, *Leviathan and the Air-Pump: Hobbes, Boyle, and the Experimental Life* (Princeton University Press, 1985)

Soresini, Franco, *Alessandro Volta* (Be-Ma Editrice, 1988).

Stahle, Nils, *Alfred Nobel and the Nobel Prizes* (The Nobel Foundation, 1978)

Sugimoto, Kenji, *Albert Einstein: A Photographic Biography* (Schocken Books, 1989)

Suplee, Curt, *Milestones of Science* (National Geographic Society, 2000)

Swenson, Loyd S., *The Ethereal Aether* (University of Texas Press, 1972)

Swenson, Loyd S., *The Ethereal Aether* (University of Texas Press, 1972)

Tolstoy, Ivan, *James Clerk Maxwell: A Biography* (Canongate, 1981)

Turing, Sarah, *Alan M. Turing* (W. Heffer and Sons, 1959).

Vitezslav, Orel, *Mendel: The First Geneticist* (Oxford University Press, 1996)

Weber, R. L. (ed.) *More Random Walks in Science* (The Institute of Physics, 1982)

Westacott, Evalyn, *Roger Bacon in Life and Legend* (New York: Philosophical Library, 1953)

Williams, Trevor (ed.), *Collins Biographical Dictionary of Scientists* (HarperCollins, 1994)

Williams, Trevor I. (ed.), *A Biographical Dictionary of Scientists* (Adam & Charles Black, 3rd edn, 1982)

Glossary

ALCHEMY/ALCHEMIST A pseudo-scientific, often spiritual and philosophical, predecessor of chemistry.

ALGEBRA The mathematics of generalized arithmetical operations.

ALGORITHM An organized procedure for performing a given type of calculation or solving a given type of problem.

ANGLE OF INCIDENCE The angle between a ray of light striking a surface (an incident ray), and an imaginary line drawn perpendicular to the surface at the point of the ray's contact.

ANNULAR ECLIPSE An **eclipse** of the sun when the moon does not completely cover the sun, so that a ring of sunlight is still seen around the circle of the moon.

ANODE The electrode connected to the positive side of the battery.

ANTHROPOLOGY The study of the physical and social characteristics of humanity encompassing its origins, its institutions, and its belief systems.

ANTIMATTER Matter that is the same as normal or conventional matter, but has an opposite electrical charge.

ARGUMENT FROM DESIGN An argument for the existence of God, which says that the universe is so complex it must have been created by an all-powerful, all-knowing being.

ARISTOTLE'S METHOD OF INDUCTION See **Induction**

ARMILLARY SPHERE Astronomical instrument consisting of a series of rings representing different aspects of the celestial sphere. It was used to determine the position of stars.

ARTIFICIAL INTELLIGENCE The ability of a machine to imitate human behavior, and an interdisciplinary field, encompassing **computer science**, **neuroscience**, philosophy, **psychology**, robotics, and linguistics, which aims to reproduce human behavior and human thought in machines. Often known by the acronym AI.

ASTROLABE An early scientific instrument that calculated the position of the sun or other celestial bodies compared to the horizon, and was used for reckoning time, for astronomical observation, and for navigation.

ASTROLOGY The theory and practice of the positions and aspects of celestial bodies in the belief that they have an influence on human affairs.

ASTRONOMY The observational study of the universe.

ASTROPHYSICS A branch of **astronomy** concerned with the physical and chemical properties, structure, and evolution of cosmic objects, including the universe itself.

ATOM The smallest chemical particle of an element (a substance composed of atoms). Atoms consist of outer **electrons** and an inner **nucleus**.

ATOMIC PHYSICS The field of **physics** that studies atoms as an isolated system of **electrons** and an atomic **nucleus**.

ATOMIC STRUCTURE The structure of an **atom**.

AXON The long projection of a nerve cell or neuron that conducts electrical impulses; axons together make up the primary transmission lines of the **nervous system**.

BACTERIA Microscopic organisms that are present everywhere, including in the human body.

BACTERIOLOGY The scientific study of **bacteria**, a branch of **microbiology**.

BEHAVIORISM In **psychology**, the view that human behavior is to be understood exclusively in terms of patterns of environmental stimuli and responses.

BEHAVIORAL PSYCHOLOGY See **Behaviorism**

BIG BANG THEORY A theory of the creation of the universe according to which the universe began about 15 billion years ago with a massive explosion, and continues to expand.

BIOCHEMISTRY The study of the **chemistry** of living organisms.

BIOLOGY The study of living organisms.

BIOMOLECULE Any **organic molecule** that is an essential part of a living organism.

BIOPHYSICS The **physics** of **biological** systems.

BIOSCIENCE Any of the branches of natural science dealing with the structure and behavior of living organisms.

BLACK HOLE An infinitely dense, infinitely small point in space caused by some dying stars (those with a mass of at least three times that of our sun) continuing to collapse in on themselves. This is one of the last possible stages in the life of a star such as the sun, when the nuclear energy source at the center of a star is exhausted and the outward energy no longer balances its **gravitational** force, so it begins to collapse under the influence of its own gravity. A black hole is the most compressed form of a dying star, and its gravity is so strong that not even light can escape from it.

BOTANY The study of plant life.

BROWNIAN MOTION The continuous random motion of molecules in a fluid or gas.

CALCINATION The change of a metal to a powder or calx (now called an oxide), such as that obtained when iron rusts.

CALCULUS A method of advanced mathematical analysis.

CATALYSIS A chemical reaction caused by the introduction of a material (a catalyst) which does not itself undergo any permanent chemical change.

CATHODE RAYS Rays emitted by the cathode or electrically negative plate of a **discharge tube**. They are now known to consist of **electrons**.

CATHODE The electrode connected to the negative side of the battery.

CELLULAR BIOLOGY The study of the structure, physiology, and life of cells. Also known as cell biology.

CHEMICAL ENGINEERING The activity of applying **chemistry** to the solution of practical problems.

CHEMICAL REVOLUTION The process in the late eighteenth century during which chemistry transformed to a modern science. It centered around **Antoine Lavoisier**'s discoveries of the law of conservation of mass and the oxygen theory of combustion, and his introduction of scientific methods and modern naming of chemical elements.

CHEMISTRY The study of the composition, structure, and properties of matter, and its interactions and changes.

CHINESE ROOM ARGUMENT A thought experiment devised by the philosopher John Searle (b. 1932) designed to show that a machine could never exhibit human understanding or **intentionality**.

CHROMOSOME The threadlike structures in the **nucleus** of a **cell** which carry **DNA**.

CIRCUMSCRIBED CIRCLE In **geometry**, a circle drawn around a polygon so that it passes through all the vertices of the polygon.

CLASSICAL (NEWTONIAN) PHYSICS Long-standing models of the universe, particularly those of **mechanics** and **electromagnetism** described by Sir **Isaac Newton** and **James Clerk Maxwell**, that developed before the rise of **quantum physics** in the twentieth century.

COMPUTER SCIENCE The systematic study of computing systems and computation.

CONCHOLOGY The study of shells or mollusks.

CONTROL THEORY A branch of **mathematics** and engineering concerned with dynamical systems.

COPENHAGEN INTERPRETATION The version of **quantum physics** put forward in 1927 by **Niels Bohr** and his colleagues at the Institute of Theoretical Physics in Copenhagen. It accepted indeterminacy and probability, and proposed that the quantum effect is ignored in physics at a macroscopic level. **Albert Einstein** and some others never really accepted this interpretation, although it became the paradigm.

COPERNICANISM The **heliocentric** theories proposed by **Nicolaus Copernicus** that the earth rotates on its axis once a day and revolves around the sun.

COSMOLOGY The study of the universe and man's place in it.

CREATIONISM The belief that the universe and all life was created by a deity.

CROOKES TUBE A type of **discharge tube**.

CRYPTANALYSIS The process of obtaining encrypted information.

CRYSTALLOGRAPHY The study of crystals.

CYBERNETICS The study of the control processes in biological and artificial systems.

CYCLOTRON Circular particle accelerator used to produce high-energy charged particles by **electromagnetism**. The particles may be used for research or for medical applications.

CYTOGENETICS The study of the structure, function, and abnormalities of human **chromosomes**.

DARWINISM Scientific school or theories influenced by the scientific ideas of **Charles Darwin**, particularly evolution through natural selection, and variation due to genetic mutation.

DECIDABILITY QUESTIONS (IN PURE MATHEMATICS) Questions about the effective method for determining membership in a set of formulas.

DECISION PROBLEM In computability theory (the study of whether problems can be solved computatioally), a problem in a formal system that can be answered with either a "yes" or a "no".

DEFRACTION The phenomenon that occurs when a wave encounters an obstacle.

DENDRITE A thread-like extension from a nerve cell that serves as an antenna to receive messages from the **axons** of other nerve cells.

DETERMINISM The view that the universe operates in causal or predetermined ways so that accurate predictions can be made about physical events.

DIFFERENTIAL EQUATION A mathematical equation involving several derivatives or differentials of a function of a variable.

DIFFERENTIAL GEARS A gear system that either receives one input, but splits the output into two, or that receives two inputs and produces an output that is the sum or difference of the two.

DISCHARGE TUBE A tube containing a vacuum in which visible and invisible discharges can be produced by the application of high voltage differences to their electrodes at the end of the tube. Lenard, Hittorf, and **Crookes tubes** are types of discharge tube.

DNA (DEOXYRIBONUCLEIC ACID) A chemical found mainly in the **nucleus** of cells that stores and transmits **genetic** information.

DOUBLE-BLIND STUDY An experiment in which neither the researchers nor the experimental subjects know who is receiving a placebo and who is receiving the active test substance.

ECLIPSE, SOLAR The phenomenon that occurs when the moon is directly between the sun and the earth, and the moon's shadow is cast upon the earth.

ECOLOGY The study of the relationship between living organisms and their environment.

ECOSYSTEM A community of organisms and the physical environment in which they interact.

ELECTRIC GENERATOR A machine that converts mechanical energy to electrical energy.

ELECTRICAL TRANSFORMER A device that transfers electrical current from circuit to circuit.

ELECTROCHEMISTRY The study of the interchange of chemical and electrical energy.

ELECTRODE An electrical conductor that makes contact with non-metallic parts of a circuit, such as an **electrolyte**.

ELECTRODYNAMICS The study of the interaction of electric currents with **magnetic fields**.

ELECTROLYSIS The decomposition of a substance by passing an electric current through a liquid.

ELECTROLYTE An electrically conductive liquid.

ELECTROMAGNETIC WAVES Waves that consist of an electric field in conjunction with a **magnetic field** oscillating with the same frequency.

ELECTROMAGNETISM The interaction of electric and magnetic forces, which are actually aspects of the same physical phenomenon.

ELECTRON DEGENERACY THEORY See **Pauli exclusion principle**

ELECTRON TUBE A device in which conduction of electricity by **electrons** takes place through a vacuum or a gas within a sealed container.

ELECTRON An elementary particle carrying a negative charge which groups around the nucleus of an atom.

ELECTRONICS The branch of technology concerned with the development and application of circuits or systems using electron devices.

EMPIRICISM A movement which regards the general notion of "experience" gained through the five senses as the central source of human knowledge. Often defined in opposition to **rationalism**.

EMPIRICAL A general term, which might apply to propositions, statements, knowledge, etc., which indicates a connection to experience. For example, an item of knowledge is empirical if the way an individual gains that knowledge depends in some way on sensory experience.

ENDOCRINOLOGY The study of endocrine glands and hormones, i.e. any of various glands that secrete hormones directly into the blood or lymph and not through a duct.

ENDOSPORES Dormant, non-reproductive, and highly resistant structures formed by some **bacteria** to preserve their **genetic** material in hostile conditions (e.g. heat, radiation, chemical agents).

ENLIGHTENMENT A movement in the eighteenth century that stressed the importance of reason and science in philosophy and the study of human society.

ENTOMOLOGY The study of insects.

ENTROPY A physical quantity that is the measurement of the amount of disorder in a system.

ENZYME A protein, made in the body, that brings about or speeds up a specific chemical reaction but is not itself changed or destroyed.

EPICYCLE In Ptolemy's geocentric model of the universe, a small circle – representing the path of the planets – rolling around a greater circle that was effectively centered on the earth.

EPIDEMIOLOGY The study of the frequency, distribution, and control of diseases in human populations.

EQUINOX When the sun is directly above the earth's equator.

ETHNOGRAPHY The scientific description and classification of the various cultural and racial groups of humankind.

ETHOLOGY/ETHOLOGIST The study of animal behavior.

EUCLIDEAN GEOMETRY Elementary geometry based on **Euclid**'s axioms.

EVOLUTION The gradual and continuous process by which the first and most primitive of living organisms have developed into the plants and animals of today.

FEMTOCHEMISTRY A branch of **chemistry** that studies chemical reactions as they occur at extremely fast timescales.

FERMENTATION In food or drink processing, the action of bacteria or yeast in breaking down sugars, releasing energy and changing the original liquid or

substance. Examples are bread-making and brewing.

FIELD THEORY A theory that looks at physical phenomenon in terms of a field or region in which the matter's properties allow physical forces to operate, and the manner in which it interacts with matter or with other fields.

FLUORESCENCE The glow induced in certain materials when exposed to light.

FUNCTIONALISM The view in philosophy that mental states are constituted by their functional role.

GALAPAGOS ISLANDS A group of islands in the Pacific Ocean inhabited by numerous endemic species.

GALVANIC BATTERY A battery consisting of a number of voltaic cells arranged in series or parallel. See also **Volta-electricity**

GALVANISM The contraction of a muscle stimulated by an electric current.

GALVANOMETER An instrument that measures electric current.

GAUSSIAN CURVATURES In differential geometry, the location of principle curvature of a given point.

GAUSSIAN PROBABILITY DISTRIBUTION The values and probabilities associated with a random event.

GENE A unit of chromosome that codes for a specific hereditary trait.

GENERAL RELATIVITY A theory of **gravitation** that describes gravity as a property of the geometry of space and time. The theory published by **Albert Einstein** in 1916.

GENETIC ENGINEERING A technology used to alter the **genetic** material of living cells in order to make them capable of producing new substances or performing new functions.

GENETICS The study of **genes**.

GENOME The hereditary information of an organism encoded in its **DNA**.

GENOMICS The study of the complete **genetic** make up of an individual.

GEOCENTRISM The belief that the earth is the center of the universe.

GEODESY The study by direct measurements of the size and shape of the earth.

GEODETIC (adj) Of or relating to the science of **geodesy**.

GEOLOGY The study of the origin, history, and structure of the earth.

GEOMAGNETISM The study of the natural magnetism of the earth.

GEOMETRY The branch of **mathematics** to do with the measurement of, the properties of, and the relationships between points, lines, angles, curves, and surfaces.

GEOPHYSICS The study of the physical properties of rocks and minerals.

GOLDEN SECTION The proportion of the two divisions of a straight line such that the smaller is to the larger as the larger is to the whole line. It is also known as the golden mean, the golden ratio and the divine proportion.

GRAVITATION The force of attraction between any two objects that have mass.

GRAVITATIONAL SINGULARITY A point in space-time in which gravitational forces are so strong as to cause matter to have an infinite density and zero volume.

GREGORIAN CALENDAR The calendar now used internationally, devised by Pope Gregory in the 1580s to replace the **Julian calendar**.

HELIOCENTRISM The theory that the sun is the center of the solar system, not the earth.

HISTOLOGY The study of the structure and behavior of cells and body tissues.

HYDROSTATICS The study of the mechanical properties of liquids which are at rest, not in a state of motion.

IDEAL GAS A hypothetical gas which obeys Boyle's law (after **Robert Boyle**) exactly at all temperatures and pressures and which has internal energy that depends only on the temperature.

IMMUNOLOGY The study of disease and the body's response to disease.

INCIDENCE, ANGLE OF See **Angle of incidence**

INDUCTION (ELECTRICITY) An electrical phenomenon whereby an electromotive force is generated in a closed circuit by a change in the flow of current.

INDUCTION COIL Any coil of wire used to introduce **induction** into a circuit.

INDUCTION, ARISTOTLE'S METHOD OF A process in which the knowledge of the truth of axioms is drawn from experience and involves a combination of **empirical** examination of examples with conceptual analysis of exemplified notions. Named after the ancient Greek philosopher **Aristotle**.

INDUCTIVE REASONING The method of reasoning used in scientific investigation where the premises of the argument offer support for the conclusion but do not entail the conclusion.

INERTIA The property of matter that causes it to resist any change in its motion.

INORGANIC CHEMISTRY The study of the compounds of all elements other than covalently bonded carbon compounds (**organic compounds**) such as hydrocarbons.

INSCRIBED CIRCLE In **geometry**, a circle drawn inside a polygon so that it touches each side of the polygon at exactly one point.

INTEGER A positive or negative whole number or zero.

INTEGRAL CALCULUS The branch of mathematics concerned with working out the area, volume, mass, displacement, or other properties of an irregular shape.

INTELLIGENT DESIGN The theory that life and the universe did not come into existence by chance but were designed and created by an intelligent being or god.

INTENTIONALITY A philosophical notion that highlights the relation between a given mental phenomenon and its content (i.e. the thing the mental phenomenon is "about").

INTERNET A computer network allowing a group of connected computers to exchange data.

ION An **atom**, or group of atoms, that has lost or gained electrons and now therefore has an electric charge, either positive or negative.

IONIZATION The process of producing **ions**.

IRRADIATION Exposure to **radiation**.

IRRATIONAL NUMBERS Numbers that cannot be expressed as the ratio between two **integers**. When written in decimal figures, an irrational number has an infinite number of decimals. An example of an irrational number is **pi**.

ISOMERS Chemical compounds that have the same chemical formula but have different structural arrangements of the **atoms** within the **molecules**.

ISOTOPES Different nuclear forms of the same element.

JULIAN CALENDAR System of dating instituted by Julius Caesar and followed from 46 BCE. See also **Gregorian calendar**

KINETIC (ENERGY) The energy of a body resulting from its motion.

LATITUDE Lines of equal distance measured north and south of the equator.

LINEAR ACCELERATOR A machine accelerating high-energy charged particles in a straight line. Many applications are medical, e.g. to create high energy **x-rays** or to deliver **radiation therapy**.

LOGARITHM A particular mathematical transformation often used to express economic variables.

LOGIC The study of the formal structure of reasoning and the principles of valid inference.

LOGICIAN A practitioner of formal **logic**.

LONGITUDE Distance east or west on the earth's surface measured from the prime **meridian** (Greenwich, England) and expressed in degrees or time.

MAGNETIC FIELD The region around a magnet that experiences the force of attraction or repulsion of that magnet.

MAGNETIC MONOPOLE One single **magnetic pole**, rather than the usual complementary pair such as the north and the south poles.

MAGNETIC POLE Region on a body where magnetic inclination is at a maximum.

MAGNETO-ELECTRICITY Electric current produced by magneto-electric **induction**.

MANHATTAN PROJECT The Allied research project developing a nuclear weapon or atomic bomb during World War II.

MATHEMATICS The study of concepts such as quantity, structure, space, and change.

MATRIX MECHANICS A mathematical approach to exploring **quantum mechanics**, based on mathematical matrices, first put forward by **Werner Heisenberg** in 1925.

MECHANICS The branch of **physics** or **mathematics** concerned with the action of forces on bodies, which includes **kinetics** and statics.

MEDICAL ETHICS The study of moral values and judgements as they apply to medicine.

MEDICINE The practice of averting unnecessary human deaths by studying, diagnosing, and treating patients.

MERIDIAN An imaginary circle around the globe that runs through both the north and the south poles. Also known as a longitude line.

METALLOGRAPHY The study of the crystalline structures of metals and alloys.

METEOROLOGY The scientific study of the atmosphere – the layer of gases surrounding the earth. Focuses on weather processes and forecasting.

METHOD OF EXHAUSTION A mathematical technique for finding the properties of circular bodies by drawing a straight-sided polygon around the outside of the circle and another inside the circle, then adding sides to the polygons until they approximate the area of the circle. The area and other properties are easier to calculate for a polygon than for a circle.

MICROBES/MICROORGANISMS Tiny forms of life, such as **bacteria**, some of which cause disease.

MICROBIOLOGY The study of **microbes**.

MIND-BODY PROBLEM The problem in philosophy concerned with the relationship between the mind and the brain.

MINERALOGY The study of minerals.

MODULATED WAVEFORM The mathematical representation of a wave, especially a graph obtained by plotting a characteristic of the wave against time.

MOLECULAR BIOLOGY The field of **biology** that studies the molecular level of organization.

MOLECULE The smallest naturally-occurring particle of a substance.

M-THEORY A "master theory" involving an 11-dimensional universe unifying all current **string** and **gravitational** theories.

NANOTECHNOLOGY Engineering or building of machines at the tiny molecular scale.

NASA The US's National Aeronautics and Space Administration.

NATURAL PHILOSOPHY Term used until the development of modern science for all investigations into nature and the physical world.

NERVOUS SYSTEM The specialized network in a body whose main components are nerve cells or **neurons**.

NEUROBIOLOGY The branch of **biology** concerned with the structure and function of **cells** of the **nervous system**.

NEUROFIBRIL Any of the long, thin, microscopic fibrils that run through the body of a **neuron** and extend into the **axon** and **dendrites**, giving the **neuron** support and shape.

NEUROLOGY The study of disorders of the **nervous system**.

NEURONS Interconnected nerve cells in a body that transmit information by electrical impulse; the core components of the brain, spinal cord/nerve cord, and peripheral nerves, that make up the **nervous system**.

NEUROPSYCHIATRY The branch of medicine dealing with mental disorders attributable to diseases of the nervous system.

NEUROSCIENCE The study of the **nervous system**.

NEUTRON A subatomic particle with no charge, part of the atom's nucleus.

NEUTRON STAR An incredibly dense, small remnant of a star consisting mainly of **neutrons**. This is one of the last possible stages in the life of a star such as the sun, when the nuclear energy source at the center of a star is exhausted and the outward energy no longer balances its **gravitational** force, so it begins to collapse under the influence of its own gravity.

NITROCELLULOSE A flammable compound made by nitrating cellulose, i.e. introducing a nitro group, through exposure to nitric acid.

NMR (NUCLEAR MAGNETIC RESONANCE) The application of a **magnetic field** to the **nucleus** of certain **atoms**, producing resonance that can be measured. NMR spectroscopy is used to study a range of chemical structures.

NON-EUCLIDEAN GEOMETRY/GEOMETRICS Geometry which does not follow **Euclid**'s fifth postulate or parallel postulate, which effectively says that a parallel line is the only one possible line that does not intersect with another line. In contrast, hyperbolic (a non-Euclidean) geometry has many non-intersecting lines and in elliptical geometry every pair of lines intersect. Although mathematicians since Euclid's time tried to prove or disprove the postulate, this was not achieved until the 1820s. Non-Euclidean geometries eventually led to **Albert Einstein**'s Theory of **General Relativity**.

NUCLEAR FISSION The splitting of a **nucleus** of an **atom** into separate parts with lighter nuclei as well as some free **neutrons**. In the process, energy is released which can be captured for nuclear energy or used as the driving force of a nuclear explosion.

NUCLEAR PHYSICS The branch of **physics** concerned with the nucleus of the **atom**.

NUCLEAR REACTOR A device to contain a nuclear chain reaction, allowing the energy to be controlled. Nuclear power stations are large examples, whereas a nuclear-powered ship runs on energy generated within a small reactor.

NUCLEUS The positively-charged center of an **atom**.

NUMBER THEORY A field of **mathematics** concerned with the study of the properties of whole numbers.

ONCOLOGY The study of cancer.

OPTICS The branch of **physics** studying the properties and behavior of light.

ORGANIC CHEMISTRY The branch of **chemistry** that studies the composition, properties, and reactions of hydrocarbon compounds.

ORTHOMOLECULAR MEDICINE The alternative **medicine** according to which larger than usual doses of certain nutrients can actually prevent or cure disease.

OSCILLATOR In **physics**, an electronic circuit that generates a specific tone or frequency.

OXIDATION The process in which an **atom**, **ion**, or **molecule** loses **electrons**.

PALEOCLIMATOLOGY The study of climatic conditions – and their causes and effects – in the geologic past, using evidence found in glacial deposits, fossils, and sediments.

PALEONTOLOGY The study of prehistoric life.

PALEOZOIC An era of geologic time from about 544 to 250 million years ago. During this period the first fishes, amphibians, reptiles, and land plants developed.

PARABOLA A curved line representing the path of a projectile.

PARALLEL COMPUTING Using a string of computers to greatly amplify their computing power.

PATHOLOGY The study of disease.

PAULI EXCLUSION PRINCIPLE First proposed in 1925 by the Austrian physicist Wolfgang Pauli (1900–58), this states that within an **atom**, no two **electrons** can occupy exactly the same **quantum** state at the same time.

PETROLOGY The geological and chemical study of rocks.

PHARMACEUTICS Medicines, drugs, or other pharmaceutical compounds.

PHARMACOLOGY The study of the effects of drugs on living organisms.

PHENOTYPE The visible physical or biological traits of an organism, such as height, hair color, observable diseases. The phenotype results from a combination of genes and external environment.

PHOSPHORESCENCE A continuing luminescence without heat.

PHOTOELECTRIC EFFECT The phenomenon that occurs when a **photon** strikes a metal surface causing an **electron** to be ejected.

PHOTON The elementary particle of light and the **quantum** unit of light energy.

PHYCOLOGY The study of algae, a type of water plant.

PHYSICS The scientific study of the natural world and the laws of the universe.

PHYSIOLOGY The study of the physical functions of living organisms.

PI (π) Pi An important constant used in **mathematics**, science, and engineering. Approximately equal to 3.14159, in Euclidean **geometry** (after **Euclid**) it represents the ratio of a circle's circumference to its diameter, which is the same as the ratio of a circle's area to the square of its radius.

PLACEBO EFFECT An apparent improvement in health when the patient believes that he or she had been given effective treatment, even if the supposed **medicine** is actually inert.

PLACE-VALUE An individual element's value, from and in relationship to, the defined set of elements and values of which it is a part.

PLANT GENETICS A branch of **genetics** concerned with modifying crops in a scientific way to select for traits that are more desired by society.

PLATE TECTONICS A **geological** theory that explains the motions of the earth's lithosphere (the solid outermost shell of the planet).

PLATONIC SOLIDS Three-dimensional shapes that have equal sides, angles and faces.

POLARIZATION The limitation of **electromagnetic** waves' oscillation to a certain plane.

POLYHEDRON A three-dimensional solid made up of plane faces.

POSITRON The antiparticle to the **electron**, with a positive charge.

PRECAMBRIAN The earliest geological era, a long period from the beginning of the formation of the earth approximately 4.5 billion years ago until about 540 million years ago. It was followed by the Cambrian period during which the first primitive organisms appeared.

PRECESSION OF THE EQUINOXES The movement of the earth's rotational axis; the very slow shift of the earth's equinoctial points (points as determined by the position of the sun relative to the earth at an **equinox**) due to rotation of the polar axe, leading to a tiny difference in the annual position of the earth compared to the **sidereal** (stellar) constellations.

PROBABILITY THEORY/PROBABILISTIC PHYSICS In **physics**, a derivation of **quantum mechanics** suggesting that at the microscopic level, particles behave in a random fashion, partly due to the very act of observation. Therefore, the universe is not **deterministic**, and descriptions of **quantum** phenomena are only probabilities, not certainties. In **mathematics**, the analysis of random phenomena.

PROJECTIVE GEOMETRY (a **non-Euclidean geometry**) The geometry of properties that remain invariant or unchanged when

projected from a point to a plane or a line.

PROTEIN Complex organic compounds composed of long chains of amino acids.

PROTON A subatomic particle with a positive charge, part of the atom's nucleus.

PSYCHIATRY A branch of **medicine** concerned with mental illnesses.

PSYCHOLOGY The study of mental processes and behavior.

PYTHAGORAS' THEOREM The theorem that in a right-angled triangle, the square of the length of the hypotenuse (i.e. the longest side) is equal to the sum of the squares of the other two sides.

QUADRATIC EQUATION An equation in which one or more of the variables is squared, but nothing is raised to a higher power.

QUANTA (SINGULAR QUANTUM) Discrete units of energy.

QUANTUM ELECTRODYNAMICS Also called quantum field theory, the study of the interaction of electrically-charged particles at the **subatomic** level and their **electromagnetic** properties.

QUANTUM FIELD THEORY See **Quantum electrodynamics**

QUANTUM MECHANICS See **Quantum theory**

QUANTUM PARTICLES Elementary particles that behave in a way that resembles both waves and conventional particles.

QUANTUM STATISTICS The statistical measurement of the microscopic **quantum** system.

QUANTUM THEORY/QUANTUM PHYSICS Sometimes described as "new physics", the model of matter and energy that applies at the **atomic** and **subatomic** level. It is usually understood to incorporate philosophical ideas that scientific measurements can only be of probabilities, not certainties, and that the act of observing affects the experiment.

RADIATION The emission and propagation of radiant energy.

RADIATION SPECTRUM The range of all possible **electromagnetic** radiation in relation to a given object.

RADIATIVE TRANSFER/ENERGY The study of the absorption, emission, and scattering of energy in a system.

RADICAL (IN CHEMISTRY) A group of **atoms** that is capable of remaining unchanged during a series of chemical reactions.

RADIOACTIVITY The breakdown of atomic **nuclei** accompanied by the release of **radiation**.

RADIUM THERAPY The use of radium in **radiation** therapy.

RATIONALISM The theory that knowledge is gained through reason.

REFRACTION The change in direction of a wave due to its speed.

REGRESSION ANALYSIS A statistical technique for making quantitative predictions of one variable from the values of another.

RELATIVITY See **general relativity** and **special relativity.**

RENAISSANCE The cultural period in European history approximately from the fourteenth through seventeenth centuries, between the Middle Ages and the beginning of the modern world.

RNA (RIBONUCLEIC ACID) A nucleic acid found in all **cells**, involved in the translation of the genetic material deoxyribonucleic acid (**DNA**) into proteins.

ROYAL INSTITUTION OF GREAT BRITAIN An organization devoted to scientific education and research, based in London.

ROYAL SOCIETY OF LONDON/ROYAL SOCIETY Full name, Royal Society of London for the Improvement of Natural Knowledge, this is the oldest existing learned society for the advancement of science, founded in 1660.

SAL AMMONIAC A rare mineral composed of ammonium chloride.

SCIENTIFIC REVOLUTION The era of European history when great scientific advances were made, usually dated as beginning in the sixteenth and seventeenth centuries with its last stages stretching into the eighteenth and nineteenth centuries.

SCOLASTIC/SCOLASTICISM A tradition in medieval universities, between 1100 and 1500, to reconcile the philosophy of the ancient classical philosophers with medieval Christian theology.

SCOTTISH ENLIGHTENMENT Era of great intellectual, cultural and scientific achievements in eighteenth-century Scotland.

SEDIMENTARY ROCK Rock that is formed from sediments.

SIDEREAL ROTATION The rotation of the sun.

SIDEREAL YEAR The time for the earth to make one complete revolution around the sun, relative to the stars.

SILICA A chemical compound found in glass.

SIMULTANEOUS LINEAR EQUATION A system of linear equations, where each term is either a constant or a product of a constant, using the same set of variables.

SINGULARITY, GRAVITATIONAL See **Gravitational singularity**

SOLAR OR TROPICAL YEAR The time it takes the earth to pass once around the sun, measured between two vernal or spring equinoxes.

SPECIAL RELATIVITY, THEORY OF A physical theory proposed in 1905 by **Albert Einstein** that generalizes Galileo's principle of relativity (that all uniform motion is relative and that there is no privileged frame of reference). It incorporates the principle that the speed of light is the same for all non-accelerating observers, and is not affected by the state of motion of the source.

SPECTRAL LINES The pattern of colored lines on a **spectrum** chart produced when **electrons** release **electromagnetic** energy.

SPECTROSCOPY The study of the theory and interpretation of **electromagnetic** radiation (light) and its relationship to matter, particularly the distribution of energy emitted by a system. A **spectrum** (plural: spectra) is the chart or graph that shows the range of intensity of light emitted.

SPECTRUM ANALYSIS The study of the chemical properties of matter and gases by analyzing the bands in their optical spectrum.

SPECTRUM, SPECTRA See **Spectroscopy**

SQUARE ROOT The number or quantity that when multiplied by itself gives a given number or quantity. For example, the square roots of 4 are 2 and −2.

STATIC ELECTRICITY An electric charge that may appear on an insulator when it is rubbed.

STELLAR DYNAMICS The distribution of matter and motion in star systems.

STRING THEORY The theory that **subatomic** particles are actually one-dimensional strings, and that spacetime consists of several dimensions.

SUBATOMIC PARTICLES The parts that make up an **atom**.

SUPERNOVA The enormously powerful explosion caused by the gaseous envelope surrounding a dying star bursting away from the star's remaining core.

SYNTHETIC CHEMISTRY A branch of **chemistry** in which chemists devise ways to make specific compounds and develop new chemical reactions.

TAXONOMY The science of naming and classifying organisms.

THEORETICAL PHYSICS The description of natural phenomena in mathematical form.

THEORY OF EVOLUTION See **Evolution**

THEORY OF RELATIVITY See **Relativity**

THERMODYNAMICS The study of the movement of heat from one body to another.

THERMOLUMINESCENCE The release of previously absorbed **radiation** upon being heated.

THOUGHT EXPERIMENT A hypothetical situation designed to help the questioner explore a subject through reason and imagination. A famous example is "Schrödinger's Cat", involving a closed box containing a cat which might be killed during the experiment, but the implications of the experiment can be discussed without physically carrying it out.

TRANSMUTATION Changing form or character.

TRANSPOSON A discrete piece of **DNA** that can insert itself into other DNA sequences within the cell.

TRIGONOMETRY The branch of **mathematics** concerned with the sides and angles of triangles.

TROPICAL YEAR See **Solar year**

UNCERTAINTY PRINCIPLE Proposed by **Werner Heisenberg** in 1927, this principle shows that some predictions can never be accurately made at the **subatomic** level, and therefore casts doubt on any certainty in scientific knowledge.

UNIFIED FIELD THEORY The attempt to describe all the major physical forces and all of matter in one, single, all-encompassing framework. No one has yet created a satisfactory unified field theory.

VARIABLE STARS Stars which brighten and dim, some with precise regularity.

VIROLOGY The study of viruses.

VIRTUAL PARTICLE A particle that exists only for an extremely brief instant in an intermediary process.

VIRUS The smallest form of **microorganisms** capable of causing disease.

VOLTA-ELECTRICITY Electric current from a voltaic pile, a number of alternating disks of two different metals separated by pads moistened with acid: in effect, a series of primary cells.

WARRING STATES PERIOD The last 500 years of the Chou Dynasty in China (700–221 BCE), a period of cultural fragmentation and civil strife.

WAVE FUNCTION A mathematical function appearing in **quantum mechanical** theory that represents a particle.

WAVE MECHANICS A theory of matter based on the concept that elementary particles possess wave properties.

WHITE DWARF A dense, bright, slowly cooling remnant of a star about the size of the earth. This is one of the last possible stages in the life of a star such as the sun, when the nuclear energy source at the center of the star is exhausted and the outward energy no longer balances its **gravitational** force, so it begins to collapse under the influence of its own gravity. Only those stars under the Chandrasekhar limit (after **Subrahmanyan Chandrasekhar**) of 1.44 times the mass of our sun will stabilize as white dwarfs.

WORLD WIDE WEB The resources on the **Internet** that use the Hypertext Transfer Protocol (HTTP), a communications protocol used to transfer or convey information on the Web.

X-RAY High energy **electromagnetic** radiation with very short wavelengths, able to penetrate many substances such as bodily tissues.

ZOOLOGY The study of animals.

Index

About the contributors

Dr Emma Campbell is a science writer and editor. She is the author of entries on: Hipparchus, Roger Bacon, and William Morris Davis.

Nicola Chalton studied philosophy at University College London and is a writer and editor for history, science, and philosophy books. She is the editor of *Philosophers* in the same series ("They Changed the World"), the project editor of this title, and the author of the entry on Nicolaus Copernicus.

Jasmine Farsarakis studied biochemistry and is a writer and editor for various medical and scientific journals. She is the author of entries on: Hippocrates, Garcia de Orta, Edward Jenner, Rudolph Virchow, Marie Curie, Edwin Hubble, and Günter Blobel.

Susan Boyd Lees is a writer for educational, science, and history related publications. She is the author of the entry on Georgius Agricola.

Elizabeth Miles has worked as an editor and author for many years; she has more than 50 published titles, including reference books on history, geography, and natural history. She is the author of entries on: Carolus Linnaeus, James Dwight Dana, Gregor Johann Mendel, Ivan Petrovich Pavlov, Karl Ferdinand Braun and Guglielmo Marconi, Albert Einstein, and Linus Pauling.

Dr Simon Riches studied philosophy at the University of Southampton and University College London, where he is now a tutor. He is the author of entries on: Robert Boyle, James Hutton, Joseph Banks, Alessandro Volta, Michael Faraday, Roderick Impey Murchison, Charles Darwin, Robert Wilhelm Bunsen, Louis Pasteur, James Clerk Maxwell, Alfred Bernhard Nobel, Wilhelm Conrad Röntgen, Henri Becquerel, Albert Abraham Michelson, Santiago Ramón y Cajal, Heinrich Hertz, Florence Bascom, Alfred Wegener, Enrico Fermi, Gregory Goodwin Pincus, Alan Turing, Stephen Hawking, Craig Venter, and Tim Berners-Lee.

Dr Vanessa Schneider studied medicine and worked in clinical medicine before taking up a career as a scientific journal editor. She is the author of entries on: Ferdinand Cohn, Robert Koch, Barbara McClintock, Francis Crick and James D. Watson, and Elias James Corey.

Hugo Simms is a writer and editor of health guides, among other titles. He is the author of entries on: Leonardo da Vinci, Benjamin Franklin, John Dalton, Amedeo Avogadro, Justus von Liebig, Ludwig Boltzmann, Thomas Edison, Jean-Baptiste Dumas, and Kurt Alder.

Meredith MacArdle studied archaeology and anthropology at the University of Cambridge; she is the author of histories of China and Canada, the co-author of a timechart history of revolutions, and has written on a variety of science topics. She is the editor of this book and the author of entries on: Gan De, Aristotle, Zhang Heng, Ptolemy, Galen, Zu Chongzhi, Aryabhata, Al-Battani, Rhazes, Shen Kuo, Ibn al-Haytham, Avicenna (Ibn Sina), Al-Zarqali, Leonardo Fibonacci, Ibn al-Baitar, Jacob ben Machir ibn Tibbon, Abraham Zacuto, Xu Guangqi, Galileo Galilei, Johannes Kepler, Antony van Leeuwenhoek, Isaac Newton, Antoine Lavoisier, Carl Friedrich Gauss, Emil Fischer, Sigmund Freud, Max Planck, Niels Bohr, Erwin Schrödinger, Megh Nad Saha, Satyendra Nath Bose, Leo Szilard, Werner Heisenberg, Paul Dirac, Zhang Yuzhe, Rachel Carson, Subrahmanyan Chandrasekhar, Jonas Salk, Richard Feynman, Paul Berg, Sylvia Earle, Sidney Altman, Richard Dawkins, and Ahmed H. Zewail.